Springer Proceedings in Mathematics & Statistics

Volume 45

For further volumes:
http://www.springer.com/series/10533

Springer Proceedings in Mathematics & Statistics

This book series features volumes composed of select contributions from workshops and conferences in all areas of current research in mathematics and statistics, including OR and optimization. In addition to an overall evaluation of the interest, scientific quality, and timeliness of each proposal at the hands of the publisher, individual contributions are all refereed to the high quality standards of leading journals in the field. Thus, this series provides the research community with well-edited, authoritative reports on developments in the most exciting areas of mathematical and statistical research today.

Oleg P. Iliev • Svetozar D. Margenov
Peter D. Minev • Panayot S. Vassilevski
Ludmil T. Zikatanov
Editors

Numerical Solution of Partial Differential Equations: Theory, Algorithms, and Their Applications

In Honor of Professor Raytcho Lazarov's 40 Years of Research in Computational Methods and Applied Mathematics

Springer

Editors
Oleg P. Iliev
Department of Flow and Material Simulation
Fraunhofer Institute for Industrial Mathematics
Kaiserslautern, Germany

Peter D. Minev
Department of Mathematical and Statistical Sciences
University of Alberta
Edmonton, AB, Canada

Ludmil T. Zikatanov
Department of Mathematics
The Pennsylvania State University
University Park, PA, USA

Svetozar D. Margenov
Institute for Parallel Processing
Bulgarian Academy of Sciences
Sofia, Bulgaria

Panayot S. Vassilevski
Center for Applied Scientific Computing
Lawrence Livermore National Laboratory
Livermore, CA, USA

ISSN 2194-1009 ISSN 2194-1017 (electronic)
ISBN 978-1-4614-7171-4 ISBN 978-1-4614-7172-1 (eBook)
DOI 10.1007/978-1-4614-7172-1
Springer New York Heidelberg Dordrecht London

Library of Congress Control Number: 2013939589

Printed on acid-free paper

Springer is part of Springer Science+Business Media (www.springer.com)

Preface

The design and implementation of numerical models that accurately capture the appropriate model features of complex physical systems described by time-dependent coupled systems of nonlinear PDEs present one of the main challenges in today's scientific computing. This volume integrates works by experts in computational mathematics, and its applications focused on the modern algorithms which are in the core of accurate modeling: adaptive finite-element methods, conservative finite-difference and finite-volume methods, and multilevel solution techniques. Fundamental theoretical results are revisited in several survey articles, and new techniques in numerical analysis are introduced. Applications showing the efficiency, reliability, and robustness of the algorithms in porous media, structural mechanics, and electromagnetism are presented.

The volume consists of papers prepared in the context of the International Symposium "Numerical Solution of Partial Differential Equations: Theory, Algorithms and their Applications" in honor of Professor Raytcho Lazarov's 40 years of research in computational methods and applied mathematics and on the occasion of his 70th birthday.

The symposium was organized and sponsored by the Institute of Information and Communication Technologies (IICT), Bulgarian Academy of Sciences (BAS), Lawrence Livermore National Laboratory (USA), and Department of Mathematics, The Pennsylvania State University (USA). Members of the program committee are Oleg Iliev (ITWM Fraunhofer, Kaiserslautern, Germany), Peter Minev (University of Alberta, Canada), Svetozar Margenov (Institute of Information and Communication Technologies, BAS), Panayot Vassilevski (Lawrence Livermore National Laboratory, USA), and Ludmil Zikatanov (The Pennsylvania State University, USA).

The list of participants who were invited to contribute and authored or coauthored a paper included in this volume is:

Owe Axelsson (Uppsala University, Sweden; KAU, Saudi Arabia; Academy of Sciences, Czech Republic)
Carsten Carstensen (Humboldt University of Berlin, Germany)

Panagiotis Chatzipantelidis (University of Crete, Greece)
Ivan Dimov (IICT, Bulgarian Academy of Sciences, Bulgaria)
Stefka Dimova (Sofia University, Bulgaria)
Oleg Iliev (ITWM Fraunhofer, Germany)
Ulrich Langer (Johannes Kepler University and RICAM, Austria)
Svetozar Margenov (IICT, Bulgarian Academy of Sciences, Bulgaria)
Peter Minev (University of Alberta, Canada)
Joseph Pasciak (Texas A&M, USA)
Petr Vabishchevich (IMM, Russian Academy of Sciences, Russia)
Panayot Vassilevski (Lawrence Livermore National Laboratory, USA)
Junping Wang (National Science Foundation, USA)
Joerg Willems (RICAM, Austrian Academy of Sciences, Austria)
Ludmil Zikatanov (The Pennsylvania State University, USA)

The editors are grateful to the Institute of Information and Communication Technologies (IICT), Bulgarian Academy of Sciences, the Lawrence Livermore National Laboratory, and the Department of Mathematics at Penn State for the support of the symposium.

On behalf of all the contributors, we dedicate this volume to our teacher, friend, and colleague Raytcho Lazarov.

Kaiserslautern, Germany — Oleg P. Iliev
Sofia, Bulgaria — Svetozar D. Margenov
Edmonton, AB, Canada — Peter D. Minev
Livermore, CA, USA — Panayot S. Vassilevski
University Park, PA, USA — Ludmil T. Zikatanov

On the Occasion of the 70th Anniversary of Raytcho Lazarov

With great pleasure we introduce this collection of papers in honor of Raytcho Lazarov, professor at the Texas A&M University and Doctor of Sciences and Doctor Honoris Causa of the "St. Kliment Ohridski" University of Sofia, Bulgaria.

Raytcho Lazarov is a computational mathematician of extraordinary depth and breadth whose work has had and continues to have exceptional impact on computational and applied mathematics. He has authored or coauthored more than 200 journal publications and 4 books spanning all major areas in computational mathematics and bridging mathematical theory and scientific computing with sciences and engineering.

Raytcho Lazarov was born in Kardzhali (Кърджали), Bulgaria, on January 23, 1943. He graduated from "St. Antim I" High School in Zlatograd (Златоград) and in 1961 went to Sofia University "St. Kliment Ohridski" to continue his studies in the Department of Mathematics (industrial profile). During his first year as a college student, Raytcho demonstrated his talent for mathematics, and his dedication to study it, and he was selected to continue his education at the University of Wroclaw in Poland in 1963. In Wroclaw Raytcho was able to interact with many distinguished mathematicians from the Polish mathematical school and received first-rate mathematical training.

In 1968 Raytcho Lazarov was admitted to the PhD program of the Moscow State University. As a graduate student in Moscow, he studied and worked under the supervision of Academician A. A. Samarskii who was one of the best contemporary computational mathematicians in the world. Lazarov's thesis work was on "Finite difference schemes for elasticity problems in curvilinear domains," among the first rigorous studies of numerical approximations of problems in structural mechanics.

After receiving his PhD degree in 1972, Raytcho Lazarov worked as a research associate and senior research associate in the Institute of Mathematics (IM) of the Bulgarian Academy of Sciences (BAS) until 1987. During this time he established himself as one of the leading experts in numerical analysis. In 1976 Raytcho Lazarov visited the Rutherford Laboratory in Didcot, UK, for one year, and this visit had notable impact on his future research. His focus shifted to the theory and applications of the finite element method (FEM) which remains to be his primary field of research to this day.

Lazarov earned the degree of doctor of sciences in June 1982 with a thesis on "Error estimates of the difference schemes for some problems of mathematical physics having generalized solutions." This thesis contained several breakthrough results, which were published in more than 10 papers and formed the basis for a research monograph that he coauthored with A. A. Samarskii and V. Makarov, Difference Schemes for Differential Equations Having Generalized Solutions, which was published in 1987.

In 1986 Lazarov's superb scientific achievements earned him the title of a professor of mathematics at the Institute of Mathematics of the Bulgarian Academy of Sciences, a position that he continues to hold to this day. His leadership ability was also recognized by his colleagues, and in 1985 he became the head of the Laboratory on Numerical Analysis, BAS, and a deputy-director of the Laboratory on Parallel Algorithms and High Performance Computer Systems, BAS. In 1986 Lazarov became deputy-director of the newly established Center for Informatics and Computer Technology (CICT) at BAS. This was one of the first interdisciplinary centers worldwide for mathematical research on advanced algorithms for the emerging parallel computer systems. He played a crucial role in hiring a cohort of the best young applied mathematicians in Bulgaria—Djidjev, Vassilevski, Margenov, Dimov, Bochev, and many more. In fact, Raytcho Lazarov's leadership was the key in making CICT one of the best places for large-scale scientific computing and parallel algorithms. In 1984 Raytcho initiated a series of international conferences on numerical methods and applications in Sofia, Bulgaria, which helped to publicize the results and achievements of the Bulgarian numerical analysts and to integrate them into the international community.

Such accomplishments were noticed by his colleagues around the world. Vidar Thomée helped Raytcho to get a visiting position at the University of Wyoming in 1987. This turned out to be a critical point in Lazarov's career. In Wyoming he met and befriended Richard Ewing who at that time was a director of the Enhanced Oil Recovery Institute (EORI) and the Institute of Scientific Computation (ISC) at the University of Wyoming. During his stay in Laramie in 1988–1992, Lazarov worked on superconvergence and local refinement techniques for mixed FE methods. During that time Raytcho initiated many collaborations and friendships with prominent mathematicians such as Jim Bramble, Joe Pasciak, Panayot Vassilevski, Junping Wang, Tom Russell, Yuri Kuznetsov, Steve McCormick, Tom Manteuffel, and Owe Axelsson. At that time Raytcho Lazarov led the development of algorithms based on the Bramble–Ewing–Pasciak–Schatz (BEPS) preconditioner and locally refined mixed FE and finite-volume methods that were also implemented in the EORI proprietary codes.

The friendship and collaboration with Dick Ewing initiated another change in Raytcho's career, and in 1992 he moved to Texas A&M University as a professor of mathematics, a position that he continues to hold now. This coincided with the establishment of the Institute of Scientific Computation (ISC) at Texas A&M under the directorship of Richard Ewing, which quickly attracted a team of world-renowned experts in this area like J. Bramble, J. Pasciak, R. Lazarov, and, more recently, Y. Efendiev, J.-L. Guermond, G. Petrova, B. Popov, W. Bangerth, and A. Bonito. The work they did in the last 20 years on computational mathematics and its applications in flows in porous media, multiphysics problems, modeling of fluids, structures and their interactions, etc. had a significant impact on these and in other research areas. Raytcho's pivotal role in this research is well known from his results on least-squares FEM; discontinuous Galerkin methods; multigrid, multilevel, and multiscale methods, mixed FEM, and more recently fractional order partial differential equations.

In recognition of his achievements Raytcho Lazarov has been awarded several honorary titles and degrees: the medal "St. Kl. Ohridski" with blue ribbon (2003–the highest honors given by Sofia University, Bulgaria, to scientists); Doctor Honoris

Causa of Sofia University "St. Kl. Ohridski" (2006); the medal of the Institute of Mathematics, Bulgarian Academy of Sciences 2008; Pichoridis Distinguished Lectureship, University of Crete, Greece (2008); and Erasmus Mundus Visiting Scholar Award, University of Kaiserslautern (2008). Most recently he was named a recipient of the medal of the Bulgarian Academy of Sciences "Marin Drinov" with ribbon (2013), which is given to scholars for outstanding contributions in the advancement of science.

During his career Lazarov has held visiting positions and contributed to advancement of research in many institutions around the globe: Joint Institute for Nuclear Research in Dubna, Russia (1980); Australian National University, Canberra (1990); Mittag Leffler Institute of Mathematics, Stockholm, Sweden (1998); University of Linz and RICAM, Austria (2005); Fraunhofer Institute of Industrial Mathematics, Kaiserslautern, Germany (2006); Lawrence Livermore National Laboratory (regularly from 1998 to 2010); and KAUST in Saudi Arabia (2008–2013). He is a member of the editorial board of five international journals and a number of conference proceedings, and he is also serving on the scientific committees of several international conferences.

Raytcho Lazarov is an outstanding scholar, and his work has had a profound impact on mathematics and other fields of science and engineering during the last four decades. His extraordinary personality, with strict academic integrity requirements for himself and his collaborators complemented by truly compassionate care about their needs, has influenced the professional and personal development of those who have had a chance to work with him. The teams which he has created over the years combined research interests, philosophy, and personal friendship, and they withstood the test of time.

We congratulate Raytcho on the occasion of his 70th birthday and wish him the best of health and enjoyment in his personal life and in continuing and expanding his successful research achievements.

Kaiserslautern, Germany — Oleg P. Iliev
Sofia, Bulgaria — Svetozar D. Margenov
Edmonton, AB, Canada — Peter D. Minev
Livermore, CA, USA — Panayot S. Vassilevski
University Park, PA, USA — Ludmil T. Zikatanov

Contents

Improving Conservation for First-Order System Least-Squares Finite-Element Methods

J.H. Adler and P.S. Vassilevski

Abstract The first-order system least-squares (FOSLS) finite element method for solving partial differential equations has many advantages, including the construction of symmetric positive definite algebraic linear systems that can be solved efficiently with multilevel iterative solvers. However, one drawback of the method is the potential lack of conservation of certain properties. One such property is conservation of mass. This paper describes a strategy for achieving mass conservation for a FOSLS system by changing the minimization process to that of a constrained minimization problem. If the space of corresponding Lagrange multipliers contains the piecewise constants, then local mass conservation is achieved similarly to the standard mixed finite-element method. To make the strategy more robust and not add too much computational overhead to solving the resulting saddle-point system, an overlapping Schwarz process is used.

Keywords Conservation • First-order system least-squares • Finite elements • Domain decomposition • Two-level

Mathematics Subject Classification (2010): 65F10, 65N20, 65N30

The work of the author "P.S. Vassilevski" was performed under the auspices of the U.S. Department of Energy by Lawrence Livermore National Laboratory under Contract DE-AC52-07NA27344

J.H. Adler
Department of Mathematics, Tufts University, Medford, MA 02155, USA
e-mail: james.adler@tufts.edu

P.S. Vassilevski (✉)
Center for Applied Scientific Computing, Lawrence Livermore National Laboratory,
P.O. Box 808, L-560, Livermore, CA 94551, USA
e-mail: panayot@llnl.gov

O.P. Iliev et al. (eds.), *Numerical Solution of Partial Differential Equations: Theory, Algorithms, and Their Applications*, Springer Proceedings in Mathematics & Statistics 45, DOI 10.1007/978-1-4614-7172-1_1, © Springer Science+Business Media New York 2013

1 Introduction

The first-order system least-squares (FOSLS) approach is a finite-element discretization, which solves a system of linear partial differential equations (PDEs) by minimizing the L^2 norm of the residual of the PDE [14–16, 30, 31]. Least-squares finite-element methods, in general, have several nice properties and have been used on a wide variety of problems, e.g., [4, 6, 7, 9, 13, 29, 34]. One advantage is that they yield symmetric positive definite (SPD) algebraic systems, which are amenable to multilevel techniques. This is true for any PDE system, including systems like Stokes where a mixed finite-element method would yield a saddle-point problem and an indefinite linear system [10]. Another advantage is that they yield sharp and reliable a posteriori estimates [3]. This is useful for implementing adaptive local refinement techniques, which allow the approximations to be resolved more accurately in regions of higher error [11, 19]. A disadvantage of the least-squares methods noted in the literature is a loss of conservation for certain properties in a given system. For instance, the Stokes' or Navier–Stokes' system contains an equation for the conservation of momentum and one for the conservation of mass [20, 21]. Since the least-squares principle minimizes both equations equally, both quantities are only conserved up to the error tolerance given for the simulation. Attempts to improve the conservation of mass would result in a loss of accuracy in the conservation of momentum. Despite this, in several applications, conservation of a certain quantity is considered essential to capturing the true physics of the system. For instance, in electromagnetic problems, such as magnetohydrodynamics (the treatment of plasmas as charged fluids), loss of accuracy in the solenoidal constraint of the magnetic field, $\nabla \cdot \mathbf{B} = 0$, can lead to instabilities in the system [2, 8].

In this paper, we consider methods for improving the conservation of a divergence constraint, such as mass conservation, in a system, using the FOSLS finite-element method. There are many ways to improve the accuracy of mass conservation in such systems, including adaptive refinement to increase the spatial resolution of the discretization [6, 7], higher temporal accuracies or higher-order elements for time-dependent problems [32], using divergence-free finite-element spaces [1, 4, 17, 18], reformulating the first-order system into a more conservative one [23], as well as using a compatible least-squares method [5], which use ideas from mixed Galerkin methods to improve the mass conservation. In addition, an alternative approach called FOSLL* [27, 28] has been developed, in which an adjoint system is considered, and the error is minimized in the L^2 norm directly. This has been shown to improve conservation in satisfying the divergence constraint in incompressible fluid flow and electromagnetic problems. In this paper, we discuss an approach that simply corrects the solution approximated by the FOSLS discretization so that it conserves the given quantity. The goal is to keep the discretization as is, preserving all of the special properties of the least-squares minimization while still obtaining the appropriate conservation. As a result, the a posteriori error estimates and the simple finite-element spaces can still be used. More specifically, the aim of this paper is to show that it is possible to conserve a certain quantity in the least-squares finite-element setting by using a local subdomain correction post-processing scheme at relatively little extra cost.

The paper is outlined as follows. In Sect. 2, we consider the FOSLS discretization applied to a Poisson problem and show how the scheme can result in a type of "mass loss." Section 3 investigates a way of transforming the minimization principle into a constrained minimization problem and investigates what types of constraints are possible. Next, in Sect. 4, a local subdomain and coarse-grid correction solver is used to make the method more robust. This uses an overlapping Schwarz (Vanka-like) smoother with a coarse-grid correction to solve the constrained problem [35–37]. Finally, concluding remarks and a discussion of future work is given in Sect. 5.

2 First-Order System Least-Squares

To illustrate the FOSLS finite-element method, consider a PDE system that is first put into a differential first-order system of equations, denoted by $Lu = f$. Here, L is a mapping from an appropriate Hilbert space, $\mathcal{V}$, to an L^2 product space. In many contexts, $\mathcal{V}$ is chosen to be an H^1 product space with appropriate boundary conditions.

This minimization is written as

$$u_* = \arg\min_{u\in\mathcal{V}} G(u;f) := \arg\min_{u\in\mathcal{V}} ||Lu - f||_0^2, \tag{1}$$

where u_* is the solution in an appropriate H^1 space. The minimization results in the weak form of the problem:

Find $u_* \in \mathcal{V}$ such that

$$\langle Lu_*, Lv\rangle = \langle f, Lv\rangle \quad \forall v \in \mathcal{V}, \tag{2}$$

where $\langle\cdot,\cdot\rangle$ is the usual L^2 inner product on the product space, $(L^2)^k$, for k equations in the linear system. If the following properties of the bilinear form $\langle Lu, Lv\rangle$ are assumed,

$\exists$ constants, c_1 and c_2, such that

$$\text{continuity} \quad \langle Lu, Lv\rangle \leq c_2||u||_{\mathcal{V}}||v||_{\mathcal{V}} \quad \forall \; u,v \in \mathcal{V}, \tag{3}$$

$$\text{coercivity} \quad \langle Lu, Lu\rangle \geq c_1||u||_{\mathcal{V}}^2 \quad \forall \; u \in \mathcal{V}, \tag{4}$$

then, by the Riesz representation theorem, this bilinear form is an inner product on $\mathcal{V}$ [26]. In addition, these properties imply the existence of a unique solution, $u_* \in \mathcal{V}$, for the weak problem (2). Here, c_1 and c_2 depend only on the operator, L, and the domain of the problem. They are independent of u and v.

Next, u_* is approximated by restricting (1) to a finite-dimensional space, $\mathscr{V}^h \subseteq \mathscr{V}$, which leads to (2) restricted to $\mathscr{V}^h$. Since $\mathscr{V}^h$ is a subspace of $\mathscr{V}$, the discrete problem is also well posed. Choosing an appropriate basis, $\mathscr{V}^h = span\{\Phi_j\}$, and restricting (2) to this basis yields an algebraic system of equations involving the matrix, A, with elements

$$(A)_{ij} = \langle L\Phi_j, L\Phi_i \rangle. \tag{5}$$

It has been shown that, in the context of a SPD H^1-equivalent bilinear form restricted to a finite-element subspace, a multilevel technique exists that yields optimal convergence to the linear system [15].

2.1 Sample Problem and Loss of Conservation

To illustrate possible losses in conservation, consider the convection–diffusion equation for unknown p in two dimensions,

$$-\nabla \cdot D\nabla p + \mathbf{r} \cdot \nabla p + cp = f, \tag{6}$$

with D an SPD matrix that could depend on the domain, $\mathbf{r}$ a vector, and c a nonnegative constant, respectively. In order to make the system first order, a new variable, $\mathbf{u} = D\nabla p$, is introduced. The resulting FOSLS system becomes

$$-\nabla \cdot \mathbf{u} + D^{-1}\mathbf{r} \cdot \mathbf{u} + cp = f, \tag{7}$$

$$\nabla \times D^{-1}\mathbf{u} = 0, \tag{8}$$

$$D^{-1/2}\mathbf{u} - D^{1/2}\nabla p = 0. \tag{9}$$

Here, a scaling on D is performed to allow the resulting discrete system to be better conditioned and, thus, more amenable to multigrid methods. Also, the extra curl equation is introduced so that the weak system is continuous and coercive and, therefore, H^1 equivalent [14, 15]. For simplicity, let $D = I$, $\mathbf{r} = \mathbf{0}$, and $c = 0$. Then, the following functional is minimized:

$$\mathscr{G} = ||\nabla \cdot \mathbf{u} + f||_0^2 + ||\nabla \times \mathbf{u}||_0^2 + ||\mathbf{u} - \nabla p||_0^2.$$

The resulting discrete system is

$$A\mathscr{U} = b,$$

where $\mathscr{U} = (\mathbf{u}, p)^T$. Here, A is the matrix as defined in (5), where L now refers to system (7)–(9). Similarly, the right-hand side vector, b, is defined as $b_i = \langle \mathbf{f}, L\Phi_i \rangle$, where $\mathbf{f} = (f, 0, 0)^T$. When minimizing this functional, equal weight is given to each term in the system. Therefore, if better accuracy is needed on a certain term, such as the divergence constraint, accuracy is lost in the other portions. In many applications, however, exact conservation of certain terms is important for developing an accurate model of a physical system. For instance, one may want to conserve the "mass" of the system. This is defined as

$$\int_{\Omega} -\nabla\cdot\mathbf{u}\,d\Omega = \int_{\Omega} f\,d\Omega. \tag{10}$$

In other words, the amount of flow in or out of the system is equal to the flow contributed by the source (this has more physical meaning in a system like Stokes, where we assume $\nabla\cdot\mathbf{u} = 0$ [20]). In fact, in many applications *local* mass conservation is desired instead, where the mass is conserved in all regions of the domain, including a single element. Mixed finite-element methods can satisfy this exactly and are commonly used in these situations. However, for the least-squares methods, since the part of the functional concerned with this property is only minimized to a certain degree (i.e., truncation error of the scheme at best), this cannot be satisfied exactly. Another issue concerns the fact that in many applications of the FOSLS finite-element method, the same order of polynomials is chosen as the basis for every unknown in the discrete space. For instance, linear functions are chosen to approximate both $\mathbf{u}$ and p. As a result, in trying to satisfy the term $\mathbf{u} - \nabla p = 0$, one is trying to match linears with the gradient of linears or constants. This is not approximated very well and accuracy is lost. As a result the conservation property is also lost. Choosing higher-order elements does remedy this to some extent, especially in two dimensions. However, using higher-order elements increases the complexity of the discrete system and the grid hierarchy in a multigrid scheme, making the systems harder to solve. In addition, the effect of higher-order elements is lessened when going to three dimensions [22, 24, 32].

To improve on this, here, the idea of adding the mass conservation as a constraint to the system is considered. Thus, instead of just minimizing the FOSLS functional, the functional is minimized subject to a constraint. This constraint enforces the desired mass conservation, while still allowing the FOSLS functional to be minimized as usual, thus retaining its nice properties. We mention that the modified method can achieve full local mass conservation, if the space of corresponding Lagrange multipliers contains the piecewise constants, similarly to the standard mixed finite-element method. Next, several approaches for implementing this constraint are described.

3 Constrained FOSLS

To enforce the constraint mentioned above, a Lagrange multiplier, λ, is introduced, and the FOSLS system is augmented as follows:

$$\begin{pmatrix} A & C^T \\ C & 0 \end{pmatrix}\begin{pmatrix} \mathscr{U} \\ \lambda \end{pmatrix} = \begin{pmatrix} b \\ g \end{pmatrix}. \tag{11}$$

Here, A and $\mathscr{U}$ are as before for the FOSLS discretization, λ is the Lagrange multiplier, and C is a finite-element assembly of the constraint; in this example, $-\nabla \cdot \mathbf{u} = f$. Two possible ways to construct C are considered. For the rest of the paper, we consider a triangulation of a mesh in two dimensions, $\mathscr{T}_h$, with grid spacing h. In addition, consider the polynomial spaces of order k defined on this triangulation as $\mathscr{P}_k$. The following notation is used for matrices and spaces:

Definition 1. Let $\Phi_j \in \left[\mathscr{P}_{k_1}\right]^2$ be a vector and let $q_i \in \mathscr{P}_{k_2}$ be a scalar. Let f be some right-hand side function as defined in (6). Then, we define the following matrices:

$$(\tilde{B})_{ij} = \langle -\nabla \cdot \Phi_j, q_i \rangle,$$

$$\Lambda \Rightarrow (\Lambda)_{ij} = \langle -\nabla \cdot \Phi_j, -\nabla \cdot \Phi_i \rangle,$$

and vectors:

$$(\tilde{g})_i = \langle f, q_i \rangle, \qquad (g)_i = \langle f, -\nabla \cdot \Phi_i \rangle.$$

3.1 "Galerkin Constraint"

Letting $C = \tilde{B}$, a standard Galerkin-type construction of the divergence constraint is obtained. It should be noted that the order of the polynomials for the constraints, k_2, can be different from the order for the FOSLS unknowns, k_1, and, in fact, should be chosen to have less degrees of freedom so as not to over-constrain the system. The pairs chosen in this paper are quadratics–linears ($\mathscr{P}_2 - \mathscr{P}_1$), quadratics–constants ($\mathscr{P}_2 - \mathscr{P}_0$), and linears–constants ($\mathscr{P}_1 - \mathscr{P}_0$). In this context, $\mathscr{U} \in \left[\mathscr{P}_{k_1}\right]^3$ and $\lambda \in \mathscr{P}_{k_2}$. The resulting system is

$$\begin{pmatrix} A & \tilde{B}^T \\ \tilde{B} & 0 \end{pmatrix} \begin{pmatrix} \mathscr{U} \\ \lambda \end{pmatrix} = \begin{pmatrix} b \\ \tilde{g} \end{pmatrix}. \tag{12}$$

3.2 "Least-Squares Constraint"

To keep faith with the FOSLS methodology, a constraint is proposed that is of the same form as that is used in the FOSLS discretization, namely, letting $C = \Lambda$. This allows the same finite-element spaces for the FOSLS unknowns to be used for the Lagrange multiplier. The system is then

$$\begin{pmatrix} A & \Lambda \\ \Lambda & 0 \end{pmatrix} \begin{pmatrix} \mathscr{U} \\ \lambda \end{pmatrix} = \begin{pmatrix} b \\ g \end{pmatrix}, \tag{13}$$

where $\mathscr{U} \in \left[\mathscr{P}_{k_1}\right]^3$ and $\lambda \in \left[\mathscr{P}_{k_1}\right]^2$.

As is shown below, the system that needs to be solved in the least-squares constraint approach may not be well conditioned. However, one can construct the constraint matrix C in such a way that it can be decomposed into a form which is much easier to solve. For instance, decompose $\Lambda = B^T B$ (see definition of B in Sect. 3.3.1), and thus, the system is rewritten as

$$\begin{pmatrix} A & B^T B \\ B^T B & 0 \end{pmatrix} \begin{pmatrix} \mathcal{U} \\ \lambda \end{pmatrix} = \begin{pmatrix} b \\ g \end{pmatrix}. \tag{14}$$

However, the construction of B is not trivial in many cases (again, see Sect. 3.3.1) and it is easier to work with $\tilde{B}$ instead. If the system in the "Galerkin" approach is taken and modified, the following is obtained:

$$\begin{pmatrix} A & \tilde{B}^T \tilde{B} \\ \tilde{B}^T \tilde{B} & 0 \end{pmatrix} \begin{pmatrix} \tilde{\mathcal{U}} \\ \tilde{\lambda} \end{pmatrix} = \begin{pmatrix} b \\ \tilde{B}^T \tilde{g} \end{pmatrix}. \tag{15}$$

As it turns out, due to the following lemma, it is reasonable to solve system (15) instead of system (14).

Lemma 1. *Consider systems* (12) *and* (15)*. Let* A, $\tilde{B}$, $\mathcal{U}$, λ, $\tilde{\lambda}$, g, *and* $\tilde{g}$ *be all defined as above in Definition 1, then,*

$$\lambda = \tilde{B}\tilde{\lambda} \text{ and } \tilde{\mathcal{U}} = \mathcal{U}.$$

Proof. First combine the two systems:

$$A\mathcal{U} + \tilde{B}^T \lambda = b, \tag{16}$$

$$\tilde{B}\mathcal{U} = \tilde{g}, \tag{17}$$

$$A\tilde{\mathcal{U}} + \tilde{B}^T \tilde{B}\tilde{\lambda} = b, \tag{18}$$

$$\tilde{B}^T \tilde{B}\tilde{\mathcal{U}} = \tilde{B}^T \tilde{g}. \tag{19}$$

Next, multiply (17) on the left by $\tilde{B}^T$ and subtract the bottom two equations from the top two. Let $e_{\mathcal{U}} = \mathcal{U} - \tilde{\mathcal{U}}$ and $e_\lambda = \lambda - \tilde{B}\tilde{\lambda}$ to obtain

$$Ae_{\mathcal{U}} + \tilde{B}^T e_\lambda = 0,$$
$$\tilde{B}^T \tilde{B} e_{\mathcal{U}} = 0.$$

Since $\tilde{B}^T$ is equivalent to a gradient operator, it can be shown that it is a one-to-one operator (since divergence and, thus, $\tilde{B}$ is onto). Therefore, $\tilde{B}e_{\mathcal{U}} = 0$ and the system becomes

$$\begin{pmatrix} A & \tilde{B}^T \\ \tilde{B} & 0 \end{pmatrix} \begin{pmatrix} e_{\mathcal{U}} \\ e_\lambda \end{pmatrix} = \begin{pmatrix} 0 \\ 0 \end{pmatrix},$$

which is the global "Galerkin" system, which is known to be invertible. As a result, $e_{\mathcal{U}} = e_\lambda = 0$ and, more importantly, $\mathcal{U} = \tilde{\mathcal{U}}$, meaning solving either system results in the same solution.

Therefore, (12) and (15) are both viable options for the constraint system. Next, each of these and some variations are tested to see which yield the best mass conservation with little extra computational work.

3.3 Solvers

To solve the constrained system, the conjugate gradient (CG) method on the Schur complement is used [33]. Solving the system in this way yields the following set of equations:

$$\mathscr{U} = A^{-1}b - A^{-1}C^T\lambda,$$
$$CA^{-1}C^T\lambda = CA^{-1}b - g.$$

The second equation is solved for λ via CG and a backsolve is used to get the original $\mathscr{U}$. For the results presented here, a direct solver is used to compute A^{-1}, but in the future a multigrid solver, or whatever is used to solve the FOSLS system itself, will be substituted instead.

For the first approach (12) and second (13), the system is solved exactly as described above. In the second approach, we consider $\Lambda = B^T B$, where $(B)_{ij}$ represents the construction of $\langle -\nabla\cdot\Phi_j, r_i\rangle$, but where $r_i = \nabla\cdot\Phi_i$ is in the divergence of the space used for A, i.e., $\nabla\cdot[\mathscr{P}_{k_1}]^2$ as opposed to the full $\mathscr{P}_{k_2}$. As a result, the Schur complement equation becomes

$$B^T BA^{-1}B^T B\lambda = B^T BA^{-1}b - g. \tag{20}$$

This is badly conditioned as the system $B^T B$ is equivalent to a $-\nabla\nabla\cdot$ (grad-div) equation. However, to remedy this, the equation is multiplied on the left by BA^{-1}, resulting in

$$(BA^{-1}B^T)(BA^{-1}B^T)B\lambda = (BA^{-1}B^T)BA^{-1}b - (BA^{-1})g.$$

Notice that BB^T is equivalent to a $-\nabla\cdot\nabla$, or Laplace system, and, thus, $BA^{-1}B^T$ is well conditioned. In addition, one only needs to solve for $B\lambda$. This system simplifies further by eliminating one of the $BA^{-1}B^T$ blocks to obtain

$$(BA^{-1}B^T)B\lambda = BA^{-1}b - (BA^{-1}B^T)^{-1}BA^{-1}g. \tag{21}$$

However, in (21), two solves of $BA^{-1}B^T$ are required, increasing the number of iterations required to solve the system.

In addition, a problem with this approach is the construction of B. A simpler way is to construct $\tilde{B}$ and use this instead to get system (15). This results in

$$\tilde{B}^T\tilde{B}A^{-1}\tilde{B}^T\tilde{B}\lambda = \tilde{B}^T\tilde{B}A^{-1}b - \tilde{B}^T\tilde{g}. \tag{22}$$

Multiplying on the left by $\tilde{B}A^{-1}$ yields

$$(\tilde{B}A^{-1}\tilde{B}^T)(\tilde{B}A^{-1}\tilde{B}^T)\tilde{B}\lambda = (\tilde{B}A^{-1}\tilde{B}^T)\tilde{B}A^{-1}b - (\tilde{B}A^{-1}\tilde{B}^T)\tilde{g}$$

$$(\tilde{B}A^{-1}\tilde{B}^T)\tilde{B}\lambda = \tilde{B}A^{-1}b - \tilde{g}. \tag{23}$$

This, however, is the same system obtained from (12) and, as shown in Lemma 1, results in the same solution for $\mathscr{U}$.

3.3.1 Construction of B

Despite being able to use the simpler construction, $\tilde{B}$, it is possible to construct B for the type of constraint considered here, $\nabla \cdot \mathbf{u} = f$. In fact, the matrix B is constructed locally using the simpler construction of $\tilde{B}$. Consider an element (triangle) T and let $[\mathscr{P}_{k_1}]^2(T)$ be the vector polynomials of degree k_1. Next, consider the "least-squares" constraint, where the space of Lagrange multipliers, λ, is $\nabla \cdot [\mathscr{P}_{k_1}]^2(T)$, which is a subspace of $[\mathscr{P}_{k_1-1}](T)$. Let $\{\varphi_s\}_{s=1}^{l}$ be the basis (restricted to T) of $[\mathscr{P}_{k_1-1}](T)$. For $k_1 = 2$, $l = 3$ (since $[\mathscr{P}_{k_1-1}](T) = [\mathscr{P}_1](T)$—the space of linears). Also, let $\{\Phi_i\}_{i=1}^{n}$ be the basis of $[\mathscr{P}_{k_1}]^2(T)$. Since $\nabla \cdot \Phi_i \in \nabla \cdot [\mathscr{P}_{k_1}]^2(T) \subset [\mathscr{P}_{k_1-1}](T)$,

$$\nabla \cdot \Phi_i = \sum_{s=1}^{l} c_{i,s}\varphi_s = [\varphi_1, \ldots, \varphi_l]\mathbf{c}_i, \tag{24}$$

for some coefficients $\mathbf{c}_i = (c_{i,s}) \in \mathbb{R}^l$. Therefore,

$$\left(\tilde{B}_T\right)_{s,i} = \langle \nabla \cdot \Phi_i, \varphi_s \rangle = [\langle \varphi_s, \varphi_1 \rangle, \ldots, \langle \varphi_s, \varphi_l \rangle]\mathbf{c}_i = \mathbf{e}_s^T M \mathbf{c}_i.$$

Here, $\mathbf{e}_s \in \mathbb{R}^l$ is the sth unit coordinate vector and $M = M_T$ is the element mass matrix coming from the space $[\mathscr{P}_{k_1-1}](T)$. In conclusion, the element matrix $\tilde{B} = \tilde{B}_T = (\langle \nabla \cdot \Phi_i, \varphi_s \rangle)_{1\le i \le n,\, 1 \le s \le l}$ admits the following form:

$$\tilde{B}_T = M_T[\mathbf{c}_1, \mathbf{c}_2, \ldots, \mathbf{c}_n].$$

For the entries $\langle \nabla \cdot \Phi_j, \nabla \cdot \Phi_i \rangle = (B_T^T B_T)_{ij}$, using the representation (24) yields

$$(B_T^T B_T)_{ij} = \langle \nabla \cdot \Phi_j, \nabla \cdot \Phi_i \rangle = \mathbf{c}_j^T \left(\langle \varphi_r, \varphi_s \rangle \right)_{r,s=1}^{l} \mathbf{c}_i = \mathbf{c}_j^T M_T \mathbf{c}_i = \left(\tilde{B}_T^T M_T^{-1} \tilde{B}_T \right)_{ij}.$$

Therefore,

$$B_T = M_T^{-\frac{1}{2}} \tilde{B}_T.$$

Thus, B is constructed relatively easily. Namely, over each element the local matrix, $\tilde{B}_T$, is built, which is the Galerkin finite-element construction of the divergence operator using $\mathscr{P}_{k_1} - \mathscr{P}_{k_1-1}$ elements. Then, $B_T = M_T^{-1/2}\tilde{B}_T$, where $M_T^{-1/2}$ is the mass matrix associated with the given element and $\mathscr{P}_{k_1-1}$.

3.4 Numerical Results

In the following numerical tests, four approaches are considered:

- Method 1: Solve the "Galerkin" constraint system (12), resulting in (23). Note that this is the same as solving system (15) and simplifying the Schur complement system.
- Method 2: Solve the "least-squares" constraint system (13), resulting in (20).
- Method 3: Solve the "least-squares" constraint system using the simpler construction, (15), resulting in (22).
- Method 4: Solve the "least-squares" constraint system (14), with the simplified Schur complement system (21).

Again, $D = I$, $\mathbf{r} = \mathbf{0}$, and $c = 0$. The right-hand side is chosen as $f = 2\pi^2 \sin(\pi x)\sin(\pi y)$ so that the true solution is $p = \sin(\pi x)\sin(\pi y)$. The problem is solved on a unit square with homogeneous Dirichlet boundary conditions for p. The system is solved using the four approaches described above for a combination of the finite-element spaces, $\mathscr{P}_2$, $\mathscr{P}_1$, and $\mathscr{P}_0$. The L^2 norms of the errors of the numerical solutions, p and $\mathbf{u} = \nabla p$, are shown in the following tables. Here, $u_{err} = ||\mathbf{u} - \mathbf{u}^*||_0/||\mathbf{u}^*||_0$ and $p_{err} = ||p - p^*||_0/||p^*||_0$ for the constrained system, where $\mathbf{u}^*$ and p^* are the true solutions. The FOSLS functional, $\mathscr{F} = ||L\mathscr{U} - f||_0$, is given for both the unconstrained system, $\mathscr{F}$, and the constrained system, $\mathscr{F}_c$. In addition, the mass conservation (or mass loss) is shown as

$$m_L = \left| \int_\Omega (\nabla \cdot \mathbf{u} + f)\, d\Omega \right|,$$

for the unconstrained FOSLS system as well as with the constraint, m_L^c. In addition, we consider local conservation of mass by integrating over each element measuring the largest mass loss over all elements in the domain,

$$\hat{m}_L = \max_T \left| \int_T (\nabla \cdot \mathbf{u} + f)\, dT \right|,$$

as this is a more practical measurement for satisfying physical conservation laws. Finally, the number of iterations needed in the CG algorithm to reduce the algebraic residual by 10^{-8} is shown (Tables 1–4).

3.5 Discussion

A couple of things to note are the fact that the first test yields some of the most optimal results. Method 2 attempts to solve the ill-conditioned $\nabla\nabla\cdot$-like system and, as is shown, requires too many iterations to be used reliably. Methods 3 and 4 improve on this; however, as they require extra matrix inversions in the solution process, they require more work than in the first case.

Table 1 (Method 1) Solve $\tilde{B}A^{-1}\tilde{B}^T\lambda = \tilde{B}A^{-1}b - \tilde{g}$

k_1	k_2	h	$\hat{m}_L$	$\hat{m}_L^c$	m_L	m_L^c	$\mathscr{F}$	$\mathscr{F}_c$	u_{err}	p_{err}	Iterations
1	0	1/16	6.9e−4	5.5e−12	2.8e−2	4.9e−11	0.90	1.12	0.181	0.015	113
1	0	1/32	6.6e−5	1.6e−12	1.0e−2	2.9e−11	0.48	0.86	0.181	0.004	232
2	0	1/16	1.7e−5	1.2e−14	3.7e−5	5.7e−14	3.7e−2	3.7e−2	1.16e−3	1.40e−4	4
2	0	1/32	1.1e−6	9.5e−16	2.4e−6	3.0e−14	9.5e−3	9.5e−3	1.82e−4	1.73e−5	2
2	1	1/16	1.7e−5	1.7e−5	3.7e−5	1.9e−13	3.7e−2	3.7e−2	1.12e−3	1.38e−4	13
2	1	1/32	1.1e−6	1.0e−6	2.4e−6	3.4e−14	9.5e−3	9.5e−3	1.81e−4	1.72e−5	7

This approach is equivalent to using the "Galerkin" approach (12) and the "least-squares" approach plus simplification of the Schur complement system on $\tilde{B}$ (15)

Table 2 (Method 2) Solve $\Lambda A^{-1}\Lambda\lambda = \Lambda A^{-1}b - g$

k_1	k_2	h	$\hat{m}_L$	$\hat{m}_L^c$	m_L	m_L^c	$\mathscr{F}$	$\mathscr{F}_c$	u_{err}	p_{err}	Iterations
1	1	1/16	6.9e−4	2.9e−12	2.8e−2	5.1e−12	0.90	1.12	0.181	0.015	1,730
1	1	1/32	6.6e−5	1.6e−11	1.0e−2	8.1e−11	0.48	0.86	0.181	0.004	20,375
2	2	1/16	1.7e−5	1.6e−13	3.7e−5	1.5e−11	3.7e−2	9.8e−2	0.012	1.38e−4	1,100
2	2	1/32	1.1e−6	9.8e−15	2.4e−6	1.7e−12	9.5e−3	4.8e−2	4.94e−3	1.72e−5	4,319

This approach is equivalent to using the "least-squares" approach, but without splitting the constraint matrix and solving the full Schur complement system (13)

Table 3 (Method 3) Solve $\tilde{B}^T\tilde{B}A^{-1}\tilde{B}^T\tilde{B}\lambda = \tilde{B}^T\tilde{B}A^{-1}b - \tilde{B}^T\tilde{g}$

k_1	k_2	h	$\hat{m}_L$	$\hat{m}_L^c$	m_L	m_L^c	$\mathscr{F}$	$\mathscr{F}_c$	u_{err}	p_{err}	Iterations
1	1	1/16	6.9e−4	2.2e−12	2.8e−2	2.8e−12	0.90	1.12	0.181	0.015	1,600
1	1	1/32	6.6e−5	1.4e−11	1.0e−2	2.9e−11	0.48	0.86	0.181	0.004	15,268
2	1	1/16	1.7e−5	9.2e−14	3.7e−5	9.3e−13	3.7e−2	3.7e−2	1.16e−3	1.40e−4	15
2	1	1/32	1.1e−6	1.2e−14	2.4e−6	7.5e−14	9.5e−3	9.5e−3	1.82e−4	1.73e−5	4
2	2	1/16	1.7e−5	1.7e−5	3.7e−5	5.9e−14	3.7e−2	3.7e−2	1.15e−3	1.38e−4	12
2	2	1/32	1.1e−6	1.0e−6	2.4e−6	7.3e−14	9.5e−3	9.5e−3	1.81e−4	1.72e−5	6

This approach is equivalent to using the "least-squares" approach with the simpler construction of the constraint, but without splitting the constraint matrix and solving the full Schur complement system (15)

Table 4 (Method 4) Solve $\Lambda A^{-1}\Lambda\lambda = \Lambda A^{-1}b - g$

k_1	k_2	h	$\hat{m}_L$	$\hat{m}_L^c$	m_L	m_L^c	$\mathscr{F}$	$\mathscr{F}_c$	u_{err}	p_{err}	Iterations
1	1	1/16	6.9e−4	3.9e−8	2.8e−2	1.3e−10	0.90	1.12	0.181	0.015	84+134
1	1	1/32	6.6e−5	4.0e−8	1.0e−2	7.8e−10	0.48	0.86	0.181	0.004	146+307
2	2	1/16	1.7e−5	9.6e−10	3.7e−5	5.1e−10	3.7e−2	9.8e−2	0.012	1.38e−4	72+101
2	2	1/32	1.1e−6	5.7e−10	2.4e−6	1.7e−9	9.5e−3	4.8e−2	7.73e−3	1.72e−5	124+198

This approach is equivalent to using the "least-squares" approach and using the simplification of the full Schur complement system using B (21). Note that since two solves of $BA^{-1}B^T$ are required, the iterations for both solves are displayed in the last column of the table

In addition, only when a stable pair of elements with the constraint is used (i.e., $\mathscr{P}_2 - \mathscr{P}_0$ or $\mathscr{P}_2 - \mathscr{P}_1$) are the optimal results obtained. This results from the fact that only for the stable combinations is there enough room to minimize the FOSLS functional. All cases yield improved conservation as this is enforced directly. However, for the unstable pairings as the constraint is enforced, only a few possible solutions are allowed and, as a result, when the FOSLS functional is minimized, there is no longer enough room to minimize certain terms in the functional any more (such as $\mathbf{u} - \nabla p = 0$). Thus, the best $\mathbf{u}$ is not found. The solution has better mass conservation, but the approximation is not necessarily capable of minimizing the FOSLS functional. This can be seen by looking at the reduction in the error of $\mathbf{u}$. In all cases, the solution, p, is approximated well and the error is reduced with h as expected. However, for the unstable pairs, the gradient, $\mathbf{u}$, is not approximated well. Thus, the functional is no longer estimating the H^1 error accurately and the a posteriori error estimator is lost. Therefore, the conclusion is that the constraint always needs to be chosen from a space which gives a stable finite-element pair with whatever unknowns from the FOSLS system that you wish to conserve. This requires considering an inf–sup condition for the FOSLS unknown and Lagrange multiplier pairs, but in many applications these pairs of spaces are well known [12, 20, 21]. In addition, it should be noted that we also obtain *local* conservation across the elements when the constraint space uses discontinuous elements (i.e., $\mathscr{P}_0$, $\nabla \cdot [\mathscr{P}_1]^2$, or $\nabla \cdot [\mathscr{P}_2]^2$). This is similar to mixed finite-element methods where $\int_T (\nabla \cdot \mathbf{u} + f)\, dT$ will be zero (or small if the system is solved only approximately) for each element T.

Alternatively, we may use for the constraints test functions from a coarse subspace of a space that generally may not provide a stable fine–grid pair. For instance, if the constraint matrix, $\tilde{B}$, is constructed using the "Galerkin-like" approach using the same polynomial space as the FOSLS system, the finite-element pairs are not stable. However, if this operator is restricted to a coarser space, H, and the Lagrange multiplier, λ_H, is chosen in that coarser space, stability is regained (assuming the coarse space is "coarse enough"). In the following results, this is tested using linears and quadratics. An interpolation operator is constructed via standard finite-element interpolation, Q_H, which takes DOF from a grid of size H and interpolates it to the fine–grid, h. Thus, the constrained system becomes

$$\begin{pmatrix} A & \tilde{B}^T Q_H \\ Q_H^T \tilde{B} & 0 \end{pmatrix} \begin{pmatrix} \mathscr{U} \\ \lambda_H \end{pmatrix} = \begin{pmatrix} b \\ Q_H^T \tilde{g} \end{pmatrix}. \tag{25}$$

As is seen in Table 5, using $\mathscr{P}_1 - \mathscr{P}_1$ and $\mathscr{P}_2 - \mathscr{P}_2$ pairs yields conservation and still allows the FOSLS functional to be minimized as expected. Thus, the solution, p, and its gradient, $\mathbf{u}$, are approximated well with only a handful of extra iterations needed. Again, if the coarse Lagrange multiplier space were discontinuous, local conservation would also be obtained over the coarse elements.

Table 5 (Alternative approach) solve (25), where $Q_H^T \tilde{B}$ is the "Galerkin" constraint on a coarser mesh

k_1	k_2	h	H	m_L	m_L^c	$\mathscr{F}$	$\mathscr{F}_c$	u_{err}	p_{err}	Iterations
1	1	1/8	1/4	5.3e−2	1.9e−12	1.65	1.70	0.222	0.056	21
1	1	1/16	1/8	2.8e−2	9.0e−12	0.90	0.91	0.132	0.013	27
1	1	1/16	1/4	2.8e−2	3.0e−11	0.90	0.90	0.134	0.015	17
1	1	1/32	1/16	1.0e−2	6.6e−13	0.48	0.48	0.053	0.003	25
1	1	1/32	1/8	1.0e−2	7.1e−12	0.48	0.48	0.053	0.004	17
1	1	1/32	1/4	1.0e−2	4.0e−12	0.48	0.48	0.053	0.004	13
2	2	1/8	1/4	5.5e−4	1.5e−11	0.14	0.15	0.008	0.001	31
2	2	1/16	1/8	3.7e−5	7.0e−13	3.7e−2	3.9e−2	1.10e−3	1.88e−4	28
2	2	1/16	1/4	3.7e−5	8.6e−13	3.7e−2	3.8e−2	1.11e−3	1.39e−4	22
2	2	1/32	1/16	2.4e−6	4.6e−13	9.5e−3	9.9e−3	1.79e−4	3.69e−5	22
2	2	1/32	1/8	2.4e−6	7.5e−14	9.5e−3	9.6e−3	1.79e−4	2.32e−5	17
2	2	1/32	1/4	2.4e−6	9.4e−14	9.5e−3	9.5e−3	1.80e−4	1.83e−5	16

4 Locally Constrained FOSLS Correction

4.1 Overlapping Schwarz Corrections

Now that it has been shown that augmenting the FOSLS system with a constraint gives better mass conservation with only a few extra iterations, a more robust local way of solving the problem is described here. An overlapping Schwarz process, as described in [37] (Sect. 9.5), is considered to break the constrained problem into smaller local problems. First consider that the FOSLS discrete system has been solved. In other words, no constraints are yet imposed. Then, the following post-processing step is performed. Let $\{\Omega_i\}_{i=1}^{N_{sd}}$ be an overlapping partition of Ω into N_{sd} mesh subdomains (i.e., each Ω_i is a union of fine–grid elements). Then, correct the current solution $\mathscr{U}$ with

$$\mathscr{U}_i \in V_h^0(\Omega_i) = \left\{ \mathbf{v} \in V_h : \text{supp}\,(\mathbf{v}_i) \subset \overline{\Omega}_i \right\},$$

by solving the locally constrained minimization problem for $\mathscr{U}_i \in V_h^0(\Omega_i)$ and $\lambda_i \in \mathscr{R}_i = \nabla \cdot V_h^0(\Omega_i)$ posed in Ω_i:

$$\begin{aligned} a(\mathscr{U} + \mathscr{U}_i,\, \mathbf{v}_i) + \langle \lambda_i,\, \nabla \cdot \mathbf{v}_i \rangle &= \langle F,\, \mathbf{v}_i \rangle, \text{ for all } \mathbf{v}_i \in V_h^0(\Omega_i), \\ \langle \nabla \cdot (\mathscr{U} + \mathscr{U}_i),\, \varphi \rangle &= \langle f,\, \varphi \rangle \text{ for } \varphi \in \mathscr{R}_i. \end{aligned}$$

Here, for the local space $\mathscr{R}_i \equiv \nabla \cdot V_h^0(\Omega_i)$, the local systems can be constructed as in Sect. 3.3.1. Likewise, a computational basis, based on QR or SVD, can be obtained as well. This is feasible if the domains Ω_i are relatively small. Next, set $\mathscr{U} := \mathscr{U} + \mathscr{U}_i$ and move onto the next subdomain Ω_{i+1}.

After several loops over the Schwarz subdomains, a global coarse-space correction is performed. For this, a coarse space, $\mathscr{R}_H \subset \nabla \cdot V_h$, is needed with an explicit locally supported basis such that the pair $(V_h, \mathscr{R}_H)$ is LBB-stable (Ladyzenskaya–Babuska–Brezzi condition) [10, 12]. Alternatively, based on a coarse space, $V_H \subset V_h$, and coarser subdomains, $\{\Omega_i^H\}$ (i.e., union of coarse elements in $\mathscr{T}_H$), for the current approximation $\mathscr{U} \in V_h$, local coarse-space corrections, $\mathscr{U}_i^H \in V_H^0(\Omega_i^H) = \{\mathbf{v}_H \in V_H : \text{ supp } (\mathbf{v}_H) \subset \overline{\Omega}_i^H\}$, are obtained by solving the local saddle-point problems for $\mathscr{U}_i^H \in V_H^0(\Omega_i)$ and $\lambda_i^H \in \mathscr{R}_i^H = \nabla \cdot V_H^0(\Omega_i^H)$ posed in Ω_i^H:

$$
\begin{aligned}
a(\mathscr{U} + \mathscr{U}_i^H, \mathbf{v}_i^H) + \langle \lambda_i^H, \nabla \cdot \mathbf{v}_i^H \rangle &= \langle F, \mathbf{v}_i^H \rangle, \text{ for all } \mathbf{v}_i^H \in V_H^0(\Omega_i^H), \\
\langle \nabla \cdot (\mathscr{U} + \mathscr{U}_i^H), \varphi \rangle &= \langle f, \varphi \rangle \text{ for } \varphi \in \mathscr{R}_i^H.
\end{aligned}
$$

Here, the coarse spaces can be constructed in a variational way by using standard interpolation and restriction operators for polynomial finite-element spaces. Finally, let $\mathscr{U} := \mathscr{U} + \mathscr{U}_i^H$ and move onto the next coarse subdomain Ω_{i+1}^H. The process can be applied recursively in a V-cycle iteration exploiting the above constrained overlapping Schwarz (Vanka-like) smoothing corrections [36]. For this paper, however, we consider only a two-level method with one global coarse space.

4.2 Numerical Results

To test the scheme described above in Sect. 4.1, the "Galerkin"-like constrained system (12) is considered on subdomains and a coarse grid. This system gave the most optimal results (fewer iterations and better mass conservation) and, therefore, appears to be the natural choice for performing the subdomain corrections. As described above, the standard FOSLS system is solved yielding, $\mathscr{U}_0$, which is used as the initial guess for the overlapping Schwarz method. Next, the finite-element triangulation of Ω is divided into overlapping subdomains, $\mathscr{T}_i$ of Ω_i. The restriction of the FOSLS system, A, and the constraint equation, $\tilde{B}$, is formed by a simple projection onto the subdomains giving, $A_i = P_i^T A P_i$ and $B_i = Q_i^T \tilde{B} P_i$. Here, P_i and Q_i are the natural injection operators of DOFs on $\mathscr{T}_i$ to the original mesh, $\mathscr{T}$, for elements of $\mathscr{P}_{k_1}$ and $\mathscr{P}_{k_2}$, respectively. Then, on each subdomain the Schur complement system of the error equations is solved as described above in Sect. 4.1. Once all corrections on subdomains are updated, the system is projected onto a coarse grid, $\mathscr{T}_H$, where an update is again solved for. We use the standard finite-element interpolation operators to move between a coarse grid of size H to a fine grid of size h. We define these as P_H for $\mathscr{P}_{k_1}$ and Q_H for $\mathscr{P}_{k_2}$. Note that P_H is a block matrix of interpolation operators for each unknown in the FOSLS system. The transposes are used as restriction operators from fine grid to coarse grid.

The algorithm is described below, letting M_s be the maximum number of subdomain smoothing steps and N_{sd} being the number of overlapping subdomains:

Solve FOSLS System: $A\mathscr{U}_0 = b$.
Compute Residuals: $r_A = b - A\mathscr{U}_0$ and $r_B = \tilde{g} - \tilde{B}\mathscr{U}_0$.
Set $\mathscr{U} = \mathscr{U}_0$ and $\lambda = 0$.
Perform Subdomain Smoothing Steps:
for $s = 1$ **to** M_s **do**
  for $i = 1$ **to** N_{sd} **do**
    Restrict Matrices and Residuals to Subdomains.

$$\text{Solve: } \begin{pmatrix} A_i & B_i^T \\ B_i & 0 \end{pmatrix} \begin{pmatrix} \mathscr{U}_i \\ \lambda_i \end{pmatrix} = \begin{pmatrix} P_i^T r_A \\ Q_i^T r_B \end{pmatrix}.$$

    Update: $\mathscr{U} = \mathscr{U} + P_i \mathscr{U}_i$ and $\lambda = \lambda + Q_i \lambda_i$.
    Recompute Residuals: $r_A = b - A\mathscr{U} - \tilde{B}^T \lambda$ and $r_B = \tilde{g} - \tilde{B}\mathscr{U}_0$.
  end
end
Perform Coarse-Grid Correction:

$$\text{Solve: } \begin{pmatrix} P_H^T A P_H & P_H^T \tilde{B}^T Q_H \\ Q_H^T \tilde{B} P_H & 0 \end{pmatrix} \begin{pmatrix} \mathscr{U}_H \\ \lambda_H \end{pmatrix} = \begin{pmatrix} P_H^T r_A \\ Q_H^T r_B \end{pmatrix}.$$

Update: $\mathscr{U} = \mathscr{U} + P_H \mathscr{U}_H$.

The results for $\mathscr{P}_2 - \mathscr{P}_0$ and $\mathscr{P}_1 - \mathscr{P}_0$ pairs of elements for the FOSLS solution and the constraint variable are given in Table 6 using various grid spacings. The first set of results is given for the original FOSLS system with no constraint correction. The FOSLS functional is reduced by h^{k_1} as expected and it gives a good approximation of the reduction in error for both **u** and p. However, the mass loss is rather large. Using quadratics improves the results but not exactly. The remaining blocks of data give the results using various numbers of smoothing steps and with or without coarse-grid corrections. In all cases, using $\mathscr{P}_2 - \mathscr{P}_0$ elements gives much better results. As seen in Tables 1 and 2, mass conservation is obtained, and the FOSLS functional is still minimized, retaining its error approximation properties. Moreover, using unstable pairs of elements can even result in the divergence of the FOSLS functional. In the context of this problem, the solution is still obtained accurately, but the gradient of the solution is not captured well. The solution process is no longer minimizing the residual in the H^1 norm.

In addition, the results show that the use of a coarse grid improves the performance of the method. The second block in Table 6 shows results for performing one smoothing step of the subdomain solver with no coarse-grid correction. This does improve the conservation results, but not significantly. Performing 100 smoothing steps of the subdomain solver with no coarse-grid correction improves the mass conservation, but of course these iterations are expensive. Finally, the fourth set shows results for using one step of the subdomain solver with one

Table 6 Mass loss, least-squares functional, and relative errors of solutions for $\mathscr{P}_1 - \mathscr{P}_0$ elements (left) and $\mathscr{P}_2 - \mathscr{P}_0$ elements (right)

	$\mathscr{P}_1 - \mathscr{P}_0$				$\mathscr{P}_2 - \mathscr{P}_0$			
	m_L	$\mathscr{F}$	u_{err}	p_{err}	m_L	$\mathscr{F}$	u_{err}	p_{err}
$1/h$	FOSLS							
8	5.3e−2	1.65	0.223	0.070	5.5e−4	0.14	8.3e−3	1.2e−3
16	2.8e−2	0.90	0.134	0.019	3.7e−5	0.04	1.1e−3	1.4e−4
32	1.0e−2	0.48	0.053	0.005	2.4e−6	0.01	1.8e−4	1.7e−5
$1/h$	$N_{sd} = 9$, $M_s = 1$, No coarse-grid correction							
8	2.6e−3	1.77	0.180	0.061	2.3e−6	0.14	8.4e−3	1.2e−3
16	1.3e−3	1.09	0.185	0.016	1.0e−7	0.04	1.2e−3	1.4e−4
32	2.9e−5	1.47	0.201	0.004	9.2e−9	0.01	1.8e−4	1.7e−5
$1/h$	$N_{sd} = 9$, $M_s = 100$, No coarse-grid correction							
8	4.2e−12	1.83	0.184	0.059	1.0e−11	0.14	8.4e−3	1.2e−3
16	4.7e−8	1.12	0.181	0.015	9.1e−11	0.04	1.2e−3	1.4e−4
32	2.2e−1	7.81	0.413	0.005	4.5e−11	0.01	1.8e−4	1.7e−5
$1/h$	$N_{sd} = 9$, $M_s = 1$, $H = 2h$							
8	8.8e−11	1.83	0.181	0.060	2.3e−13	0.14	8.3e−3	1.2e−3
16	1.3e−12	1.92	0.188	0.015	1.2e−13	0.04	1.2e−3	1.4e−4
32	1.1e−10	10.09	0.209	0.004	2.5e−14	0.01	1.8e−4	1.7e−5
$1/h$	$N_{sd} = 9$, $M_s = 1$, $H = 4h$							
8	5.3e−3	1.84	0.191	0.060	1.2e−6	0.14	8.3e−3	1.2e−3
16	1.2e−3	1.20	0.186	0.016	1.8e−8	0.04	1.2e−3	1.4e−4
32	2.8e−4	2.63	0.200	0.004	3.0e−9	0.01	1.8e−4	1.7e−5
$1/h$	$N_{sd} = 9$, $M_s = 10$, $H = 4h$							
8	3.9e−4	1.83	0.195	0.060	1.1e−7	0.14	8.3e−3	1.2e−3
16	4.0e−3	1.29	0.192	0.015	7.8e−9	0.04	1.2e−3	1.4e−4
32	4.3e−2	9.93	0.363	0.007	6.7e−10	0.01	1.8e−4	1.7e−5

solve on a coarse grid. The mass conservation is retained and not much work is needed. Combining with the results from Table 1, this process requires around four iterations of MINRES for each local subdomain and for the coarse grid. Each of these subdomains has less DOFs, and therefore, the work required to solve the constrained system is a fraction of the cost of solving the original FOSLS system.

5 Conclusions

In summary, the results of this paper have shown that properties such as mass conservation can be obtained using the least-squares finite-element method and a post-process subdomain correction method. There are many other methods, as mentioned in the Introduction (Sect. 1), that also improve conservation properties for least-squares problems. These may involve reformulating the system or choosing better finite-element spaces for the original FOSLS system. For instance, nonconforming elements can be used that satisfy the mass conservation across interfaces much better than the standard polynomial spaces used here [1, 17, 18, 25]. The goal of our approach in this paper is to show that the system can be solved as is, with no alterations to the original FOSLS method. Thus, it should be considered a robust finite-element method for such systems which obtains physically accurate solutions efficiently. Care needs to be given in choosing the right spaces for the constraint system, so that a stable method is obtained and the FOSLS functional retains its important a posteriori error estimator properties. This includes considering discontinuous spaces, in order to ensure *local* conservation across smaller regions of the domain. However, since this post-processing is done on local subdomains and/or on coarse grids, only a fractional amount of computational cost is added to the solution process. Future work involves implementing the above algorithms in a multilevel way and including the coarse-space constraints in the local subdomain process. Also, other applications such as Stokes flow and magnetohydrodynamics are worth considering.

References

1. Baker, G.A., Jureidini, W.N., Karakashian, O.A.: Piecewise solenoidal vector-fields and the Stokes problem. SIAM J. Numer. Anal. **27**(6), 1466–1485 (1990)
2. Barth, T.: On the role of involutions in the discontinuous Galerkin discretization of Maxwell and magnetohydrodynamic systems. In: Arnold, D.N., Bochev, P.B., Lehoucq, R.B., Nicolaides, R.A., Shashkov, M. (eds.) Compatible Spatial Discretizations, pp. 69–88. Springer, New York (2006)
3. Berndt, M., Manteuffel, T.A., McCormick, S.F.: Local error estimates and adaptive refinement for first-order system least squares (FOSLS). Electron. Trans. Numer. Anal. **6**, 35–43 (1997)
4. Bochev, P., Gunzburger, M.D.: Analysis of least-squares finite-element methods for the Stokes equations. Math. Comput. **63**(208), 479–506 (1994)
5. Bochev, P.B., Gunzburger, M.D.: A locally conservative least-squares method for Darcy flows. Commun. Numer. Meth. Eng. **24**(2), 97–110 (2008)
6. Bochev, P., Cai, Z., Manteuffel, T.A., McCormick, S.F.: Analysis of velocity-flux first-order system least-squares principles for the Navier-Stokes equations: part I. SIAM J. Numer. Anal. **35**(3), 990–1009 (1998)
7. Bochev, P., Manteuffel, T.A., McCormick, S.F.: Analysis of velocity-flux least-squares principles for the Navier-Stokes equations: part II. SIAM J. Numer. Anal. **36**(4), 1125–1144 (1999)
8. Brackbill, J., Barnes, D.C.: The effect of nonzero $\nabla \cdot \mathbf{B}$ on the numerical solution of the magnetohydrodynamic equations. J. Comput. Phys. **35**(3), 426–430 (1980)
9. Bramble, J., Kolev, T., Pasciak, J.: A least-squares approximation method for the time-harmonic Maxwell equations. J. Numer. Math. **13**(4), 237 (2005)
10. Brenner, S.C., Scott, L.R.: Mathematical Theory of Finite Element Methods, 2nd edn. Springer, New York (2002)
11. Brezina, M., Garcia, J., Manteuffel, T., McCormick, S., Ruge, J., Tang, L.: Parallel adaptive mesh refinement for first-order system least squares. Numer. Lin. Algebra Appl. **19**, 343–366 (2012)
12. Brezzi, F., Fortin, M.: Mixed and Hybrid Finite Elements Methods. Springer Series in Computational Mathematics. Springer, Berlin (1991)

13. Cai, Z., Starke, G.: Least-squares methods for linear elasticity. SIAM J. Numer. Anal. **42**(2), 826–842 (electronic) (2004). doi:10.1137/S0036142902418357. http://dx.doi.org.ezproxy.library.tufts.edu/10.1137/S0036142902418357
14. Cai, Z., Lazarov, R., Manteuffel, T.A., McCormick, S.F.: First-order system least squares for second-order partial differential equations: part I. SIAM J. Numer. Anal. **31**, 1785–1799 (1994)
15. Cai, Z., Manteuffel, T.A., McCormick, S.F.: First-order system least squares for second-order partial differential equations 2. SIAM J. Numer. Anal. **34**(2), 425–454 (1997)
16. Carey, G.F., Pehlivanov, A.I., Vassilevski, P.S.: Least-squares mixed finite element methods for non-selfadjoint elliptic problems II: performance of block-ILU factorization methods. SIAM J. Sci. Comput. **16**, 1126–1136 (1995)
17. Cockburn, B., Kanschat, G., Schotzau, D.: A locally conservative LDG method for the incompressible Navier-Stokes equations. Math. Comput. **74**(251), 1067–1095 (2005)
18. Cockburn, B., Kanschat, G., Schötzau, D.: A note on discontinuous Galerkin divergence-free solutions of the Navier-Stokes equations. J. Sci. Comput. **31**, 61–73 (2007)
19. De Sterck, H., Manteuffel, T., McCormick, S., Nolting, J., Ruge, J., Tang, L.: Efficiency-based h- and hp-refinement strategies for finite element methods. Numer. Lin. Algebra Appl. **15**, 89–114 (2008)
20. Girault, V., Raviart, P.A.: Finite Element Approximation of the Navier-Stokes Equations, revised edn. Springer, Berlin (1979)
21. Girault, V., Raviart, P.A.: Finite Element Methods for Navier-Stokes Equations: Theory and Algorithms (Springer Series in Computational Mathematics). Springer, Berlin (1986)
22. Heys, J.J., Lee, E., Manteuffel, T.A., McCormick, S.F.: On mass-conserving least-squares methods. SIAM J. Sci. Comput. **28**(5), 1675–1693 (2006)
23. Heys, J.J., Lee, E., Manteuffel, T.A., McCormick, S.F.: An alternative least-squares formulation of the Navier-Stokes equations with improved mass conservation. J. Comput. Phys. **226**(1), 994–1006 (2007)
24. Heys, J.J., Lee, E., Manteuffel, T.A., McCormick, S.F., Ruge, J.W.: Enhanced mass conservation in least-squares methods for Navier-Stokes equations. SIAM J. Sci. Comput. **31**(3), 2303–2321 (2009)
25. Karakashian, O., Jureidini, W.: A nonconforming finite element method for the stationary Navier-Stokes equations. SIAM J. Numer. Anal. **35**(1), 93–120 (1998)
26. Lax, P.D.: Functional analysis. Wiley-Interscience, New York (2002)
27. Lee, E., Manteuffel, T.A.: FOSLL* method for the eddy current problem with three-dimensional edge singularities. SIAM J. Numer. Anal. **45**(2), 787–809 (2007)
28. Manteuffel, T.A., McCormick, S.F., Ruge, J., Schmidt, J.G.: First-order system LL* (FOSLL*) for general scalar elliptic problems in the plane. SIAM J. Numer. Anal. **43**(5), 2098–2120 (2006)
29. Münzenmaier, S., Starke, G.: First-order system least squares for coupled Stokes-Darcy flow. SIAM J. Numer. Anal. **49**(1), 387–404 (2011). doi:10.1137/100805108. http://dx.doi.org.ezproxy.library.tufts.edu/10.1137/100805108
30. Pehlivanov, A.I., Carey, G.F.: Error-estimates for least-squares mixed finite-elements. ESAIM Math. Model. Numer. Anal. **28**(5), 499–516 (1994)
31. Pehlivanov, A.I., Carey, G.F., Vassilevski, P.S.: Least-squares mixed finite element methods for non-selfadjoint elliptic problems I: error analysis. Numer. Math. **72**, 501–522 (1996)
32. Pontaza, J.P., Reddy, J.N.: Space-time coupled spectral/hp least-squares finite element formulation for the incompressible Navier-Stokes equations. J. Comput. Phys. **197**(2), 418–459 (2004)
33. Saad, Y.: Iterative Methods for Sparse Linear Systems. Society for Industrial & Applied Mathematics, Philadelphia (2003)
34. Starke, G.: A first-order system least squares finite element method for the shallow water equations. SIAM J. Numer. Anal. **42**(6), 2387–2407 (electronic) (2005). doi:10.1137/S0036142903438124. http://dx.doi.org.ezproxy.library.tufts.edu/10.1137/S0036142903438124

35. Toselli, A., Widlund, O.B.: Domain Decomposition Methods—Algorithms and Theory. Springer, New York (2005)
36. Vanka, S.P.: Block-implicit multigrid solution of Navier-Stokes equations in primitive variables. J. Comput. Phys. **65**(1), 138–158 (1986)
37. Vassilevski, P.S.: Multilevel Block Factorization Preconditioners. Matrix-based Analysis and Algorithms for Solving Finite Element Equations. Springer, New York (2008)

Multiscale Coarsening for Linear Elasticity by Energy Minimization

Marco Buck, Oleg Iliev, and Heiko Andrä

Abstract In this work, we construct energy-minimizing coarse spaces for the finite element discretization of mixed boundary value problems for displacements in compressible linear elasticity. Motivated from the multiscale analysis of highly heterogeneous composite materials, basis functions on a triangular coarse mesh are constructed, obeying a minimal energy property subject to global pointwise constraints. These constraints allow that the coarse space exactly contains the rigid body translations, while rigid body rotations are preserved approximately. The application is twofold. Resolving the heterogeneities on the finest scale, we utilize the energy-minimizing coarse space for the construction of robust two-level overlapping domain decomposition preconditioners. Thereby, we do not assume that coefficient jumps are resolved by the coarse grid, nor do we impose assumptions on the alignment of material jumps and the coarse triangulation. We only assume that the size of the inclusions is small compared to the coarse mesh diameter. Our numerical tests show uniform convergence rates independent of the contrast in the Young's modulus within the heterogeneous material. Furthermore, we numerically observe the properties of the energy-minimizing coarse space in an upscaling framework. Therefore, we present numerical results showing the approximation errors of the energy-minimizing coarse space w.r.t. the fine-scale solution.

Keywords Linear elasticity • Domain decomposition • Robust coarse spaces • Energy-minimizing shape functions

Mathematics Subject Classification (2010): 35R05, 65F10, 65F10, 65N22, 65N55, 74B05, 74S05

M. Buck (✉) • O. Iliev • H. Andrä
Fraunhofer Institute for Industrial Mathematics (ITWM), Fraunhofer-Platz 1, 67663 Kaiserslautern, Germany
e-mail: marco.buck@itwm.fraunhofer.de; oleg.iliev@itwm.fraunhofer.de; heiko.andrae@itwm.fraunhofer.de

O.P. Iliev et al. (eds.), *Numerical Solution of Partial Differential Equations: Theory, Algorithms, and Their Applications*, Springer Proceedings in Mathematics & Statistics 45, DOI 10.1007/978-1-4614-7172-1_2,

1 Introduction

Constantly rising demands on the range of application of today's industrial products require the development of innovative, highly effective composite materials, specifically adapted to their field of application. Virtual material design provides essential support in the development process of new materials as it substantially reduces costs and time for the construction of prototypes and performing measurements on their properties. Of special interest is the multiscale analysis of particle-reinforced composites. They combine positive features of their components such as light weight and high stiffness.

Due to large variations in the material parameters, the linear system arising from the finite element discretization of the linear elasticity PDE on such heterogeneous materials is in general very ill-conditioned. Our goal is to develop two-level domain decomposition preconditioners which are robust w.r.t. the jumps in the material coefficients of the PDE. Two-level overlapping domain decomposition preconditioners for the equations of linear elasticity are presented in several papers [9, 19, 22]. Under certain conditions on the alignment of the material jumps with the coarse grid, the aggregation-based method in [19] (see also [26] in the context of AMG) promises mesh and coefficient independent condition number bounds. These methods might not be fully robust when variations in the coefficients appear on a very small scale where the coefficients cannot be resolved by a coarse mesh. A more recent approach in [23] guarantees robustness w.r.t. arbitrary coefficient variations by solving generalized eigenvalue problems in the overlapping regions of the coarse basis functions. The dimension of the resulting coarse space strongly depends on the coefficient distribution. This approach is a variation of the method in [7, 30], where it is applied to abstract symmetric positive definite operators in a multiscale framework.

Further robust methods for solving linear elasticity problems include multilevel methods studied in [14] and further developed in [11] and [12]. A purely algebraic multigrid method for linear elasticity problems is constructed, based on computational molecules, a new variant of AMGe [3]. Such an approach has been studied earlier for scalar elliptic PDEs in [15]. Classical AMG methods for linear elasticity problems are presented in [1, 5] and the references therein.

In this paper, we construct coarse basis functions with a minimal energy property subject to the constraints that the coarse space exactly contains the rigid body translations, while the rigid body rotations are preserved approximately. Energy-minimizing methods have been proposed in [29] and [16] and were further studied in [25, 31]. In [17], such an approach is generalized and applied to non-Hermitian matrices. The approach was motivated in [29] from experimental results of one-dimensional problems. It is based on improving the approximation properties of the coarse space by reducing its dependence on the PDE coefficients. In [25], energy-minimizing coarse spaces were motivated from developments in the convergence theory for two-level Schwarz methods of scalar elliptic PDEs in [8]. In [16], energy-minimizing coarse spaces are presented also for isotropic linear elasticity, in the context of smoothed aggregation. The novel part in the paper at hand is the application to the multiscale framework. The construction on a coarse tetrahedral mesh allows large overlaps in the supports of the basis functions and the coarse space promises good upscaling properties.

An interesting method proposed in [20] constructs basis functions by minimizing their energy subject to a set of functional rather than pointwise constraints. This approach is applied to scalar elliptic PDEs. Similar to the method in [7], the objective is to prove the approximation property in a weighted Poincaré inequality. By a proper choice of the functional constraints, mesh and coefficient independent convergence rates can be obtained. Further variants of coarse spaces with a minimal energy property, including local variants, can be found in [6, 10, 13, 28].

The outline of the paper is as follows. We proceed with the continuous formulation of the governing PDE system and the discretization on the fine grid in Sect. 2. In Sect. 3 we shortly recapitulate the two-level additive Schwarz method, followed by introducing the precise structure of the underlying fine and coarse grid in three spatial dimensions. In Sect. 4, we present a detailed construction of the energy-minimizing basis. Section 5 is devoted to numerical results, a short discussion follows in Sect. 6.

2 Governing Equations and Their Discretization

2.1 The Equations of Linear Elasticity

For the sake of simplicity, let $\Omega \subset \mathbb{R}^3$ be a Lipschitz domain. We shall assume that $\Gamma = \partial\Omega$ admits the decomposition into two disjoint subsets Γ_{D_i} and Γ_{N_i}, $\Gamma = \overline{\Gamma_{D_i} \cup \Gamma_{N_i}}$ and $\operatorname{meas}(\Gamma_{D_i}) > 0$ for $i \in \{1,2,3\}$. We consider a solid body in Ω, deformed under the influence of volume forces $\boldsymbol{f}$ and traction forces $\boldsymbol{t}$. Assuming a linear elastic material behavior, the displacement field $\boldsymbol{u}$ of the body is governed by the mixed b.v.p. [2]

$$-\operatorname{div} \boldsymbol{\sigma}(\boldsymbol{u}) = \boldsymbol{f} \ \text{ in } \Omega, \tag{1}$$

$$\boldsymbol{\sigma}(\boldsymbol{u}) = \boldsymbol{C} : \boldsymbol{\varepsilon}(\boldsymbol{u}) \ \text{ in } \Omega, \tag{2}$$

$$u_i = g_i \ \text{ on } \Gamma_{D_i}, \quad i = 1,2,3,$$

$$\sigma_{ij} n_j = t_i \ \text{ on } \Gamma_{N_i}, \quad i = 1,2,3,$$

where $\boldsymbol{\sigma}$ is the stress tensor, the strain tensor $\boldsymbol{\varepsilon}$ is given by the symmetric part of the deformation gradient,

$$\boldsymbol{\varepsilon}(\boldsymbol{u}) = \frac{1}{2}\left(\nabla \boldsymbol{u} + \nabla \boldsymbol{u}^T\right)$$

and $\boldsymbol{n}$ is the unit outer normal vector on Γ and $\sigma_{ij} n_j = (\boldsymbol{\sigma} \cdot \boldsymbol{n})_i$. The fourth-order elasticity tensor $\boldsymbol{C} = \boldsymbol{C}(x), x \in \Omega$ describes the elastic stiffness of the material under mechanical load. The coefficients $c_{ijkl}, 1 \leq i,j,k,l \leq 3$ may contain large jumps within the domain Ω. They depend on the parameters of the particular materials which are enclosed in the composite. The boundary conditions are imposed separately for each component $u_i, i = 1,2,3$ of the vector-field $\boldsymbol{u} = (u_1,u_2,u_3)^T : \bar{\Omega} \to \mathbb{R}^3$.

Equation (1) is the general form of the PDE system for anisotropic linear elasticity, which simplifies when the solid body consists of one or more isotropic materials. In this case, (2) can be expressed in terms of the Lamé coefficients $\lambda \in \mathbb{R}$ and $\mu > 0$, which are characteristic constants of the specific material. The stiffness tensor of an isotropic material is given by $c_{ijkl} = \lambda \delta_{ij}\delta_{kl} + \mu(\delta_{ik}\delta_{jl} + \delta_{il}\delta_{jk})$, and the stress is $\boldsymbol{\sigma}(\boldsymbol{u}) = \lambda \mathrm{tr}(\boldsymbol{\varepsilon}(\boldsymbol{u}))\boldsymbol{I} + 2\mu\boldsymbol{\varepsilon}(\boldsymbol{u})$.

2.2 Weak Formulation

Consider the Sobolev space $\mathscr{V} := [H^1(\Omega)]^3$ of vector-valued functions whose components are square-integrable with weak first-order partial derivatives in the Lebesgue space $L^2(\Omega)$. We define the subspace $\mathscr{V}_0 \subset \mathscr{V}$,

$$\mathscr{V}_0 := \left\{\boldsymbol{v} \in [H^1(\Omega)]^3 : v_i = 0 \text{ on } \Gamma_{D_i}, i = 1,2,3\right\}. \tag{3}$$

Additionally, we define the manifold

$$\mathscr{V}_g := \left\{\boldsymbol{v} \in [H^1(\Omega)]^3 : v_i = g_i \text{ on } \Gamma_{D_i}, i = 1,2,3\right\}. \tag{4}$$

The Sobolev space $\mathscr{V}$ inherits its scalar product from $H^1(\Omega)$; it is given by

$$(\boldsymbol{u},\boldsymbol{v})_{[H^1(\Omega)]^3} := \sum_{i=1}^{3} (u_i,v_i)_{H^1(\Omega)}.$$

We assume $\boldsymbol{f} \in \mathscr{V}_0'$ to be in the dual space of $\mathscr{V}_0$, $\boldsymbol{t} \in [H^{-\frac{1}{2}}(\Gamma_N)]^3$ is in the trace space, and $c_{ijkl} \in L^\infty(\Omega)$ to be uniformly bounded. Additionally, we require the stiffness tensor $\boldsymbol{C}$ to be positive definite, i.e., it holds $(\boldsymbol{C} : \boldsymbol{\varepsilon}(\boldsymbol{v})) : \boldsymbol{\varepsilon}(\boldsymbol{v}) \geq C_0\, \boldsymbol{\varepsilon}(\boldsymbol{v}) : \boldsymbol{\varepsilon}(\boldsymbol{v})$ for a constant $C_0 > 0$. Note that for an isotropic material with the parameters λ and μ, this condition holds when $C_0/2 < \mu < \infty$ and $C_0 \leq 2\mu + 3\lambda < \infty$. We define the bilinear form $a : \mathscr{V} \times \mathscr{V} \to \mathbb{R}$,

$$a(\boldsymbol{u},\boldsymbol{v}) := \int_\Omega (\boldsymbol{C} : \boldsymbol{\varepsilon}(\boldsymbol{u})) : \boldsymbol{\varepsilon}(\boldsymbol{v})\, dx. \tag{5}$$

This form is symmetric, continuous, and coercive. The coercivity, i.e.,

$$\exists\, c_0 > 0\ :\ a(\boldsymbol{v},\boldsymbol{v}) \geq c_0 \left\|\boldsymbol{v}\right\|_{[H^1(\Omega)]^3}\ \forall \boldsymbol{v} \in \mathscr{V}_0,$$

can be shown by using Korn's inequality (cf. [2]). Furthermore, we define the continuous linear form $F : \mathscr{V} \to \mathbb{R}$,

$$F(\boldsymbol{v}) := \int_{\Omega} \boldsymbol{f} \cdot \boldsymbol{v} dx + \int_{\Gamma_N} \boldsymbol{t} \cdot \boldsymbol{v} ds.$$

The weak solution of (1) is then given in terms of $a(\cdot,\cdot)$ and $F(\cdot)$ by $\boldsymbol{u} \in \mathscr{V}_g$, such that

$$a(\boldsymbol{u},\boldsymbol{v}) = F(\boldsymbol{v}) \ \forall \boldsymbol{v} \in \mathscr{V}_0. \tag{6}$$

Under the assumptions above, a unique solution of the weak formulation in (6) is guaranteed by the Lax–Milgram lemma [2].

2.3 Finite Element Discretization

We want to approximate the solution of (6) in a finite dimensional subspace $\mathscr{V}^h \subset \mathscr{V}$. Therefore, let $\mathscr{T}_h$ be a quasi-uniform triangulation of $\boldsymbol{\Omega} \subset \mathbb{R}^3$ into tetrahedral finite elements with mesh parameter h, and let $\bar{\Sigma}_h$ be the set of vertices of $\mathscr{T}_h$ contained in $\bar{\boldsymbol{\Omega}}$. Furthermore, let $\bar{\mathscr{N}_h}$ denote the corresponding index set of nodes in $\bar{\Sigma}_h$. We denote the number of grid points in $\bar{\Sigma}_h$ by n_p. In Sect. 3, the regular grid and its triangulation are introduced in more detail. Let

$$\mathscr{V}^h := \mathrm{span}\{\varphi_k^{j,\mathrm{h}} : \bar{\boldsymbol{\Omega}} \to \mathbb{R}^3, j \in \bar{\mathscr{N}_h}, k = 1,2,3\}$$

be the space of continuous piecewise linear vector-valued functions on $\mathscr{T}_h$. Each such basis function is of the form

$$\varphi_k^{j,\mathrm{h}} = (\varphi_{k1}^{j,\mathrm{h}}, \varphi_{k2}^{j,\mathrm{h}}, \varphi_{k3}^{j,\mathrm{h}})^T, \quad \varphi_{kl}^{j,\mathrm{h}}(x^i) = \delta_{ij}\delta_{kl}, \ x^i \in \bar{\Sigma}_h, \ l \in \{1,2,3\},$$

where δ_{ij} denotes the Kronecker delta. For the sake of simplifying the notation, we assume a fixed numbering of the basis functions to be given. To be more specific, we assume that there exists a suitable surjective mapping $\{\varphi_k^{j,\mathrm{h}}\} \to \{1,\ldots,n_d\}$, $\varphi_k^{j,\mathrm{h}} \mapsto (j,k)$. Here, $n_d = 3n_p$ denotes the total number of degrees of freedom (DOFs) of $\mathscr{V}^h$. Note that this mapping automatically introduces a renumbering from $\{1,\ldots,n_p\} \times \{1,2,3\} \to \{1,\ldots,n_d\}$. We introduce the discrete analogies to the space in (3) and the manifold in (4) by

$$\mathscr{V}_0^h := \left\{\boldsymbol{v}^h \in \mathscr{V}^h : v_i^h = 0 \text{ on } \Gamma_{D_i}, i = 1,2,3\right\}, \tag{7}$$

$$\mathscr{V}_g^h := \left\{\boldsymbol{v}^h \in \mathscr{V}^h : v_i^h = g_i \text{ on } \Gamma_{D_i}, i = 1,2,3\right\}. \tag{8}$$

We want to find $\boldsymbol{u}^h \in \mathscr{V}_g^h$, where $\boldsymbol{u}^h = \boldsymbol{w}^h + \boldsymbol{g}^h$, with $\boldsymbol{w}^h \in \mathscr{V}_0^h$ and $\boldsymbol{g}^h \in \mathscr{V}_g^h$. More precisely, we seek $\boldsymbol{u}^h = (u_1^h, u_2^h, u_3^h)^T$ with

$$u_k^h = \sum_{j=1}^{n_p} \mathrm{u}_{(j,k)} \varphi_k^{j,h}, \; k = 1,2,3,$$

such that

$$a(\boldsymbol{w}^h, \boldsymbol{v}^h) = F(\boldsymbol{v}^h) - a(\boldsymbol{g}^h, \boldsymbol{v}^h) \;\; \forall \boldsymbol{v}^h \in \mathscr{V}_0^h.$$

We define the index set of DOFs of $\mathscr{V}^h$ by $\mathscr{D}^h = \{1, \ldots, n_d\}$ and introduce the subset

$$\mathscr{D}_0^h := \left\{(i,k) \in \mathbb{N} : i \in \bar{\mathscr{N}_h}, x^i \notin \Gamma_{D_k}\right\}.$$

Furthermore, we may introduce $\mathscr{D}_{\Gamma_D}^h := \mathscr{D}^h \backslash \mathscr{D}_0^h \neq \emptyset$. The bilinear form in (5) applied to the basis functions of $\mathscr{V}^h$ reads

$$a(\varphi_m^{i,h}, \varphi_k^{j,h}) = \int_\Omega \boldsymbol{\varepsilon}(\varphi_m^{i,h}) : \boldsymbol{C} : \boldsymbol{\varepsilon}(\varphi_k^{j,h}) \, \mathrm{d}x. \tag{9}$$

We define $A \in \mathbb{R}^{n_d \times n_d}$, $f \in \mathbb{R}^{n_d}$ by

$$A_{(i,m)(j,k)} = \begin{cases} a(\varphi_m^{i,h}, \varphi_k^{j,h}) & \text{if } (i,m) \in \mathscr{D}_0^h, (j,k) \in \mathscr{D}_0^h, \\ a(\varphi_m^{i,h}, \varphi_k^{j,h}) & \text{if } (i,m) = (j,k) \in \mathscr{D}_{\Gamma_D}^h, \\ 0 & \text{otherwise} \end{cases}$$

and

$$f_{(j,k)} = \begin{cases} F(\varphi_k^{j,h}) - \sum_{(i,m) \in \mathscr{D}_{\Gamma_D}^h} a(\varphi_m^{i,h}, \varphi_k^{j,h}) g_m(x^i) & \text{if } (j,k) \in \mathscr{D}_0^h, \\ F(\varphi_k^{j,h}) = a(\varphi_k^{j,h}, \varphi_k^{j,h}) g_k(x^j) & \text{if } (j,k) \in \mathscr{D}_{\Gamma_D}^h. \end{cases}$$

Observe that common supports of basis functions $\varphi_m^{i,h}$ and $\varphi_k^{j,h}$ with $(i,m) \in \mathscr{D}_0^h$, $(j,k) \in \mathscr{D}_{\Gamma_D}^h$ do not have a contribution to the entries in A. They only contribute to the loadvector $\mathbf{f}$. This leads to the sparse linear system

$$\mathbf{A}\mathbf{u} = \mathbf{f} \tag{10}$$

with the symmetric positive definite (spd) stiffness matrix A. The symmetry of A is inherited from the symmetry of $a(\cdot,\cdot)$, while the positive definiteness is a direct consequence of the coercivity of the bilinear form. Note that in the construction above, the essential DOFs in $\mathscr{D}_{\Gamma_D}^h$ are not eliminated from the linear system. Degrees of freedom related to Dirichlet boundary values are contained in A by strictly imposing $u_i^h = g_i^h$ on $\Gamma_{D_i}, i \in \{1,2,3\}$, i.e., any row in A related to a Dirichlet DOFs contains only a nonzero entry on the diagonal. The remaining Dirichlet DOFs in the columns of A vanish as they are transferred to the right-hand side in (10).

3 The Two-Level Method

We are interested in solving the linear system (10) iteratively and the construction of preconditioners for A which remove the ill-conditioning due to (i) mesh parameters and (ii) variations in the PDE coefficients. Such preconditioners involve corrections on local subdomains as well as a global solve on a coarse grid. Specifically, we apply the two-level additive Schwarz preconditioner, which we shortly recapitulate in this section. Furthermore, we precisely introduce the fine and coarse triangulation on a structured grid. The structure is such that the coarse elements can be formed by an agglomeration of fine elements.

3.1 Two-Level Additive Schwarz

Let $\{\Omega_i, i=1,\dots,N\}$ be an overlapping covering of $\bar{\Omega}$, such that $\Omega_i \setminus \partial\Omega$ is open for $i \in \{1,\dots,N\}$. $\Omega_i \setminus \partial\Omega$ is assumed to consist of the interior of a union of fine elements $\tau \in \mathscr{T}_h$. The part of Ω_i which is overlapped with its neighbors should be of uniform width $\delta_i > 0$. We define the local submatrices of A corresponding to the subdomains $\Omega_i \subset \bar{\Omega}$ by $A_i = R_i A R_i^T$. Roughly speaking, R_i is the restriction matrix of a vector defined in Ω to Ω_i (more details can be found in [24]).

Additionally to the local subdomains, we need a coarse triangulation $\mathscr{T}_H$ of $\bar{\Omega}$ into coarse elements. Here, we assume again that each coarse element T consists of a union of fine elements $\tau \in \mathscr{T}_h$ of the fine triangulation. We will construct a coarse basis whose values are determined on the coarse grid points in $\bar{\Omega}$ (excluding coarse DOFs on the Dirichlet boundaries), given by the vertices of the coarse elements in $\mathscr{T}_H$. The coarse space $\mathscr{V}_0^H \subset \mathscr{V}_0^h$ is constructed such that it is a subspace of the vector-field of piecewise linear basis functions on the fine grid. That is, each function $\phi^H \in \mathscr{V}_0^H$ omits a complete representation w.r.t. the fine-scale basis. The *restriction matrix* R_H describes a mapping from the coarse to the fine space and contains the corresponding coefficient vectors of the coarse basis functions by row. The coarse grid stiffness matrix is then defined as the Galerkin product $A_H := R_H A R_H^T$. With these tools in hand, the action of the two-level additive Schwarz preconditioner M_{AS}^{-1} is defined implicitly by

$$M_{\mathrm{AS}}^{-1} = R_H^T A_H^{-1} R_H + \sum_{i=1}^{N} R_i^T A_i^{-1} R_i.$$

In the following, we write A_0 and R_0 instead of A_H and R_H. The following two theorems are basic results in domain decomposition theory. Proofs can be found in [24]. Theorem 1 also states a reasonable assumption on the choice of the overlapping subdomains.

Theorem 1 (Finite Covering). *The set of overlapping subspaces* $\{\Omega_i, i=1,\dots,N\}$ *can be colored by* $N_C \leq N$ *different colors such that if two subspaces* Ω_i *and* Ω_j *have the same color, it holds* $\Omega_i \cap \Omega_j = \emptyset$. *For the smallest possible number* N_C, *the largest eigenvalue of the two-level preconditioned Schwarz linear system is bounded by*

$$\lambda_{\max}(M_{\mathrm{AS}}^{-1}A) \leq N_C + 1$$

Theorem 2 (Stable Decomposition). *Suppose there exists a number* $C_1 \geq 1$, *such that for every* $\boldsymbol{u}^h \in \mathcal{V}_0^h$, *there exists a decomposition* $\boldsymbol{u}^h = \sum_{i=0}^N \boldsymbol{u}^i$ *with* $\boldsymbol{u}^0 \in \mathcal{V}_0^H$ *and* $\boldsymbol{u}^i \in \mathcal{V}^h(\Omega_i)$, $i=1,\dots,N$ *such that*

$$\sum_{i=0}^{N} a(\boldsymbol{u}^i, \boldsymbol{u}^i) \leq C_1^2\, a(\boldsymbol{u}^h, \boldsymbol{u}^h).$$

Then, it holds

$$\lambda_{\min}(M_{\mathrm{AS}}^{-1}A) \geq C_1^{-2}.$$

As we can see, the choice of the coarse space has no influence on the estimate of the largest eigenvalue of the preconditioned system. However, it is crucial for obtaining a small constant C_1 in the estimate of the smallest eigenvalue in Theorem 2. We continue with introducing the structured fine and coarse grid.

3.2 *Fine and Coarse Triangulation*

The Fine Grid

Let the domain Ω be a 3D cube, i.e., $\bar{\Omega} = [0, L_x] \times [0, L_y] \times [0, L_z] \subset \mathbb{R}^3$ for given $L_x, L_y, L_z > 0$. The fine grid is constructed from an initial voxel structure which is further decomposed into tetrahedral finite elements [21]. More precisely, the set of grid points in $\bar{\Omega}$ is given by

$$\bar{\Sigma}_h := \big\{(x_i, y_j, z_k)^T \,:\, x_i = ih_x, y_j = jh_y, z_k = kh_z, \qquad (11)$$
$$i=0,\dots,n_x,\ j=0,\dots,n_y,\ k=0,\dots,n_z\big\}$$

where $n_x = L_x/h_x,$, $n_y = L_y/h_y$, $n_z = L_z/h_z$. For simplicity, we may assume that $L := L_x = L_y = L_z$ and $h := h_x = h_y = h_z$, and thus $n_h := n_x = n_y = n_z$. That is, the fine grid can be decomposed into $n_h \times n_h \times n_h$ grid blocks of size $h \times h \times h$. We denote such a fine grid block by $\square_h^{ijk}$, $1 \leq i,j,k \leq n_h$. The triple (i,j,k) uniquely determines the position of the corresponding block in $\bar{\Omega}$. Each block is further decomposed into 5 tetrahedral elements. The decomposition depends on the position of the specific grid block. To identify them, we introduce the notation $s^{ijk} := s(\square_h^{ijk}) = i+j+k$. We distinguish between two different decompositions, depending on the value of

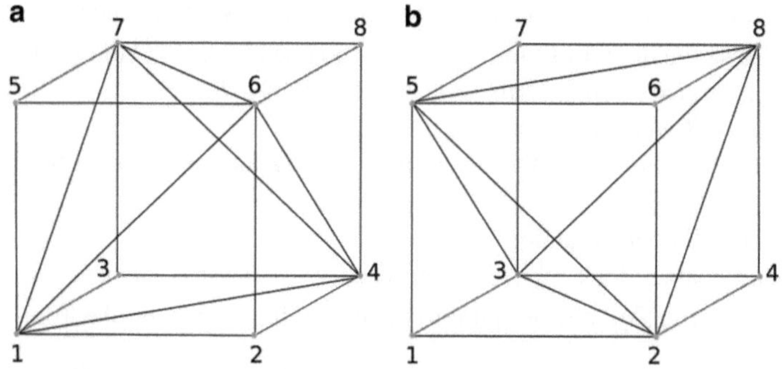

Fig. 1 Decomposition of grid block into 5 tetrahedral elements

s^{ijk} mod 2. We follow the numbering of the 8 vertices of a block as given in Fig. 1. If s^{ijk} is even (see Fig. 1a), block $\square_h^{ijk}$ is decomposed into 5 tetrahedra which are defined by the set of their four vertices within each block,

$$\{\{1,2,4,6\},\{1,3,4,7\},\{1,5,6,7\},\{4,6,7,8\},\{1,4,6,7\}\}.$$

If s^{ijk} is odd (see Fig. 1b), the decomposition of block $\square_h^{ijk}$ into the tetrahedra is done such that their vertices are given by

$$\{\{1,2,3,5\},\{2,3,4,8\},\{2,5,6,8\},\{3,5,7,8\},\{2,3,5,8\}\}.$$

With the given decomposition, a conformal triangulation of Ω into tetrahedral elements is uniquely defined, we denote this partition by $\mathcal{T}_h$. $\mathcal{T}_h$ is referred to as the fine grid triangulation, whereas the coarse grid triangulation, introduced in the following, is denoted by $\mathcal{T}_H$.

Forming Coarse Elements by Agglomeration

The coarse elements $T \in \mathcal{T}_H$ are constructed by an agglomeration of the fine elements. We construct a set of agglomerated elements $\{T\} = \mathcal{T}_H$ such that each $T = \bigcup_{i=1}^{n_T} \tau_i$, $\tau_i \in \mathcal{T}_h$ is a simply connected union of fine grid elements. Thus, for any two $\tau_i, \tau_j \in \mathcal{T}_h$, there exists a connecting path of elements $\{\tau_k\}_k \subset T$ beginning in τ_i and ending in τ_j. Each fine grid element τ should belong to exactly one agglomerated element T. Due to the regular structure of the underlying grid, the agglomeration is done such that the coarse elements have the same tetrahedral form as the fine elements, and automatically form a coarser grid of equal structure. The table *AE_element* (cf. [27]) is used to store the fine elements which belong to an agglomerated (coarse) element. Given the fine triangulation $\mathcal{T}_h$ of Ω, the agglomeration process proceeds as follows:

1. Given a fixed coarsening-factor c_f, compute the position of the coarse nodes to decompose the domain Ω into imaginary coarse blocks $\square_H^{ijk}$ of size $H \times H \times H$, where $1 \le i, j, k \le n_H \in \mathbb{N}$, $n_H = n_h/c_f$, and $H = c_f h$.
2. Build the *CB_element* table:
   For each $\tau \in \mathscr{T}_h$, obtain the position of τ in Ω and assign it to the belonging coarse block $\square_H^{ijk}$.
3. Build the *AE_element* table:
   For each coarse block $\square_H^{ijk} \subset \bar{\Omega}$ and each $\tau \subset \square_H^{ijk}$ (*CB_element*), measure the position of τ in $\square_H^{ijk}$ and assign it to the belonging coarse tetrahedron.

In step 3 of the agglomeration process, we use again the mapping $s^{ijk} := s(\square_H^{ijk}) = i + j + k$ to identify the coarse tetrahedra into which a given block is decomposed. This partition automatically defines a set of coarse grid points, given by the vertices of the coarse elements. It remains to show that a straightforward decomposition of a coarse block into coarse tetrahedral elements leads to the same result as forming the coarse tetrahedra by agglomerating fine elements. The proof of this concept is given in Lemma 1.

Lemma 1 (Mesh Alignment). *The meshes $\mathscr{T}_h$ and $\mathscr{T}_H$ are aligned.*

Proof. Let $\square_h^{ijk} \subset \bar{\Omega}$ be a fine grid block. We introduce the four vectors $n^1 = (-1,1,1)^T, n^2 = (1,-1,1)^T, n^3 = (1,1,-1)^T$, and $n^4 = (-1,-1,-1)^T$. If s_h^{ijk} is odd (see Fig. 1a), they form the inner normal vectors on the four faces of the tetrahedron which is centered in the interior of the block $\square_h^{ijk}$; if s_h^{ijk} is even (see Fig. 1b), they form the outer normal vectors on the faces of the tetrahedron in the center of $\square_h^{ijk}$. The given normal vectors n^ℓ, $\ell = 1,\dots,4$, characterize the four families of planes $\Xi_\ell^h := \{n^\ell \cdot x = 2zh, x \in \bar{\Omega}, z \in \mathbb{Z}\}$. We want to show that these families induce the splitting of any fine voxel $\square_h^{ijk} \subset \bar{\Omega}$ into the five tetrahedra by their intersection with $\square_h^{ijk}$. To see this, let us first assume that s_h^{ijk} is odd, that is, the fine voxel is decomposed according to the splitting in Fig. 1a. We denote by $\mathscr{F}_\ell(\square_h^{ijk})$ the face of the tetrahedra in $\square_h^{ijk}$ which is normal to n^ℓ, $\ell \in \{1,\dots,4\}$. Moreover, let $x^{i'j'k'} = (i'h, j'h, k'h)^T$ be the vertex of $\square_h^{ijk}$ which is closest to the origin (node 1 in Fig. 1a), that is, $(i', j', k') = (i-1, j-1, k-1)$. Then it holds indeed that $(n^\ell \cdot x)/h \mod 2 = (i'+j'+k') \mod 2$ for all $x \in \mathscr{F}_\ell(\square_h^{ijk})$, $\ell = 1,\dots,4$. Since $i+j+k$ is odd by assumption, we have that $(i'+j'+k') \mod 2 = 0$. Hence, it holds $\mathscr{F}_\ell(\square_h^{ijk}) = \Xi_\ell^h \cap \square_h^{ijk}$, and the decomposition of $\square_h^{ijk}$ into tetrahedra is induced by the families Ξ^ℓ, $\ell = 1,\dots,4$. Assuming now that s_h^{ijk} is even, the fine voxel is decomposed according to the splitting in Fig. 1b. For $\ell = 1,\dots,4$, let $\mathscr{F}_\ell(\square_h^{ijk})$ denote the angular face of the tetrahedra in $\square_h^{ijk}$ to which n^ℓ is normal. We denote by $x^{ijk} = (ih, jh, kh)^T$ the vertex of $\square_h^{ijk}$ which is most distant form the origin (node 8 in Fig. 1b). It holds for all $x \in \mathscr{F}_\ell(\square_h^{ijk})$, $\ell \in \{1,\dots,4\}$, that $(n^\ell \cdot x)/h \mod 2 = (i+j+k) \mod 2$. Since $i+j+k$ is even by assumption, we conclude again that $\Xi_\ell^h \cap \square_h^{ijk}$ defines the decomposition of $\square_h^{ijk}$ into tetrahedra. The same arguments can be applied to show that for $\ell \in \{1,\ldots,4\}$, the sets $\Xi_\ell^H := \{n^\ell \cdot x = 2zH, x \in \bar{\Omega}, z \in \mathbb{Z}\}$ form the family of planes which induce the decomposition of the coarse blocks into tetrahedra. Since the families Ξ_ℓ^h and Ξ_ℓ^H, $\ell = 1,\ldots,4$, intersect in the origin and due to $H = c_f h$ for some $c_f \in \mathbb{N}$, the coarse grid family of planes is a subset of the fine ones which shows that fine and coarse meshes are aligned.

3.3 Abstract Multiscale Coarse Space

In Sect. 3.2, we introduced the structured fine and coarse mesh which will be used in our numerical tests. For the construction of the basis functions, the assumptions on $\mathcal{T}_H$ can be slightly weakened. In general, we require that $\mathcal{T}_H$ is a conforming tetrahedral coarse mesh, such that each $T \in \mathcal{T}_H$ consists of a union of fine elements $\tau \in \mathcal{T}_h$ with $\mathcal{T}_H$ being shape-regular w.r.t. $H := \max_{T\in\mathcal{T}_H} H_T$, $H_T = \mathrm{diam}(T)$. Let $\bar{\Sigma}_H$ be the set of coarse nodes of $\mathcal{T}_H$ in $\bar{\Omega}$. We denote the index set of coarse nodes of $\mathcal{T}_H$ on $\bar{\Omega}$ by $\bar{\mathcal{N}_H}$. For each coarse grid point $x^p \in \bar{\Sigma}_H$, we introduce the set

$$\omega_p := \text{interior}\left(\bigcup_{\{T\in\mathcal{T}_H : x^p \in T\}} T \right), \tag{12}$$

given by the interior of the union of coarse elements which are attached to node x^p. We will construct a coarse vector-valued basis whose values are determined on the coarse grid points in $\bar{\Omega}$, given by the vertices of the coarse elements in $\mathcal{T}_H$. The coarse basis functions are constructed such that they can be represented w.r.t. the vector-field of piecewise linear basis functions $\mathcal{V}^h$ on the fine grid. Given the coarse basis functions, we introduce the coarse space in abstract form by

$$\mathcal{V}^H := \mathrm{span}\{\phi_m^{p,\mathrm{H}}, p \in \bar{\mathcal{N}_H}, m = 1,2,3\}. \tag{13}$$

This space can be viewed as a generalization of the space of piecewise linear vector-fields on $\mathcal{T}_H$. The coarse basis functions are constructed to have the following form.

Assumption 3.1 (Abstract Coarse Space).

(C1) $\phi_m^{p,\mathrm{H}} = (\phi_{m1}^{p,\mathrm{H}}, \phi_{m2}^{p,\mathrm{H}}, \phi_{m3}^{p,\mathrm{H}})^T$, $\phi_{mk}^{p,\mathrm{H}}(x^q) = \delta_{pq}\delta_{mk}$, $p \in \bar{\mathcal{N}_H}, k \in \{1,2,3\}$,

(C2) $\mathrm{supp}\,\phi_m^{p,\mathrm{H}} \subset \bar{\omega}_p$,

(C3) $\|\phi_{mk}^{p,\mathrm{H}}\|_{L^\infty(\Omega)} \le C, k \in \{1,2,3\}$,

(C4) $\sum_{p\in\bar{\mathcal{N}_H}} \phi_{mk}^{p,\mathrm{H}}(x) = \delta_{mk}$, $x \in \bar{\Omega}, k \in \{1,2,3\}$,

Assumption ***(C4)*** implies that the rigid body translations are globally contained in the coarse space. Additionally, we might require that the coarse space also contains the rigid body rotations, and thus,

(C5) $\mathcal{RBM} \subset \mathrm{span}\{\phi_m^{p,\mathrm{H}} : p \in \bar{\mathcal{N}_H}, k \in \{1,2,3\}\}$, where the space $\mathscr{RBM}$ of rigid body modes in $\bar{\Omega}$ is defined by

$$\mathscr{RBM} = \{\boldsymbol{v} \in [L^2(\bar{\Omega})]^3 : \boldsymbol{v} = \boldsymbol{a} + \boldsymbol{b} \times x, \boldsymbol{a}, \boldsymbol{b} \in \mathbb{R}^3, x \in \bar{\Omega}\}.$$

It is shown in [4] that multiscale finite element coarse spaces for linear elasticity with vector-valued linear boundary conditions contain the rigid body modes globally. Although the construction of the energy-minimizing coarse space which we present in Sect. 4 does not guarantee that the three rigid body rotations are globally contained in the coarse space, the numerical tests in Sect. 5 validate the robustness of the method for problems where the boundary conditions prohibit global rotations, i.e., $\text{meas}(\Gamma_{D_i}) > c_0$, $i = 1,2,3$, with $c_0 > 0$.

4 Energy Minimization for the Elasticity System

In this section we present the construction of the energy-minimizing coarse space for the 3D system of linear elasticity. We start with the definition of the basis and the corresponding coarse space $\mathscr{V}^H = \mathscr{V}^{\text{EMin}}$, followed by some details of its properties. Furthermore, we provide a precise definition of the interpolation operators which are determined by the coarse basis and show how these basis functions can be computed efficiently.

4.1 The Energy-Minimizing Coarse Space

We construct the energy-minimizing coarse space $\mathscr{V}^H$ on $\mathscr{T}_H$ according to assumption 3.1. We denote by $|\cdot|_{a,\Omega} := a(\cdot,\cdot)^{1/2}$ the semi-norm on $[H^1(\Omega)]^3$, induced by the bilinear form in (5). For $m = 1,2,3$ and each $p \in \bar{\mathscr{N}}_H$, we construct a basis function

$$\phi_m^{p,\text{EMin}} : \omega_p \to \mathbb{R}^3.$$

Ensuring that the three translations are exactly contained in the coarse space, the construction is done separately for $m \in \{1,2,3\}$, such that

$$\sum_{p \in \bar{\mathscr{N}}_H} |\phi_m^{p,\text{EMin}}|^2_{a,\Omega} \to \min \tag{14}$$

$$\text{subject to} \quad \sum_{p \in \bar{\mathscr{N}}_H} \phi_{mk}^{p,\text{EMin}} = \delta_{mk} \quad k = 1,2,3, \ \text{in } \Omega. \tag{15}$$

Thus, the basis is constructed such that the coarse basis preserves the three translations exactly. The rigid body rotations are contained only approximately. The basis satisfies Assumption 3.1 ***(C1)–(C4)***. Hence, the given functions are linearly independent and the energy-minimizing coarse space is defined as in (13). Note that we define the subspace $\mathscr{V}_0^{\mathrm{EMin}} \subset \mathscr{V}^{\mathrm{EMin}}$ as the subspace which contains only basis functions which correspond to coarse nodes $x^p \in \bar{\Sigma}_H$ which do not touch the global Dirichlet boundary. Furthermore, we exclude any fine grid DOFs on the boundary $\Gamma_{D_i}, i = 1,2,3$ when constructing the interpolation operator. More details are given in Sect. 4.3. In the following, we give a constructive proof for the existence and uniqueness of the solution of the minimization problem in (14) and (15). Therefore, we denote by $\bar{A} \in \mathbb{R}^{n_d \times n_d}$ the global stiffness matrix where no essential boundary conditions are applied. The entries of $\bar{A}$ are determined by (9). Furthermore, we denote by R_p the matrix describing the restriction to ω_p of a vector which corresponds to DOFs on $\mathscr{V}^h$ in $\bar{\Omega}$. The principal submatrix of $\bar{A}$ is then given by $\bar{A}_p = R_p \bar{A} R_p^T$. Note that $\bar{A}_p$ is non-singular for any suitable R_p. Furthermore, let $\mathbf{1}^m \in \mathbb{R}^{n_d}$ be the coefficient vector which represents a rigid body translation in the component $m \in \{1,2,3\}$ in terms of the fine-scale basis of $\mathscr{V}^h$.

Theorem 3. *The solution of the minimization problem in* (14) *and* (15) *on the space* $\mathscr{V}^h$ *is given by*

$$\Phi_m^{p,\mathrm{EMin}} = R_p^T \bar{A}_p^{-1} R_p \Lambda_m, \tag{16}$$

where $\Lambda_m \in \mathbb{R}^{n_d}$ *is the vector of Lagrange multipliers, which satisfies*

$$\sum_{p \in \bar{\mathscr{N}}_H} R_p^T \bar{A}_p^{-1} R_p \Lambda_m = \mathbf{1}^m.$$

Proof. The minimization problem couples the quadratic objective function in (14) with linear constraints, given in (15). Introducing the *Lagrange multiplier* Λ_m, a solution can be found by the extrema of the quadratic Lagrange functional

$$\mathscr{L}_m\Big(\{\Phi_m^{p,\mathrm{EMin}}\}, \Lambda_m\Big) = \frac{1}{2} \sum_{p \in \bar{\mathscr{N}}_H} \Phi_m^{p,\mathrm{EMin}^T} \bar{A}\, \Phi_m^{p,\mathrm{EMin}} - \Lambda_m{}^T \Big(\sum_{p \in \bar{\mathscr{N}}_H} \Phi_m^{p,\mathrm{EMin}} - \mathbf{1}^m \Big).$$

We enforce an additional constraint on the support of the basis functions by substituting $\Phi_m^{p,\mathrm{EMin}} = R_p^T \hat{\Phi}_m^{p,\mathrm{EMin}}$. The vector $\hat{\Phi}_m^{p,\mathrm{EMin}}$ can be considered as the local representation of $\Phi_m^{p,\mathrm{EMin}}$ on its support ω_p w.r.t. the basis of $\mathscr{V}^h(\omega_p)$. To find the critical point of this functional, we impose $\nabla_{\Lambda_m} \mathscr{L}_m = 0$ and $\nabla_{\hat{\Phi}_m^{p,\mathrm{EMin}}} \mathscr{L}_m = 0$, which results in the saddle point problem

$$\bar{A}_p \hat{\Phi}_m^{p,\mathrm{EMin}} - R_p \Lambda_m = 0 \quad \forall p \in \bar{\mathscr{N}}_H, \tag{17}$$

$$\sum_{p \in \bar{\mathscr{N}}_H} R_p^T \hat{\Phi}_m^{p,\mathrm{EMin}} - \mathbf{1}^m = 0. \tag{18}$$

From (17), we conclude

$$\hat{\Phi}_m^{p,\mathrm{EMin}} = \bar{A}_p^{-1} R_p \Lambda_m \quad \forall p \in \bar{\mathscr{N}}_H. \tag{19}$$

Substituting (19) into (18) yields

$$\mathbf{1}^m = \sum_{p \in \bar{\mathcal{N}}_H} R_p^T \bar{A}_p^{-1} R_p \Lambda_m.$$

We introduce $L := \sum_{p \in \bar{\mathcal{N}}_H} R_p^T \bar{A}_p^{-1} R_p$ and obtain for $m \in \{1,2,3\}$,

$$\Lambda_m = L^{-1} \mathbf{1}^m. \tag{20}$$

□

Thus, to compute the basis, we have to solve the global *Lagrange multiplier system* in (20) for each $m \in \{1,2,3\}$ and solve local subproblems in (19) to compute the particular basis functions.

4.2 Properties of the Energy-Minimizing Coarse Space

As we can conclude from the construction, the coarse space contains the three rigid body translations globally in $\bar{\Omega}$. However, it is not clear how well this coarse space approximates the set of rigid body rotations. The rotations are, in general, not exactly contained in $\mathcal{V}^H$. The energy-minimizing construction of the basis functions allows quite general supports, and the method is easily applicable to unstructured meshes. If we denote by $\omega_p^{\text{int}} := \{x \in \omega_p : x \notin \omega_q \text{ for any } q \neq p\}$ the subset of ω_p which is not overlapped with the support of any other basis function, it is clear that rigid body rotations cannot be globally contained in the coarse space as long as $\text{meas}(\omega_p^{\text{int}}) > 0$. Thus, to ensure that the presented construction of the coarse space allows an adequate approximation of the rigid body rotations, a necessary requirement needs to be stated on the supports of the basis functions. Defining the coarse basis functions on the coarse mesh $\mathcal{T}_H$ as introduced before yields large overlaps in the supports of neighboring basis functions. It holds $\omega_p^{\text{int}} = \{x^p\}$, and thus, we obtain $\text{meas}(\omega_p^{\text{int}}) = 0$. However, this requirement is not sufficient to ensure that all the rigid body rotations are preserved exactly by the coarse space.

An important property, showing the multiscale character of the presented energy-minimizing coarse space, is summarized in the following. We show that the Lagrange multipliers $\Lambda_m, m = 1,2,3$, are supported on the coarse element boundaries, and thus, the energy-minimizing basis functions are given by a discrete PDE-harmonic extension of local boundary data. Before proving this statement, we introduce the following notation. For $T \in \mathcal{T}_H$, let

$$\text{range}(T) := \bigcap_{p \in \bar{\mathcal{N}}_H(T)} \text{range}(R_p^T)$$

be the set of vectors in $\mathbb{R}^{n_d}$ which correspond to functions in $\mathcal{V}^h$ which are supported in the interior of T. We show that the Lagrange multiplier Λ_m, $m = 1,2,3$, has nonzero values only in a set which is complementary to $\{\text{range}(T) : T \in \mathscr{T}_H\}$. The nonzero entries correspond to fine basis functions which are supported on the boundaries of the coarse elements $T \in \mathscr{T}_H$.

Lemma 2. *Let $m \in \{1,2,3\}$ be fixed and let $\Lambda_m = L^{-1}\mathbf{1}^m$. Then for each $T \in \mathscr{T}_H$, we have*

$$\xi^T \Lambda_m = 0 \quad \forall \xi \in \text{range}(T).$$

Proof. Let $n_T = \#\{p \in \bar{\mathscr{N}}_H(T)\}$ be the number of vertices of T. For $m \in \{1,2,3\}$, it holds

$$\sum_{p \in \bar{\mathscr{N}}_H(T)} \Phi_m^{p,\text{EMin}} = \mathbf{1}^m \quad \text{on } T.$$

For each $\xi \in \text{range}(T)$, let $\hat{\xi}_p := R_p \xi$, $p \in \bar{\mathscr{N}}_H(T)$ be the local representation of ξ in $\omega_p \subset \Omega$. Note that it also holds $R_p^T \hat{\xi}_p = \xi$ since ξ_p is supported in range(R_p^T) by assumption. We have by (17),

$$n_T\, \xi^T \Lambda_m = \sum_{p \in \bar{\mathscr{N}}_H(T)} \hat{\xi}_p^T R_p \Lambda_m = \sum_{p \in \bar{\mathscr{N}}_H(T)} \hat{\xi}_p^T \bar{A}_p \hat{\Phi}_m^{p,\text{EMin}} = \xi^T \bar{A} \mathbf{1}^m = 0,$$

where we used $\xi \in \text{range}(T)$ twice. The last equality follows since $\mathbf{1}^m \in \text{Ker}(\bar{A})$. □

This shows that the basis functions are locally PDE-harmonic, a well-known property (cf. [31]) of the energy-minimizing basis. From the solution of the Lagrange multiplier system, optimal boundary conditions for the local basis functions are extracted on $\{\partial T, T \in \mathscr{T}_H\}$. It is obvious that the energy-minimizing basis functions are continuous along the boundaries of the coarse elements and lead to a conforming coarse space.

4.3 The Interpolation Operator

In the following, we construct the interpolation operator which is given by the energy-minimizing coarse space. Let us first summarize some notations. The number of grid points in $\bar{\Omega}$ on the fine grid is denoted by n_p; the number of grid points on the coarse grid is denoted by N_p. To each grid point, fine or coarse, we associate a vector-field $u = (u_1, u_2, u_3)^T : \bar{\Omega} \to \mathbb{R}^3$ of displacements. We denote the corresponding components $u_i, i = 1,2,3$ of the vector-field by *unknowns*. The number of fine and coarse DOFs on the fine and coarse triangulation (in $\bar{\Omega}$) is given by $n_d = 3n_p$, $N_d = 3N_p$, respectively. Furthermore, for $\beta \in \{h, H\}$, the set $\mathscr{D}^\beta = \mathscr{D}^\beta(\bar{\Omega})$ denotes the index set of fine ($\beta = h$), respectively, coarse ($\beta = H$) DOFs of $\mathscr{V}^\beta$. For any subset $W \subset \bar{\Omega}$, let $\mathscr{D}^\beta(W) \subset \mathscr{D}^\beta(\bar{\Omega})$ be the restriction of $\mathscr{D}^\beta$ to the local set of DOFs in W, given in a local numbering. To keep the notation with indices more intuitive for the reader, we use the following convention. To indicate DOFs in $\mathscr{D}^h$, we use (i,k) or (j,l) to indicate DOFs, while the index (p,m) or (q,r) are used for the indication of a coarse degree of freedom in $\mathscr{D}^H$. We use the fine-scale representation of a coarse basis function $\phi_m^{p,\text{EMin}}$ to define the interpolation operator, respectively, the restriction operator. Each energy-minimizing basis function omits the representation

$$\phi_m^{p,\text{EMin}} = \sum_{k=1}^{3} \sum_{i=1}^{n_p} \bar{r}_{(p,m),(i,k)} \varphi_k^{i,h}. \tag{21}$$

This representation defines a matrix $\bar{R} \in \mathbb{R}^{N_d \times n_d}$ which contains the coefficient vectors, representing a coarse basis function in terms of the fine-scale basis, by rows. Note that $\bar{R}$ does not define the final restriction operator used in the additive Schwarz setting. The restriction operator R_H, which we use in the additive Schwarz algorithm is then constructed as a submatrix of $\bar{R}$, which contains only the rows corresponding to coarse basis functions of $\mathscr{V}_0^H$. Thus, it contains the rows related to coarse basis functions which vanish on the global Dirichlet boundaries Γ_{D_i}, $i = 1,2,3$ and do not contain any fine DOFs on the global Dirichlet boundary. Denoting the entries of R_H by $(r_{p',j'})_{p',j'}$, we define

$$r_{p',j'} = \begin{cases} \bar{R}_{p',j'} & \text{if } p' \in \mathscr{D}^H(\Omega^*), \; j' \in \mathscr{D}_0^h(\bar{\Omega}), \\ 0 & \text{if } p' \in \mathscr{D}^H(\Omega^*), \; j' \in \mathscr{D}_{\Gamma_D}^h(\bar{\Omega}), \end{cases}$$

where $\mathscr{D}^H(\Omega^*)$, $\Omega^* := \bar{\Omega} \backslash (\cup^i \Gamma_{D_i})$ denotes the coarse interior DOFs in Ω^*. The matrix representing the interpolation from the coarse space $\mathscr{V}_0^H$ to the fine space $\mathscr{V}_0^h$ is simply given by the transposed, R_H^T. The coarse stiffness matrix can be computed by the Galerkin product $A^H = R_H A R_H^T$.

5 Numerical Experiments

In this section, we give a series of examples involving binary media, showing the performance of the energy-minimizing preconditioner under variations of the mesh parameters as well as the material coefficients. In addition to that, we measure the approximation error of the energy-minimizing coarse space to a fine-scale solution. In each numerical test, we compare the energy-minimizing coarse space with a standard linear coarse space. We perform our simulations on the domain $\bar{\Omega} = [0,1] \times [0,1] \times [0,L], L > 0$, with fine and coarse mesh as introduced in Sect. 3.2. Dirichlet conditions in the first unknown are given on $\Gamma_1 = \{(x,y,z)^T \in \partial\Omega : x = 0, x = 1\}$, in the second unknown on $\Gamma_2 = \{(x,y,z)^T \in \partial\Omega : y = 0, y = 1\}$, and in the third unknown on $\Gamma_3 = \{(x,y,z)^T \in \partial\Omega : z = 0, z = L\}$. For the numerical tests,

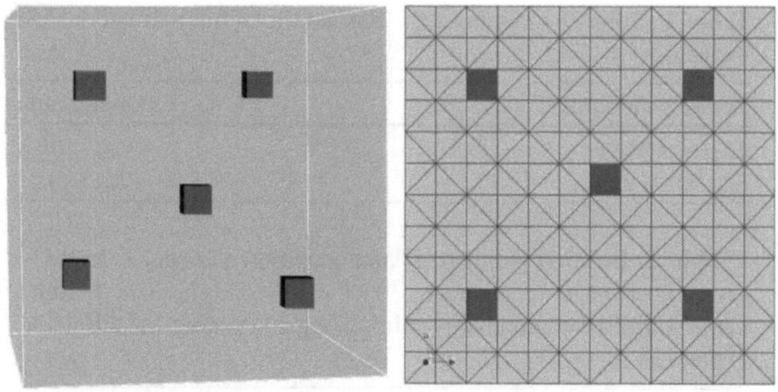

Fig. 2 Medium 1: binary composite; matrix material and $1 \times 1 \times 1$ inclusions; discretization in $12 \times 12 \times 12$ voxels; each voxel is decomposed in 5 tetrahedra; inclusions lie in the interior of a coarse tetrahedral element; 3D view (*left*) and 2D projection with fine mesh, showing the position of the inclusions (*right*)

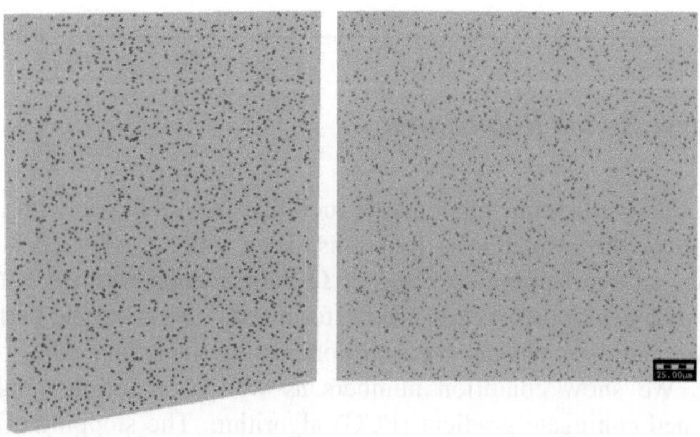

Fig. 3 Medium 2: binary composite: discretization in $240 \times 240 \times 12$ voxels; matrix material and $1 \times 1 \times 1$ inclusions identically distributed; 3D view (*left*) and 2D projection (*right*)

we consider different heterogeneous media. First, we assume that the discontinuities are isolated, that is, the material jumps occur only in the interior of coarse elements. Figure 2 shows such a binary medium with one inclusion inside each coarse tetrahedral element.

For a second medium, we do not impose any restriction on the position of the small inclusions. More precisely, we generate a binary medium whose inclusions are identically distributed. An example of such a medium is given in Fig. 3.

In the following, we refer to the binary medium where inclusions are isolated in the interior of coarse elements as medium 1, while the medium with identically distributed inclusions is referred to as medium 2. For each medium, the Young's modulus E as well as Poisson ratio ν for matrix material and inclusions are given in Table 1. The contrast $\Delta_E := E_{inc}/E_{mat}$ may vary over several orders of magnitude.

Table 1 Young's modulus and Poisson ratio of matrix material and inclusions

Young's modulus	Poisson ratio
$E_{mat} = 1$ MPa	$\nu_{mat} = 0.2$
$E_{inc} = \Delta_E E_{mat}$	$\nu_{inc} = 0.2$

Table 2 Iteration numbers n_{it} and condition numbers $\kappa(M_{AS}^{-1}A)$ for experiment 1; geometry: $1/h \times 1/h \times H/h$, $h = 1/240$, $H = 12h$; linear and energy-minimizing coarsening for different contrasts $\Delta_E \geq 1$

	Lin		EMin	
Δ_E	n_{it}	$\kappa(M_{AS}^{-1}A)$	n_{it}	$\kappa(M_{AS}^{-1}A)$
10^0	13	4.4	14	4.9
10^3	21	18.7	14	5.0
10^6	25	109.0	14	5.0
10^9	25	109.0	14	5.0

5.1 Coarse Space Robustness

We choose the overlapping subdomains such that they coincide with the supports $\bar{\omega}_p$, $p \in \bar{\mathcal{N}}_H$ of the coarse basis functions. Then, $\{\Omega_i, i = 1,\ldots,N\} = \{\omega_p, p \in \bar{\mathcal{N}}_H\}$ defines an overlapping covering of $\bar{\Omega}$ with overlap width $\delta = O(H)$, often referred to as a *generous overlap*. We perform tests observing the performance of the two-level additive Schwarz preconditioner using linear and energy-minimizing coarsening. We show condition numbers as well as iteration numbers of the preconditioned conjugate gradient (PCG) algorithm. The stopping criterion is to reduce the preconditioned initial residual by six orders of magnitude, i.e., $\|r\|_{M_{AS}^{-1}} \leq 10^{-6}\|r_0\|_{M_{AS}^{-1}}$. For the construction of the energy-minimizing basis functions, the Lagrange multiplier systems are solved using the CG algorithm; the initial residual is reduced by three orders of magnitude. The estimated condition numbers of $\kappa(M_{AS}^{-1}A)$ are computed based on the three-term recurrence which is implicitly formed by the coefficients within the PCG algorithm (cf. [18]).

In a first experiment (1), we test the robustness of the method on medium 1 for fixed mesh parameters under the variation of the contrast Δ_E. Tables 2 and 3 show the corresponding condition numbers and iteration numbers having stiff ($\Delta_E > 1$) and soft ($\Delta_E < 1$) inclusions. In the former case, robustness is achieved only for the energy-minimizing coarse space, while linear coarsening leads to nonuniform convergence results.

In experiment 2, performed on medium 1, we measure the condition numbers and iteration numbers under variation of the mesh parameters, while the PDE coefficients remain fixed. We observe similar results as in experiment 1.

Table 4 shows the condition numbers for linear and energy-minimizing coarsening. For the linear coarse space, the condition number shows a linear dependence

Table 3 Iteration numbers n_{it} and condition numbers $\kappa(M_{AS}^{-1}A)$ for experiment 1; geometry: $1/h \times 1/h \times H/h$, $h = 1/240$, $H = 12h$; linear and energy-minimizing coarsening for different contrasts $\Delta_E \leq 1$

	Lin		EMin	
Δ_E	n_{it}	$\kappa(M_{AS}^{-1}A)$	n_{it}	$\kappa(M_{AS}^{-1}A)$
10^{-0}	13	4.4	13	4.9
10^{-3}	13	4.4	13	5.0
10^{-6}	13	4.4	13	5.0
10^{-9}	13	4.4	13	5.0

Table 4 Iteration numbers n_{it} and condition numbers $\kappa(M_{AS}^{-1}A)$ for experiment 2; geometry: $1/h \times 1/h \times H/h$; $H = 12h$; linear and energy-minimizing coarsening for different h; contrast: $\Delta_E = 10^6$

	Lin		EMin	
h	n_{it}	$\kappa(M_{AS}^{-1}A)$	n_{it}	$\kappa(M_{AS}^{-1}A)$
1/60	14	7.9	13	4.4
1/120	17	28.1	14	5.0
1/180	21	61.8	14	4.9
1/240	25	109.0	14	5.0

Table 5 Iteration numbers n_{it} and condition numbers $\kappa(M_{AS}^{-1}A)$ for experiment 1 on medium 2; geometry: $1/h \times 1/h \times H/h$, $h = 1/240$, $H = 12h$; linear and energy-minimizing coarsening for different contrasts $\Delta_E \geq 1$

	Lin		EMin	
Δ_E	n_{it}	$\kappa(M_{AS}^{-1}A)$	n_{it}	$\kappa(M_{AS}^{-1}A)$
10^0	13	4.4	14	4.9
10^3	27	19.3	14	4.9
10^6	66	414	14	5.0
10^9	68	427	14	5.0

on the number of subdomains, while the condition number for energy-minimizing coarsening is uniformly bounded.

In the experiment above, we obtained coefficient independent convergence rates of the energy-minimizing coarse space on medium 1. In a second part, we test the performance of the method on medium 2, where the small inclusions are identically distributed. This is what we see in Tables 5 and 6 for experiment 1 on medium 2: For fixed mesh parameters under the variation of the contrast Δ_E, they show the corresponding condition numbers and iteration numbers having stiff ($\Delta_E > 1$) and soft ($\Delta_E < 1$) inclusions. Robustness for the linear coarse space is only achieved in the later case where soft inclusions are considered. For stiff inclusions, the linear coarsening strategy leads to iteration numbers and condition numbers which strongly depend on the contrast in the medium. The energy-minimizing coarse space is fully robust w.r.t. coefficient variations.

Table 6 Iteration numbers n_{it} and condition numbers $\kappa(M_{AS}^{-1}A)$ for experiment 1 on medium 2; geometry: $1/h \times 1/h \times H/h$, $h = 1/240$, $H = 12h$; linear and energy-minimizing coarsening for different contrasts $\Delta_E \leq 1$

	Lin		EMin	
Δ_E	n_{it}	$\kappa(M_{AS}^{-1}A)$	n_{it}	$\kappa(M_{AS}^{-1}A)$
10^{-0}	13	4.4	14	4.9
10^{-3}	13	4.4	14	5.0
10^{-6}	13	4.4	14	5.0
10^{-9}	13	4.4	14	5.0

Table 7 Iteration numbers n_{it} and condition numbers $\kappa(M_{AS}^{-1}A)$ for experiment 2 on medium 2; geometry: $1/h \times 1/h \times H/h$; $H=12h$; linear and energy-minimizing coarsening for different h; contrast: $\Delta_E = 10^6$

	Lin		EMin	
h	n_{it}	$\kappa(M_{AS}^{-1}A)$	n_{it}	$\kappa(M_{AS}^{-1}A)$
1/60	26	39.2	13	4.4
1/120	48	154	14	5.0
1/180	52	261	14	4.9
1/240	66	414	14	5.0

Now, we perform experiment 2 on medium 2 and measure the condition numbers and iteration numbers under variation of the mesh parameters and fixed PDE coefficients. Table 7 shows iteration and condition numbers for linear and energy-minimizing coarsening. Mesh independent bounds are achieved for the energy-minimizing coarse space, while for the linear coarse space, iteration numbers as well as condition numbers grow with the number of subdomains.

5.2 Coarse Space Approximation

In a second set of experiments, we test the approximation properties of the energy-minimizing coarse space. The domain $\bar{\Omega} = [0,1] \times [0,1] \times [0,L]$ contains a binary medium with small inclusions. Again, we distinguish between medium 1 (Fig. 2: inclusions in the interior of each coarse element) and medium 2 (Fig. 3: identically distributed inclusions). We solve the linear system $-\mathrm{div}\boldsymbol{\sigma}(\boldsymbol{u}) = \boldsymbol{f}$ in $\bar{\Omega} \setminus \Gamma_D$ with a constant volume force $\boldsymbol{f} = (1,1,0)^T$ in the x- and y-component. Homogeneous Dirichlet and Neumann boundary conditions are applied on the boundary $\partial\Omega$.

Let $\boldsymbol{u}^h$ denote the approximate solution on a fine mesh $\mathscr{T}_h$. With the bilinear form defined in (6) and the space $\mathscr{V}_0^h$ of piecewise linear vector-valued basis functions as defined in (7), it holds $a(\boldsymbol{u}^h, \boldsymbol{v}^h) = F(\boldsymbol{v}^h)\ \forall \boldsymbol{v}^h \in \mathscr{V}_0^h$. This formulation leads to the linear system $A\mathbf{u}^h = \mathbf{f}^h$. Let $\mathscr{V}_0^H$ be the space of energy-minimizing basis functions

Table 8 Approximation of fine-scale solution by linear and energy-minimizing coarse space for medium 1; geometry: $1/h \times 1/h \times H/h$, $h = 1/120$, $H = 12h$

	$\frac{\|\mathbf{u}^h - \mathbf{u}^c\|_{l_2}}{\|\mathbf{u}^h\|_{l_2}}$		$\frac{\|\mathbf{u}^h - \mathbf{u}^c\|_A}{\|\mathbf{u}^h\|_A}$	
Δ_E	Lin	EMin	Lin	EMin
10^{-9}	8.63×10^{-3}	1.09×10^{-1}	8.92×10^{-2}	3.32×10^{-1}
10^{-6}	8.63×10^{-3}	1.09×10^{-1}	8.92×10^{-2}	3.32×10^{-1}
10^{-3}	8.63×10^{-3}	1.09×10^{-1}	8.91×10^{-2}	3.32×10^{-1}
10^{0}	8.09×10^{-3}	1.09×10^{-1}	8.53×10^{-2}	3.31×10^{-1}
10^{3}	7.39×10^{-1}	1.07×10^{-1}	8.60×10^{-1}	3.28×10^{-1}
10^{6}	9.97×10^{-1}	1.07×10^{-1}	9.99×10^{-1}	3.28×10^{-1}
10^{9}	9.97×10^{-1}	1.07×10^{-1}	9.99×10^{-1}	3.28×10^{-1}

on the coarse triangulation $\mathscr{T}_H$ which vanish on the Dirichlet boundary $\Gamma_i, i = 1,2,3$ (see Sect. 4.3). The energy-minimizing solution is given by $\boldsymbol{u}^{\mathrm{EMin}} \in \mathscr{V}_0^H$, such that $a(\boldsymbol{u}^{\mathrm{EMin}}, \boldsymbol{v}^H) = F(\boldsymbol{v}^H)\ \forall \boldsymbol{v}^H \in \mathscr{V}_0^H$. Using the fine-scale representation of an energy-minimizing basis function as defined in (21), the equivalent linear system reads $A_H \mathbf{u}^H = \mathbf{f}^H$. Here, $A_H = R_H A R_H^T$ is the coarse stiffness matrix, $\mathbf{f}^H = R_H \mathbf{f}^h$, and $\mathbf{u}^{\mathrm{EMin}} = R_H^T \mathbf{u}^H$ is the vector whose entries define the fine-scale representation of $\boldsymbol{u}^{\mathrm{EMin}}$ in terms of the basis of $\mathscr{V}_0^h$.

For fixed mesh parameters h and H, under the variation of the contrast Δ_E, Tables 8 and 9 show the relative approximation errors $\|\mathbf{u}^h - \mathbf{u}^c\|$ in l_2 and in the "energy"-norm for linear (c=Lin) and energy-minimizing (c=EMin) coarse space for medium 1 and medium 2, respectively.

The fine solution $\mathbf{u}^h$ is computed approximately within the PCG algorithm by reducing the initial preconditioned residual by 12 orders of magnitude. The coarse solution $\mathbf{u}^H$ is computed exactly by a sparse direct solve of the coarse linear system. For both media, the energy-minimizing coarse space gives stable approximation errors, only slightly varying with the contrast. The linear coarse space only shows a poor approximation of the fine-scale solution for high contrasts $\Delta_E \gg 1$. The explanation is that for $\Delta_E \gg 1$, the fine-scale solution is contained in a space which is nearly A-orthogonal to the space spanned by the linear coarse basis functions. Note that this is in agreement with the results presented in Table 4, where the condition number grows almost linearly with the number of subdomains.

We also observe from Tables 8 and 9 that for soft inclusions ($\Delta_E \leq 1$), the approximation error is smaller by the linear coarse space than by the energy-minimizing coarse space. The latter is due to the circumstance that the vector-valued energy-minimizing basis is, even for homogeneous coefficients, not piecewise linear on the coarse triangulation. It is known that the shape of the energy-minimizing basis functions is in general mesh dependent, e.g., for the discretization of the scalar Poisson problem on a regular mesh in 2D, an energy-minimizing basis is observed to be piecewise linear in [29] (see also [25]). However, for the vector-valued problem considered here with the mesh as in Sect. 3.2, the vector-valued energy-minimizing

Table 9 Approximation of fine-scale solution by linear and energy-minimizing coarse space for medium 2; geometry: $1/h \times 1/h \times H/h$, $h = 1/120$, $H = 12h$

	$\frac{\|\mathbf{u}^h-\mathbf{u}^c\|_{l_2}}{\|\mathbf{u}^h\|_{l_2}}$		$\frac{\|\mathbf{u}^h-\mathbf{u}^c\|_A}{\|\mathbf{u}^h\|_A}$	
Δ_E	Lin	EMin	Lin	EMin
10^{-9}	8.60×10^{-3}	1.09×10^{-1}	8.90×10^{-2}	3.32×10^{-1}
10^{-6}	8.60×10^{-3}	1.09×10^{-1}	8.90×10^{-2}	3.32×10^{-1}
10^{-3}	8.60×10^{-3}	1.09×10^{-1}	8.90×10^{-2}	3.32×10^{-1}
10^{0}	8.09×10^{-3}	1.09×10^{-1}	8.53×10^{-2}	3.31×10^{-1}
10^{3}	7.01×10^{-1}	1.15×10^{-1}	8.37×10^{-1}	3.40×10^{-1}
10^{6}	9.99×10^{-1}	1.12×10^{-1}	1.00×10^{-0}	3.36×10^{-1}
10^{9}	1.00×10^{-0}	1.12×10^{-1}	1.00×10^{-0}	3.36×10^{-1}

basis is not piecewise linear on the coarse mesh for reasonable mesh sizes $H > h > 0$. The latter also implies that the rigid body rotations are only approximated globally.

We can summarize the numerical results obtained in this section as follows. The energy-minimizing construction allows a low-energy approximation of the basis functions, independently of the Young's modulus of the inclusions. We considered different media where the discontinuities are either isolated in the interior of coarse elements or randomly distributed. Using an energy-minimizing coarse space, our experiments show uniform condition number bounds w.r.t. both, coefficient variations in the Young's modulus and the mesh size. In contrast, robustness is not achieved with the linear coarse space. The linear basis function cannot capture the smallest eigenvalues associated to the discontinuities in the material parameters. The energy of the basis function strongly depends on the Young's modulus of the inclusion. As the experiments show, no uniform iteration number or condition number bounds are achieved. This observation holds for all considered media.

6 Discussion

We constructed energy-minimizing coarse spaces for microstructural problems in 3D linear elasticity. The coarse basis is such that it contains the rigid body translations exactly, while the rigid body rotations are preserved approximately. We used the coarse basis for the construction of two-level overlapping domain decomposition preconditioners in the additive version and performed experiments on binary media. For the class of problems which excludes pure traction boundary values, the results show uniform condition number bounds w.r.t. both, coefficient variations in the Young's modulus and the mesh size. Furthermore, we tested the fine-scale approximation of the energy-minimizing coarse space and observed uniform results, independent of the contrast in the composite material.

Acknowledgements The authors would like to thank Dr. Panayot Vassilevski and Prof. Ludmil Zikatanov for many fruitful discussions and their valuable comments on the subject of this paper.

References

1. Baker, A.H., Kolev, T., Yang, U.M.: Improving algebraic multigrid interpolation operators for linear elasticity problems. Numer. Lin. Algebra Appl. **17**, 495–517 (2010)
2. Braess, D.: Finite Elements: Theory, Fast Solvers and Applications in Solid Mechanics, 3rd edn. Cambridge University Press, Cambridge (2007)
3. Brezina, M., Cleary, A.J., Falgout, R.D., Henson, V.E., Jones, J.E., Manteuffel, T.A., Mccormick, S.F., Ruge, J.W.: Algebraic multigrid based on element interpolation (AMGe). SIAM J. Sci. Comput. **22**, 1570–1592 (2000)
4. Buck, M., Iliev, O., Andrä, H.: Multiscale finite element coarse spaces for the application to linear elasticity. Cent. Eur. J. Math. **11**(4), 680–701 (2013)
5. Clees, T.: AMG strategies for PDE systems with applications in industrial semiconductor simulation. Thesis, Faculty of Mathematics, University of Cologne (2005)
6. Dohrmann, C., Klawonn, A., Widlund, O.: A family of energy minimizing coarse spaces for overlapping Schwarz preconditioners. Domain Decomposition Methods in Science and Engineering XVII, Springer, 247–254 (2008)
7. Efendiev, Y., Galvis, J., Lazarov, R., Willems, J.: Robust domain decomposition preconditioners for abstract symmetric positive definite bilinear forms. ESAIM Math. Model. Numer. Anal. **46**, 1175–1199 (2012)
8. Graham, I.G., Lechner, P.O., Scheichl, R.: Domain decomposition for multiscale PDEs. Numer. Math. **106**, 589–626 (2007)
9. Janka, A.: Algebraic domain decomposition solver for linear elasticity. Appl. Math. **44**, 435–458 (1999)
10. Jones, J., Vassilevski, P.S.: AMGe based on element agglomeration. SIAM J. Sci. Comput. **23**, 109–133 (2001)
11. Karer, E.: Subspace correction methods for linear elasticity. Thesis, University of Linz (2011)
12. Karer, E., Kraus, J.K.: Algebraic multigrid for finite element elasticity equations: determination of nodal dependence via edge matrices and two-level convergence. Int. J. Numer. Meth. Eng. **83**, 642–670 (2010)
13. Kolev, T.V., Vassilevski, P.S.: AMG by element agglomeration and constrained energy minimization interpolation. Numer. Lin. Algebra Appl. **13**, 771–788 (2006)
14. Kraus, J.K.: Algebraic multigrid based on computational molecules, II: linear elasticity problems. SIAM J. Sci. Comput. **30**, 505–524 (2008)
15. Kraus, J.K., Schicho, J.: Algebraic multigrid based on computational molecules I: scalar elliptic problems. Computing **77**, 57–75 (2006)
16. Mandel, J., Brezina, M., Vaněk, P.: Energy optimization of algebraic multigrid bases. Computing **62**, 205–228 (1999)
17. Olson, L.N., Schroder, J.B., Tuminaro, R.S.: A general interpolation strategy for algebraic multigrid using energy minimization. SIAM J. Sci. Comput. **33**, 966–991 (2011)
18. Saad, Y.: Iterative Methods for Sparse Linear Systems, 2nd edn. SIAM, Philadelphia (2003)
19. Sarkis, M.: Partition of unity coarse spaces: enhanced versions, discontinuous coefficients and applications to elasticity. In: Domain Decomposition Methods in Science and Engineering XIV., Natl. Auton. Univ. Mex., Mexico, 149–158 (2003)
20. Scheichl, R., Vassilevski, P.S., Zikatanov, L.T.: Weak approximation properties of elliptic projections with functional constraints. Multiscale Model. Simul. **9**, 1677–1699 (2011)
21. Schulz, V., Andrä, H., Schmidt, K.: Robuste Netzgenerierung zur Mikro-FE-Analyse mikrostrukturierter Materialien. In: NAFEMS Magazin, vol. 2, pp. 28–30 (2007)

22. Smith, B.F.: Domain decomposition algorithms for the partial differential equations of linear elasticity. Thesis, Courant Institute of Mathematical Sciences, New York University (1990)
23. Spillane, N., Dolean, V., Hauret, P., Nataf, F., Pechstein, C., Scheichl, R.: Abstract robust coarse spaces for systems of PDEs via generalized eigenproblems in the overlaps. Technical Report 2011–07, University of Linz, Institute of Computational Mathematics (2011)
24. Toselli, A., Widlund, O.: Domain Decomposition Methods, Algorithms and Theory. Springer, Berlin (2005)
25. Van lent, J., Scheichl, R., Graham, I.G.: Energy-minimizing coarse spaces for two-level Schwarz methods for multiscale PDEs. Numer. Lin. Algebra Appl. **16**, 775–799 (2009)
26. Vaněk, P., Brezina, M., Tezaur, R.: Two-grid method for linear elasticity on unstructured meshes. SIAM J. Sci. Comput. **21**, 900–923 (1999)
27. Vassilevski, P.S.: Multilevel Block Factorization Preconditioners: Matrix-Based Analysis and Algorithms for Solving Finite Element Equations. Springer, New York (2008)
28. Vassilevski, P.S.: General constrained energy minimizing interpolation mappings for AMG. SIAM J. Sci. Comput. **32**, 1–13 (2010)
29. Wan, W., Chan, T.F., Smith, B.: An energy-minimizing interpolation for robust multigrid methods. SIAM J. Sci. Comput. **21**, 1632–1649 (2000)
30. Willems, J.: Robust multilevel methods for general symmetric positive definite operators. Technical Report 2012–06, RICAM Institute for Computational and Applied Mathematics (2012)
31. Xu, J., Zikatanov, L.T.: On an energy minimizing basis in algebraic multigrid methods. Comput. Vis. Sci. **7**, 121–127 (2004)

Preconditioners for Some Matrices of Two-by-Two Block Form, with Applications, I

Owe Axelsson

Abstract Matrices of two-by-two block form with matrix blocks of equal order arise in various important applications, such as when solving complex-valued systems in real arithmetics, in linearized forms of the Cahn–Hilliard diffusive phase-field differential equation model and in constrained partial differential equations with distributed control. It is shown how an efficient preconditioner can be constructed which, under certain conditions, has a resulting spectral condition number of about 2. The preconditioner avoids the use of Schur complement matrices and needs only solutions with matrices that are linear combinations of the matrices appearing in each block row of the given matrix and for which often efficient preconditioners are already available.

Keywords Two-by-two block-structured matrices • Preconditioning • Complex-valued system • Cahn–Hilliard phase-field model • Optimal control • Distributed control

Mathematics Subject Classification (2010): 65F10, 65F35, 76T10, 49J20

1 Introduction

To motivate the study, we give first some examples of two-by-two block matrices where blocks of equal order, i.e. square blocks, appear. Although the matrices are of special type, as we shall see there are several important applications where

O. Axelsson (✉)
IT4 Innovations Department, Institute of Geonics AS CR, Ostrava, Czech Republic

King Abdulaziz University, Jeddah, Saudi Arabia
e-mail: owe.axelsson@it.uu.se

O.P. Iliev et al. (eds.), *Numerical Solution of Partial Differential Equations: Theory, Algorithms, and Their Applications*, Springer Proceedings in Mathematics & Statistics 45,
DOI 10.1007/978-1-4614-7172-1_3,

they arise. One such example is related to the solution of systems with complex-valued matrices. Complex-valued systems arise, for instance, when solving certain partial differential equations (PDE) appearing in electromagnetics and wave propagation; see [1]. Complex arithmetics requires more memory storage and may require more involved implementation. Therefore it is desirable to rewrite a complex-valued matrix system in a form that can be handled using real arithmetics.

Using straightforward derivations, for a complex-valued matrix $A+iB$, where A and B are real and A is nonsingular, it holds

$$(A+iB)(I-iA^{-1}B)=A+BA^{-1}B$$

so

$$(A+iB)^{-1}=(I-iA^{-1}B)(A+BA^{-1}B)^{-1}.$$

It follows that a complex-valued system

$$(A+iB)(\mathbf{x}+i\mathbf{y})=\mathbf{f}+i\mathbf{g},$$

where $\mathbf{x},\mathbf{y},\mathbf{f},\mathbf{g}$ are real vectors, can be solved by solving two real-valued systems with matrix $A+BA^{-1}B$ with right-hand sides $\mathbf{f}$ and $\mathbf{g}$ respectively, in addition to a matrix vector multiplication with B and two solutions of systems with the matrix A.

In many applications, $A+BA^{-1}B$ can be ill conditioned and costly to construct and solve systems with, in particular as it involves solutions with inner systems with the matrix A. Therefore, this approach is normally less efficient.

As has been shown in [2] (see also [1,3]), it may be better to rewrite the equation in real-valued form

$$\begin{bmatrix} A & -B \\ B & A \end{bmatrix} \begin{bmatrix} \mathbf{x} \\ \mathbf{y} \end{bmatrix} = \begin{bmatrix} \mathbf{f} \\ \mathbf{g} \end{bmatrix}. \tag{1}$$

A matrix factorization shows that

$$\begin{bmatrix} A & 0 \\ B & A+BA^{-1}B \end{bmatrix} \begin{bmatrix} I & -A^{-1}B \\ 0 & I \end{bmatrix} \begin{bmatrix} \mathbf{x} \\ \mathbf{y} \end{bmatrix} = \begin{bmatrix} \mathbf{f} \\ \mathbf{g} \end{bmatrix},$$

where I is the identity matrix. It is seen that here it suffices with one solution with matrix $A+BA^{-1}B$, in addition to two solves with A. However, we will show that the form (1) allows for an alternative solution method based on iteration and the construction of an efficient preconditioner that involves only two systems with matrices that are linear combinations of matrices A and B and that a corresponding iterative solution of (1) can substantially lower the computational expense. We shall show that such a preconditioner can be constructed for a matrix in the more general form

$$\mathcal{A}=\begin{bmatrix} A & -B^T \\ \beta^2 B & \alpha^2 A \end{bmatrix}, \tag{2}$$

where α,β are positive numbers. By the introduction of a new, scaled second variable vector $\mathbf{y} := \frac{1}{\alpha^2}\mathbf{y}$, the systems transform into the alternative form

$$\mathcal{A} = \begin{bmatrix} A & -aB^T \\ bB & A \end{bmatrix}, \tag{3}$$

where $a = \frac{1}{\alpha^2}, b = \beta^2$. This form arises in the two-phase version of the Cahn–Hilliard equation used to track interfaces between two fluids with different densities using a stationary grid; see [4, 5].

As we shall see in the sequel, a matrix in the form (2), with $\beta = 1$ arises also in optimization problems for PDE, with a distributed control function, that is, a control function defined in the whole domain of definition of the PDE. For an introduction to such problems, see [6, 7].

Problems of this kind appear in various applications in engineering and geosciences but also in medicine [8] and finance [9]. As a preamble to this topic, we recall that the standard form of a constrained optimization problem with a quadratic function takes the form

$$\min_{\mathbf{u}} \left\{ \frac{1}{2}\mathbf{u}^T A \mathbf{u} - \mathbf{u}^T \mathbf{f} \right\}$$

subject to the constraint $B\mathbf{u} = \mathbf{g}$. Here, $\mathbf{u}, \mathbf{f} \in \Re^n, \mathbf{g} \in \Re^m$, and A is a symmetric and positive definite (spd) matrix of order $n \times n$ and B has order $m \times n$, $m \leq n$. For the existence of a solution, if $m = n$ we must assume that $\dim \Re(B) < m$, where $\Re(B)$ denotes the range of B. The corresponding Lagrangian function with multiplier $\mathbf{p}$ and regularization term $-\alpha \mathbf{p}^T C \mathbf{p}$, where α is a small positive number and C is spd, takes the form

$$\mathfrak{L}(\mathbf{u}, \mathbf{p}) = \frac{1}{2}\mathbf{u}^T A \mathbf{u} - \mathbf{u}^T \mathbf{f} + \mathbf{p}^T (B\mathbf{u} - \mathbf{g}) - \frac{1}{2}\alpha \mathbf{p}^T C \mathbf{p}.$$

By the addition of the regularization term, the Lagrange multiplier vector $\mathbf{p}$ becomes unique.

The necessary first-order conditions for an optimal, saddle point solution lead to

$$\begin{bmatrix} A & B^T \\ B & -\alpha C \end{bmatrix} \begin{bmatrix} \mathbf{u} \\ \mathbf{p} \end{bmatrix} = \begin{bmatrix} \mathbf{f} \\ \mathbf{g} \end{bmatrix}. \tag{4}$$

Here, we can extend the matrix B with $n - m$ zero rows and the vector $\mathbf{g}$ with $n - m$ zero components, to make B of the same order as A. Similarly, C is extended. It is possible to let $C = A$. (Then the $n - m$ correspondingly added components of $\mathbf{p}$ become zero.) As we shall see, in optimal control problems with a distributed control, we get such a form with no need to add zero rows to B.

If we change the sign of $\mathbf{p}$, the corresponding matrix takes the form $\begin{bmatrix} A & -B^T \\ B & A \end{bmatrix}$, i.e. the same form as in (1). The matrix in (4) is indefinite. It can be preconditioned with a block-diagonal matrix, but it leads to eigenvalues on both sides of the origin, which slows down the convergence of the corresponding iterative acceleration method, typically of a conjugate gradient type, such as MINRES in [10]. In this paper we show that much faster convergence can be achieved if instead we precondition $\mathcal{A}$ with a matrix that is a particular perturbation of it, since this leads to positive eigenvalues and no Schur complements need to be handled. We consider then preconditioning of matrices of the form (2) or (3). Thereby we assume that A is symmetric and positive definite, or at least positive semidefinite and $ker(A) \cap ker(B) = \{\emptyset\}$, which will be shown to guarantee that $\mathcal{A}$ is nonsingular.

In Sect. 2 we present a preconditioner to this matrix, but given in the still more general form

$$\mathcal{A} = \begin{bmatrix} A & -aB_2 \\ bB_1 & A \end{bmatrix}, \tag{5}$$

where it is assumed that $H_i = A + \sqrt{ab}B_i$, $i = 1,2$ are regular. It involves only solutions with the matrices H_1 and H_2. Hence, no Schur complements needed to be handled arise here.

In Sect. 3 we perform an eigenvalue analysis of the preconditioning method. This result extends the applicability of the previous results, e.g. in [2] and [4]. Furthermore, the present proofs are sharper and more condensed.

In Sect. 4 we show that certain constrained optimal control problems for PDE with a distributed control can be written in the above two-by-two block form. The results in that section extend related presentations in [7].

Further development of the methods and numerical tests will be devoted to part II of this paper.

The notation $A \leq B$ for symmetric matrices A, B means that $A - B$ is positive semidefinite.

2 The Preconditioner and Its Implementation

Given a matrix in the form (2), we consider first a preconditioner to $\mathcal{A}$ in the form

$$\mathcal{B} = \begin{bmatrix} A & 0 \\ \beta^2 B & \tilde{\alpha}A + \beta B \end{bmatrix} \begin{bmatrix} A^{-1} & 0 \\ 0 & A^{-1} \end{bmatrix} \begin{bmatrix} A & -B^T \\ 0 & \tilde{\alpha}A + \beta B^T \end{bmatrix} \tag{6}$$

where $\tilde{\alpha}$ is a positive preconditioning method parameter to be chosen. A computation shows that

$$\mathcal{B} = \mathcal{A} + \begin{bmatrix} 0 & 0 \\ 0 & (\tilde{\alpha}^2 - \alpha^2)A + \tilde{\alpha}\beta(B + B^T) \end{bmatrix}$$

We show now that an action of its inverse requires little computational work.

Proposition 1. *An action of the inverse of the form of the matrix $\mathcal{B}$ in (6) requires one solution of each of the matrices $A, \tilde{\alpha}A + \beta B$ and $A, \tilde{\alpha}A + \beta B^T$, in this order.*

Proof. To solve a system

$$\mathcal{B}\begin{bmatrix}\mathbf{x}\\ \mathbf{y}\end{bmatrix}=\begin{bmatrix}\mathbf{f}\\ \mathbf{g}\end{bmatrix},$$

solve first

$$\begin{bmatrix}A & 0\\ \beta^2 B & \tilde{\alpha}A+\beta B\end{bmatrix}\begin{bmatrix}\tilde{\mathbf{x}}\\ \tilde{\mathbf{y}}\end{bmatrix}=\begin{bmatrix}\mathbf{f}\\ \mathbf{g}\end{bmatrix},$$

which requires a solution with A and $\tilde{\alpha}A+\beta B$. Solve then

$$\begin{bmatrix}A & -B^T\\ 0 & \tilde{\alpha}A+\beta B^T\end{bmatrix}\begin{bmatrix}\mathbf{x}\\ \mathbf{y}\end{bmatrix}=\begin{bmatrix}A\tilde{\mathbf{x}}\\ A\tilde{\mathbf{y}}\end{bmatrix}=\begin{bmatrix}\mathbf{f}\\ A\tilde{\mathbf{y}}\end{bmatrix}$$

by solving

$$(\tilde{\alpha}A+\beta B)\mathbf{y}=A\tilde{\mathbf{y}},$$

$$\mathbf{z}:=A^{-1}B^T\mathbf{y} \quad \text{as}$$

$$\mathbf{z}=\frac{1}{\beta}(\tilde{\mathbf{y}}-\tilde{\alpha}\mathbf{y})$$

to finally obtain

$$\mathbf{x}=\tilde{\mathbf{x}}+\mathbf{z}.$$ ■

In applications, often A is a mass matrix and B is a stiffness matrix. When A depends on heterogeneous material coefficients, the matrices $\tilde{\alpha}A+\beta B$ and $\tilde{\alpha}A+\beta B^T$ can be better conditioned than A. We show now that by applying the explicit expression for $\mathcal{B}^{-1}$, the separate solution with A in (6) can be avoided.

We find it convenient to show this first for preconditioners $\mathcal{B}$ applied to the matrix $\mathcal{A}$ in the form (3). Here,

$$\mathcal{B}=\begin{bmatrix}A & -aB^T\\ bB & A+\sqrt{ab}(B+B^T)\end{bmatrix}. \tag{7}$$

For its inverse the following proposition holds. For its proof, we assume first that A is spd.

Proposition 2. *Let A be spd. Then*

$$\begin{aligned}\mathcal{B}^{-1}&=\begin{bmatrix}A & -aB^T\\ bB & A+\sqrt{ab}(B+B^T)\end{bmatrix}^{-1}\\ &=\begin{bmatrix}H^{-1}+H^{-T}-H^{-T}AH^{-1} & \sqrt{\frac{a}{b}}(I-H^{-T}A)H^{-1}\\ -\sqrt{\frac{b}{a}}H^{-T}(I-AH^{-1}) & H^{-T}AH^{-1}\end{bmatrix},\end{aligned}$$

where $H=A+\sqrt{ab}B$, which is assumed to be nonsingular.

Proof. For the derivation of the expression for the inverse we use the form of the inverse of a general matrix in two-by-two block form. (However, clearly we can verify the correctness of the expression directly by computation of the matrix times its inverse. An alternative derivation can be based on the Schur–Banachiewicz form of the inverse.) Assume that A_{ii}, $i=1,2$ are nonsingular. Then

$$\begin{bmatrix} A_{11} & A_{12} \\ A_{21} & A_{22} \end{bmatrix}^{-1} = \begin{bmatrix} S_1^{-1} & -A_{11}^{-1}A_{12}S_2^{-1} \\ -S_2^{-1}A_{21}A_{11}^{-1} & S_2^{-1} \end{bmatrix}.$$

Here, the Schur complements $S_i, i=1,2$ equal

$$S_i = A_{ii} - A_{ij}A_{jj}^{-1}A_{ji},\ i,j=1,2,\ i \neq j.$$

Further, $S_2^{-1}A_{21}A_{11}^{-1} = A_{22}^{-1}A_{21}S_1^{-1}$.

For the given matrix it holds

$$\begin{aligned} S_2 &= A + \sqrt{ab}(B+B^T) + abBA^{-1}B^T \\ &= (A+\sqrt{ab}B)A^{-1}(A+\sqrt{ab}B^T). \end{aligned}$$

Further,

$$\begin{aligned} -A_{11}^{-1}A_{12}S_2^{-1} &= aA^{-1}B^T(A+\sqrt{ab}B^T)^{-1}A(A+\sqrt{ab}B)^{-1} \\ &= \sqrt{\frac{a}{b}}A^{-1}((\sqrt{ab}B^T+A)-A)(A+\sqrt{ab}B^T)^{-1}A(A+\sqrt{ab}B)^{-1} \\ &= \sqrt{\frac{a}{b}}(H^{-1}-H^TAH^{-1}) = \sqrt{\frac{a}{b}}(I-H^{-T}A)H^{-1}. \end{aligned}$$

Similarly,

$$-A_{22}^{-1}A_{21}S_1^{-1} = -\sqrt{\frac{b}{a}}H^{-T}(I-AH^{-1}).$$

Finally, since the pivot block in the inverse matrix equals the inverse of the Schur complement, the corresponding equality holds for the pivot block in the matrix itself, that is,

$$A_{11} = (S_1^{-1} - A_{11}^{-1}A_{12}S_2^{-1}A_{21}A_{11}^{-1})^{-1}. \tag{8}$$

Therefore,

$$\begin{aligned} S_1^{-1} &= A_{11}^{-1} + A_{11}^{-1}A_{12}S_2^{-1}A_{21}A_{11}^{-1} \\ &= A^{-1}[A-(I-AH^{-T})A(I-H^{-1}A)]A^{-1} \\ &= H^{-1}+H^{-T}-H^{-T}AH^{-1} \end{aligned}$$

■

Remark 1. Incidently, relation (8) can be seen as a proof of the familiar Sherman–Morrison–Woodbury formula.

We show now that Proposition 2 implies that an action of the matrix $\mathcal{B}^{-1}$ needs only a solution with each of the matrices H and H^T. This result has appeared previously in [4], but the present proof is more condensed and more generally applicable. We will then show it for a matrix in the general form (5).

Guided by the result in Proposition 2, we give now the expression for the inverse of the preconditioner to a matrix in the form (5).

Proposition 3. *Let*

$$\mathcal{B} = \begin{bmatrix} A & -aB_2 \\ bB_1 & A+\sqrt{ab}(B_1+B_2) \end{bmatrix}$$

then

$$\mathcal{B}^{-1} = \begin{bmatrix} H_1^{-1}+H_2^{-1}-H_2^{-1}AH_1^{-1} & \sqrt{\frac{a}{b}}(I-H_2^{-1}A)H_1^{-1} \\ -\sqrt{\frac{b}{a}}H_2^{-1}(I-AH_1^{-1}) & H_2^{-1}AH_1^{-1} \end{bmatrix}$$

where $H_i = A+\sqrt{ab}B_i, i=1,2$, *which are assumed to be nonsingular.*

Proof. We show first that $\mathcal{B}$ is nonsingular. If

$$\mathcal{B}\begin{bmatrix} \mathbf{x} \\ \mathbf{y} \end{bmatrix} = \begin{bmatrix} 0 \\ 0 \end{bmatrix}. \tag{9}$$

then $A\mathbf{x} = aB_2\mathbf{y}$ and

$$A\mathbf{y}+bB_1\mathbf{x}+\sqrt{ab}(B_1+B_2)\mathbf{y} = 0.$$

Then

$$(A+\sqrt{ab}B_1)\mathbf{y}+\sqrt{\frac{b}{a}}(\sqrt{ab}B_1\mathbf{x}+aB_2\mathbf{y}) = 0$$

or

$$(A+\sqrt{ab}B_1)(\sqrt{\frac{b}{a}}\mathbf{x}+\mathbf{y}) = 0.$$

Hence, $\mathbf{x} = -\sqrt{\frac{a}{b}}\mathbf{y}$, so $\sqrt{\frac{a}{b}}(A+\sqrt{ab}B_2)\mathbf{y} = 0$ or $\mathbf{y} = 0$, so (9) has only the trivial solution. The expression for $\mathcal{B}^{-1}$ follows by direct inspection. ∎

Proposition 4. *Assume that* $A+\sqrt{ab}B_i, i=1,2$ *are nonsingular. Then* $\mathcal{B}$ *is nonsingular and a linear system with the preconditioner* $\mathcal{B}$,

$$\begin{bmatrix} A & -aB_2 \\ bB_1 & A+\sqrt{ab}(B_1+B_2) \end{bmatrix}\begin{bmatrix} \mathbf{x} \\ \mathbf{y} \end{bmatrix} = \begin{bmatrix} \mathbf{f}_1 \\ \mathbf{f}_2 \end{bmatrix}$$

can be solved with only one solution with $A+\sqrt{ab}B_1$ *and one with* $A+\sqrt{ab}B_2$.

Proof. It follows form Proposition 3 that an action of the inverse of $\mathcal{B}$ can be written in the form

$$\begin{bmatrix} A & -aB_2 \\ bB_1 & A+\sqrt{ab}(B_1+B_2) \end{bmatrix}^{-1} \begin{bmatrix} \mathbf{f}_1 \\ \mathbf{f}_2 \end{bmatrix} =$$

$$= \begin{bmatrix} H_1^{-1}\mathbf{f}_1 + H_2^{-1}\mathbf{f}_1 - H_2^{-1}AH_1^{-1}\mathbf{f}_1 + \sqrt{\frac{a}{b}}(I-H_2^{-1}A)H_1^{-1}\mathbf{f}_2 \\ -\sqrt{\frac{b}{a}}H_2^{-1}(I-AH_1^{-1})\mathbf{f}_1 + H_2^{-1}AH_1^{-1}\mathbf{f}_2 \end{bmatrix}$$

$$= \begin{bmatrix} H_2^{-1}\mathbf{f}_1 + \mathbf{g} - H_2^{-1}A\mathbf{g} \\ -\sqrt{\frac{b}{a}}H_2^{-1}\mathbf{f}_1 + \sqrt{\frac{b}{a}}H_2^{-1}A\mathbf{g} \end{bmatrix}$$

$$= \begin{bmatrix} \mathbf{g} + H_2^{-1}(\mathbf{f}_1 - A\mathbf{g}) \\ -\sqrt{\frac{b}{a}}H_2^{-1}(\mathbf{f}_1 - A\mathbf{g}) \end{bmatrix} = \begin{bmatrix} \mathbf{g}+\mathbf{h} \\ -\sqrt{\frac{b}{a}}\mathbf{h} \end{bmatrix}$$

where

$$\mathbf{g} = H_1^{-1}(\mathbf{f}_1 + \sqrt{\frac{a}{b}}\mathbf{f}_2), \mathbf{h} = H_2^{-1}(\mathbf{f}_1 - A\mathbf{g}).$$

The computation can take place in the following order:

(i) Solve $H_1\mathbf{g} = \mathbf{f}_1 + \sqrt{\frac{a}{b}}\mathbf{f}_2$.
(ii) Compute $A\mathbf{g}$ and $\mathbf{f}_1 - A\mathbf{g}$.
(iii) Solve $H_2\mathbf{h} = \mathbf{f}_1 - A\mathbf{g}$.
(iv) Compute $\mathbf{x} = \mathbf{g}+\mathbf{h}$ and $\mathbf{y} = -\sqrt{\frac{b}{a}}\mathbf{h}$. ■

Remark 2. In some applications $H_1 = A + \sqrt{ab}B_1$, and $H_2 = A + \sqrt{ab}B_2$ may be better conditioned than A itself. Even if it is not, often software for these combinations exists.

3 Condition Number Bounds

To derive condition number bounds for the preconditioned matrix $\mathcal{B}^{-1}\mathcal{A}$, we consider two cases:

(i) $B_1 = B, B_2 = B^T, A$ is symmetric, A and $B+B^T$ are positive semidefinite, and

$$ker(A) \cap ker(B_i) = \{\emptyset\}, i = 1,2$$

(ii) A is symmetric and positive definite and certain conditions, to be specified later, hold for B_1 and B_2.

3.1 A Is Symmetric and Positive Semidefinite

Assume that conditions (i) hold. Then it follows that $A+\sqrt{ab}B$ and $A+\sqrt{ab}B^T$, and hence also $\mathcal{B}$, are nonsingular. We show first that then $\mathcal{A}$ is also nonsingular.

Proposition 5. *Let condition (i) hold. Then $\mathcal{A}$ is nonsingular.*

Proof. If

$$\begin{bmatrix} A & -aB^T \\ bB & A \end{bmatrix} \begin{bmatrix} \mathbf{x} \\ \mathbf{y} \end{bmatrix} = \begin{bmatrix} \mathbf{0} \\ \mathbf{0} \end{bmatrix}$$

then

$$\mathbf{x}^* A\mathbf{x} - a\mathbf{x}^* B^T \mathbf{y} = \mathbf{0},$$
$$b\mathbf{y}^* B\mathbf{x} + \mathbf{y}^* A\mathbf{y} = \mathbf{0}$$

so $\frac{1}{a}\mathbf{x}^* A\mathbf{x} + \frac{1}{b}\mathbf{y}^* A\mathbf{y} = 0$, where $\mathbf{x}^*$, $\mathbf{y}^*$ denote the complex conjugate vector.

Since A is positive semidefinite, it follows that $\mathbf{x}, \mathbf{y} \in kerA$. But then $B^T\mathbf{y} = \mathbf{0}$ and $B\mathbf{x} = \mathbf{0}$, implying that $\mathbf{x}, \mathbf{y} \in kerB$, so $\mathcal{A}\begin{bmatrix} \mathbf{x} \\ \mathbf{y} \end{bmatrix} = \begin{bmatrix} \mathbf{0} \\ \mathbf{0} \end{bmatrix}$ has only the trivial solution. ■

Proposition 6. *Let $\mathcal{A} = \begin{bmatrix} A & aB^T \\ -bB & A \end{bmatrix}$, where a, b are nonzero and have the same sign and let $\mathcal{B} = \begin{bmatrix} A & aB^T \\ -bB & A+\sqrt{ab}(B+B^T) \end{bmatrix}$. If conditions (i) hold, then the eigenvalues of $\mathcal{B}^{-1}\mathcal{A}$, are contained in the interval $[\frac{1}{2}, 1]$.*

Proof. For the generalized eigenvalue problem

$$\lambda \mathcal{B} \begin{bmatrix} \mathbf{x} \\ \mathbf{y} \end{bmatrix} = \mathcal{A} \begin{bmatrix} \mathbf{x} \\ \mathbf{y} \end{bmatrix}$$

it follows from Proposition 5 that $\lambda \neq 0$. It holds

$$\left(\frac{1}{\lambda} - 1\right) \mathcal{A} \begin{bmatrix} \mathbf{x} \\ \mathbf{y} \end{bmatrix} = \begin{bmatrix} 0 \\ \sqrt{ab}\,(B+B^T)\mathbf{y} \end{bmatrix}$$

Here, $\lambda = 1$ if $\mathbf{y} \in ker(B+B^T)$. If $\lambda \neq 1$, then

$$A\mathbf{x} = -aB^T\mathbf{y}$$

and

$$\left(\frac{1}{\lambda} - 1\right)(\mathbf{y}^* A\mathbf{y} - b\mathbf{y}^* B\mathbf{x}) = \sqrt{ab}\,\mathbf{y}^*(B+B^T)\mathbf{y}$$

or

$$\left(\frac{1}{\lambda} - 1\right)(\mathbf{y}^* A\mathbf{y} + \frac{b}{a}\mathbf{x}^* A\mathbf{x}) = \sqrt{ab}\,\mathbf{y}^*(B+B^T)\mathbf{y}.$$

Since both A and $B+B^T$ are positive semidefinite, it follows that $\lambda \leq 1$.

Further it holds,

$$-\mathbf{y}^* A\mathbf{x} = a\mathbf{y}^* B^T \mathbf{y}$$

so

$$\left(\frac{1}{\lambda}-1\right)(a\mathbf{y}^* B^T \mathbf{y} + b\mathbf{x}^* B\mathbf{x}) = -\sqrt{ab}\,\mathbf{x}^*(B+B^T)\mathbf{y}$$

or

$$\left(\frac{1}{\lambda}-1\right)(a\mathbf{y}^*(B+B^T)\mathbf{y} + b\mathbf{x}^*(B+B^T)\mathbf{x}) = -2\sqrt{ab}\,\mathbf{x}^*(B+B^T)\mathbf{y}.$$

Since $B+B^T$ is positive semidefinite, $|\mathbf{x}| + |\mathbf{y}| \neq 0$, and a and b have the same sign, it follows that

$$\frac{1}{\lambda} - 1 \leq \frac{2\sqrt{ab}\,|\mathbf{x}^*(B+B^T)\mathbf{y}|}{|a|\,\mathbf{y}^*(B+B^T)\mathbf{y} + |\mathbf{b}|\,\mathbf{x}^*(B+B^T)\mathbf{x}} \leq 1,$$

that is, $\lambda \geq \frac{1}{2}$. ■

3.2 A Is Symmetric and Positive Definite

Assume now that A is symmetric and positive definite. Let $\mathcal{A}$ be defined in (5) and let $\tilde{B}_i = \sqrt{ab}A^{-1/2}B_iA^{-1/2}$, $i=1,2$. Assume that the eigenvalues of the generalized eigenvalue problem,

$$\mu(I+\tilde{B}_1\tilde{B}_2)\mathbf{z} = (\tilde{B}_1+\tilde{B}_2)\mathbf{z},\ \mathbf{z}\neq 0 \tag{10}$$

are real and $\mu_{\max} \geq \mu \geq \mu_{\min} > -1$.

Proposition 7. *Let $\mathcal{A}$ be defined in (5), let $\tilde{B}_i = \sqrt{ab}A^{-1/2}B_iA^{-1/2}$, $i=1,2$, and assume that $\tilde{B}_1+\tilde{B}_2$ is spd and (10) holds. Then the eigenvalues of $\mathcal{B}^{-1}\mathcal{A}$ are contained in the interval* $\left[\frac{1}{1+\mu_{\max}}, \frac{1}{1+\mu_{\min}}\right]$.

Proof. $\lambda\mathcal{B}\begin{bmatrix}\mathbf{x}\\ \mathbf{y}\end{bmatrix} = \mathcal{A}\begin{bmatrix}\mathbf{x}\\ \mathbf{y}\end{bmatrix}$ implies

$$(\lambda-1)\begin{bmatrix}A\mathbf{x} - aB_2\mathbf{y}\\ A\mathbf{y} + bB_1\mathbf{x} + \sqrt{ab}(B_1+B_2)\mathbf{y}\end{bmatrix} = \begin{bmatrix}0\\ \sqrt{ab}(B_1+B_2)\mathbf{y}\end{bmatrix}.$$

Hence, a block-diagonal transformation with $\begin{bmatrix}A^{-1/2} & 0\\ 0 & A^{-1/2}\end{bmatrix}$ shows that

$$(\lambda-1)\begin{bmatrix}\tilde{\mathbf{x}}-\sqrt{\frac{a}{b}}\tilde{B}_2\tilde{\mathbf{y}}\\ \tilde{\mathbf{y}}+\sqrt{\frac{b}{a}}\tilde{B}_1\tilde{\mathbf{x}}+(\tilde{B}_1+\tilde{B}_2)\tilde{\mathbf{y}}\end{bmatrix}=\begin{bmatrix}0\\ -(\tilde{B}_1+\tilde{B}_2)\tilde{\mathbf{y}}\end{bmatrix},$$

where $\tilde{\mathbf{x}}=A^{1/2}\mathbf{x}, \tilde{\mathbf{y}}=A^{1/2}\mathbf{y}$.

If $\lambda\neq 1$, then

$$(1-\lambda)\left[I+\tilde{B}_1\tilde{B}_2\right]\tilde{\mathbf{y}}=\lambda(\tilde{B}_1+\tilde{B}_2)\tilde{\mathbf{y}},$$

Hence, by (10),

$$\frac{1}{\lambda}-1=\mu \quad \text{or} \lambda=\frac{1}{1+\mu},$$

which implies the stated eigenvalue bounds. ∎

Corollary 1. *If $B_1=B, B_2=B^T$, and $I+\tilde{B}$ is nonsingular, then*

$$\frac{1}{2}\leq\lambda\leq\frac{1}{1+\mu_{\min}},$$

where $\mu_{\min}>-1$. If the symmetric part of B is positive semidefinite, then

$$\frac{1}{2}\leq\lambda\leq 1.$$

Proof. Since

$$(I-\tilde{B})(I-\tilde{B}^T)\geq 0$$

it follows that

$$I+\tilde{B}\tilde{B}^T\geq\tilde{B}+\tilde{B}^T$$

which implies $\mu\leq 1$ in (10). Similarly,

$$(I+\tilde{B})(I+\tilde{B}^T)\geq 0,$$

that is,

$$I+\tilde{B}\tilde{B}^T\geq-(\tilde{B}+\tilde{B}^T)$$

implies $\mu_{\min}\geq -1$. But $\mu_{\min}>-1$ since $I+\tilde{B}$, and hence $I+\tilde{B}^T$, are nonsingular. If $B+B^T\geq 0$, then $\mu_{\min}=0$. ∎

Corollary 2. *If $B_1=B$, $B_2=B-\delta/\sqrt{ab}A$ for some real number δ, where B is spd and $2B>\delta/\sqrt{ab}A$, that is, $B_1+B_2=2B-\delta/\sqrt{ab}A$ is spd, then*

$$\frac{\sqrt{4-\delta^2}}{2+\sqrt{4-\delta^2}}\leq\lambda\leq 1.$$

Proof. Here, (10) takes the form

$$\mu(I+\tilde{B}^2-\delta\tilde{B})\tilde{\mathbf{z}}=(2\tilde{B}-\delta I)\tilde{\mathbf{z}},$$

where $\tilde{B}=\sqrt{ab}A^{-1/2}BA^{-1/2}$. Let β be an eigenvalue of $\tilde{B}$.
Then

$$\mu=\frac{2\beta-\delta}{1+\beta^2-\beta\delta}.$$

Since $\delta<2\beta$, it follows that $\mu>0$, that is, $\lambda\leq 1$. Further, a computation shows that μ takes its largest value when

$$(2\beta-\delta)^2=2(1+\beta^2-\beta\delta)$$

or

$$(2\beta-\delta)^2=2+\frac{1}{2}(2\beta-\delta)^2-\frac{\delta^2}{2},$$

that is when

$$2\beta-\delta=\sqrt{4-\delta^2}.$$

Then $\mu=2/\sqrt{4-\delta^2}$ and the statement follows from $\lambda=1/(1+\mu)$. ■

Remark 3. Matrices in the form as given in Corollary 2 appear in phase-field models; see, e.g. [4, 5]. For complex-valued systems, normally the coefficients are $a=b=1$. In other applications, such as those in Sects. 4.1 and 4.2, a form such as in Proposition 1 arises. One can readily transform from one form into the other.

Propositions 6 and 7 show that if A is spd and $B+B^T$ is positive semidefinite, then the condition number of the preconditioned matrix satisfies

$$\mathcal{K}(\mathcal{B}^{-1}\mathcal{A})\leq 1+\mu_{\max}\leq 2.$$

Using a preconditioning parameter, as in (6), we derive now a further-improved condition number bound under the assumption that matrix B is symmetric. We consider then the form (2) of matrix $\mathcal{A}$.

Proposition 8. *Let* $\mathcal{A}=\begin{bmatrix}A & -B^T\\ \beta^2 B & \alpha^2 A\end{bmatrix}$, *where* $\alpha>0,\beta>0$, *and let* $\mathcal{B}$ *be defined in (6). Assume that A and B are symmetric and that A is positive definite. Let* $\tilde{B}=\beta A^{-1/2}BA^{-1/2}$ *and assume that* $\tilde{B}$ *has eigenvalues* μ *in the interval* $[\mu_{\min},\mu_{\max}]$, *where* $0\leq|\mu_{\min}|<\mu_{\max}$, *and that* $\frac{\tilde{\alpha}}{\alpha}=|\tilde{\mu}_{\min}|+\sqrt{1+\tilde{\mu}_{\min}^2}$ *where* $\tilde{\mu}_{\min}=\mu_{\min}/\alpha$, $\tilde{\mu}_{\max}=\mu_{\max}/\alpha$. *Then the eigenvalues of* $\mathcal{B}^{-1}\mathcal{A}$ *satisfy*

$$\lambda(\mathcal{B}^{-1}\mathcal{A})=\frac{\alpha^2+\mu^2}{(\tilde{\alpha}+\mu)^2}.$$

For its condition number it holds

$$\min_{\tilde{\alpha}} \kappa(\mathcal{B}^{-1}\mathcal{A}) = \left(\frac{1-\delta}{1+\gamma}\right)^2 + (1+\tilde{\mu}_{\max}^2)\left(\frac{\gamma+\delta}{1+\delta}\right)^2,$$

where $\delta = |\mu_{\min}| / \mu_{\max}$ *and* $\gamma = \sqrt{(1+\tilde{\mu}_{\min}^2)/(1+\tilde{\mu}_{\max}^2)}$. *Here it holds*

$$\frac{\tilde{\alpha}}{\alpha} = \frac{\tilde{\alpha}_{opt}}{\alpha} = \frac{|\tilde{\mu}_{\min}| + \gamma\tilde{\mu}_{\max}^2}{1-\gamma}.$$

If B is positive semidefinite, then

$$\kappa(\mathcal{B}^{-1}A) \le 1 + 1 / \left(1 + \frac{1}{\sqrt{1+\tilde{\mu}_{\max}^2}}\right)^2,$$

where the upper bound is taken for

$$\frac{\tilde{\alpha}}{\alpha} = \frac{1}{\tilde{\mu}_{\max}} + \sqrt{1 + \frac{1}{\tilde{\mu}_{\max}^2}}.$$

Proof. Since both $\mathcal{A}$ and $\mathcal{B}$ are nonsingular, the eigenvalues λ of the generalized eigenvalue problem,

$$\lambda \mathcal{B} \begin{bmatrix} \mathbf{x} \\ \mathbf{y} \end{bmatrix} = \mathcal{A} \begin{bmatrix} \mathbf{x} \\ \mathbf{y} \end{bmatrix}$$

are nonzero. Using (2) and (6), we find

$$\left(\frac{1}{\lambda} - 1\right) \mathcal{A} \begin{bmatrix} \mathbf{x} \\ \mathbf{y} \end{bmatrix} = \begin{bmatrix} 0 \\ [(\tilde{\alpha}^2 - \alpha^2)A + \tilde{\alpha}\beta(B+B^T)]\,\mathbf{y} \end{bmatrix}.$$

If $\mathbf{y} = \mathbf{0}$, then for all $\mathbf{x} \neq \mathbf{0}$ it follows that $\lambda = 1$. For $\lambda \neq 1$, it follows that $A\mathbf{x} = B^T\mathbf{y}$ and, since A is spd,

$$\left(\frac{1}{\lambda} - 1\right)(\beta^2 BA^{-1}B^T + \alpha^2 A)\,\mathbf{y} = [(\tilde{\alpha}^2 - \alpha^2)A + \tilde{\alpha}\beta(B+B^T)]\,\mathbf{y},$$

or

$$\frac{1}{\lambda}(\tilde{B}\tilde{B}^T + \alpha^2 I)\,\tilde{\mathbf{y}} = (\tilde{\alpha}^2 I + \tilde{B}\tilde{B}^T + \tilde{\alpha}(\tilde{B}+\tilde{B}^T))\,\tilde{\mathbf{y}},$$

where $\tilde{B} = \beta A^{-1/2} B A^{-1/2}$ and $\tilde{\mathbf{y}} = A^{1/2}\mathbf{y}$. Since $\tilde{B}$ is symmetric, if $\tilde{B}\tilde{\mathbf{y}} = \mu\tilde{\mathbf{y}}$, $\tilde{\mathbf{y}} \neq 0$, i.e. μ is an eigenvalue of $\tilde{B}$, it follows that μ is real and

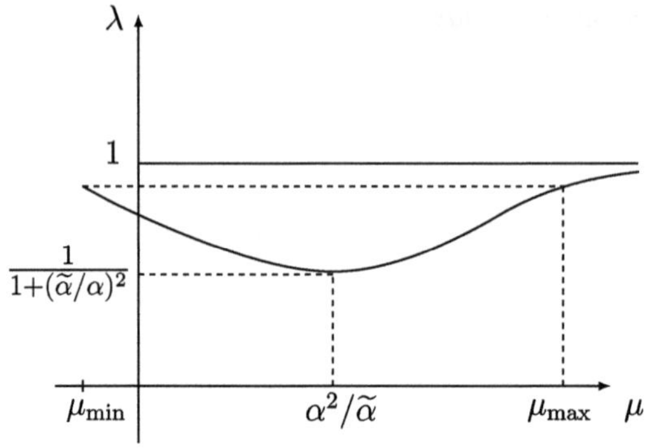

Fig. 1 $\lambda(\mu) = (\alpha^2+\mu^2)/(\tilde{\alpha}+\mu)^2$

$$\lambda = \lambda(\mu) = \frac{\alpha^2+\mu^2}{\tilde{\alpha}^2+\mu^2+2\tilde{\alpha}\mu} = \frac{\alpha^2+\mu^2}{(\tilde{\alpha}+\mu)^2}.$$

The eigenvalues vary as indicated in Fig. 1.

Consider first the case where there exists negative eigenvalues. To get $\lambda < 1$ for negative values of μ, we must choose $(\tilde{\alpha}+\mu)^2 > \alpha^2+\mu^2$, i.e. $\tilde{\alpha}^2+2\tilde{\alpha}\mu-\alpha^2>0$, that is,

$$\tilde{\alpha} > |\,\mu\,| + \sqrt{\mu^2+\alpha^2} \qquad \text{or}$$

$$\frac{\tilde{\alpha}}{\alpha} > |\,\tilde{\mu}_{\min}\,| + \sqrt{1+\tilde{\mu}_{\min}^2}.$$

The minimum value of $\lambda(\mu)$ can be found from

$$\lambda'(\mu) = \frac{2}{(\tilde{\alpha}+\mu)^3}\,(\tilde{\alpha}\mu-\alpha^2) = 0,$$

that is,

$$\min\lambda(\mu) = \lambda_{\min} = \lambda(\alpha^2/\tilde{\alpha}) = \frac{\alpha^2+\alpha^4/\tilde{\alpha}^2}{(\tilde{\alpha}+\alpha^2/\tilde{\alpha})^2} = \frac{1}{1+(\tilde{\alpha}/\alpha)^2}$$

To minimize the condition number, it can be seen (cf. Fig. 1) that we must choose $\tilde{\alpha}$ such that

$$\lambda(\mu_{\min}) = \lambda(\mu_{\max}),$$

that is,

$$\lambda_{\max} = \frac{\alpha^2+\mu_{\min}^2}{(\tilde{\alpha}-|\mu_{\min}|)^2} = \frac{\alpha^2+\mu_{\max}^2}{(\tilde{\alpha}+\mu_{\max})^2},$$

or

$$\frac{\tilde{\alpha}/\alpha - \tilde{\mu}_{\min}}{\tilde{\alpha}/\alpha + \tilde{\mu}_{\max}} = \gamma := \left(\frac{1+\tilde{\mu}_{\min}^2}{1+\tilde{\mu}_{\max}^2}\right)^{1/2}.$$

Here $\gamma < 1$, since by assumption $\mu_{\max} > | \mu_{\min} |$. Hence,

$$\frac{\tilde{\alpha}}{\alpha} = \frac{\tilde{\alpha}_{opt}}{\alpha} = \frac{| \tilde{\mu}_{\min} | + \gamma\tilde{\mu}_{\max}}{1-\gamma}$$

Then

$$\kappa(\mathcal{B}^{-1}\mathcal{A}) = \frac{\lambda_{\max}}{\lambda_{\min}} = \frac{1+\tilde{\mu}_{\max}^2}{\left(\frac{|\tilde{\mu}_{\min}|+\gamma\tilde{\mu}_{max}}{1-\gamma} + \tilde{\mu}_{\max}\right)^2}[1+\left(\frac{| \tilde{\mu}_{\min} | + \gamma\tilde{\mu}_{\max}}{1-\gamma}\right)^2]$$

$$= \frac{1+\tilde{\mu}_{\max}^2}{(| \tilde{\mu}_{\min} | + \tilde{\mu}_{\max})^2}\left[(1-\gamma)^2 + (| (\tilde{\mu}_{\min} | + \gamma\tilde{\mu}_{\max})^2\right].$$

It holds

$$(1-\gamma)^2 = \left(\frac{1-\gamma^2}{1+\gamma}\right)^2 = \frac{(\tilde{\mu}_{\max}^2 - \tilde{\mu}_{\min}^2)^2}{(1+\tilde{\mu}_{\max}^2)(1+\gamma)^2} = \frac{(\tilde{\mu}_{\max} + | \tilde{\mu}_{\min} |)^2(\tilde{\mu}_{\max} + \tilde{\mu}_{\min} |)^2}{(1+\tilde{\mu}_{\max}^2)(1+\gamma)^2}.$$

Hence,

$$\kappa(\mathcal{B}^{-1}\mathcal{A}) = \left(\frac{1-\delta}{1+\gamma}\right)^2 + (1+\tilde{\mu}_{\max}^2)\left(\frac{\gamma+\delta}{1+\delta}\right)^2.$$

If B is positive semidefinite, then we let $\mu_{\min} = 0$ so $\delta = 0, \gamma = 1/\sqrt{1+\tilde{\mu}_{\max}^2}$ and

$$\kappa(\mathcal{B}^{-1}\mathcal{A}) \le 1 + \frac{1}{(1+\frac{1}{\sqrt{1+\tilde{\mu}_{\max}^2}})^2}$$

which is taken for

$$\frac{\tilde{\alpha}}{\alpha} = \frac{1}{\tilde{\mu}_{\max}} + \sqrt{1+\frac{1}{\tilde{\mu}_{\max}^2}} \qquad \blacksquare$$

Remark 4. If $\mu_{\min} = 0$ then $\kappa(\mathcal{B}^{-1}\mathcal{A}) < 2$ and if $\mu_{\max} \to \infty$ then $\tilde{\alpha} \to \alpha$ and $\kappa(\mathcal{B}^{-1}\mathcal{A}) \to 2$. If $\tilde{\mu}_{\max} = 1$ then $\tilde{\alpha}/\alpha = 1+\sqrt{2}$ and

$$\kappa(\mathcal{B}^{-1}\mathcal{A}) \le 1 + \frac{1}{(1+\frac{1}{\sqrt{2}})^2} \approx 1.34.$$

Remark 5. As is well known, when eigenvalue bounds of a preconditioned matrix, as in the case with $\mathcal{B}^{-1}\mathcal{A}$, are known, then one can replace the conjugate gradient (CG) with a Chebyshev acceleration method. This can be important, for instance, if one uses some domain decomposition method for massively parallel computations, as it avoids the global communication of inner products used in CG methods.

4 Distributed Optimal Control of Elliptic and Oseen Equations

Let Ω be a bounded domain in $\Re^d$, $d = 1,2$ or 3, and let $\partial\Omega$ be its boundary which is assumed to be sufficiently smooth. Let $L^2(\Omega), H^1(\Omega)$ and $H_0^1(\Omega)$ denote the standard Lebesgue and Sobolev spaces of functions in Ω, where $H_0^1(\Omega)$ denotes functions with homogeneous Dirichlet boundary values at $\Gamma_0 \subset \partial\Omega$ where Γ_0 has a nonzero measure. Further, let $(\cdot,\cdot)$ and $\|\cdot\|$ denote the inner product and norm, respectively, in $L^2(\Omega)$, both for scalar and vector functions. Extending, but following [7], and based on [6], we consider now two optimal control problems. In [7] a block-diagonal preconditioner is used. Here we apply instead the preconditioner presented in Sect. 2.

4.1 An Elliptic State Equation

The problem is to find the state $u \in H_0^1(\Omega)$ and the control function $y \in L^2(\Omega)$ that minimizes the *cost function*

$$J(u,y) = \frac{1}{2} \| u - u_d \|^2 + \frac{\alpha}{2} \| y \|^2$$

subject to the *state equation*

$$\begin{cases} -\Delta u + (\mathbf{b} \cdot \nabla) u = y \text{ in } \Omega \\ \text{with boundary conditions} \\ u = 0 \text{ on } \Gamma_0;\ \nabla u \cdot \mathbf{n} = 0 \text{ on } \Gamma_1 = \partial\Omega \setminus \Gamma_0. \end{cases} \tag{11}$$

Here $\mathbf{b}$ is a given, smooth vector. For simplicity, assume that $\mathbf{b} \cdot \mathbf{n} |_{\Gamma_1} = 0$. Further, u_d denotes a given, desired state (possibly obtained by measurements at some discrete points and then interpolated to the whole of Ω). The forcing term y acts as a control of the solution to the state equation. By including the control in the cost functional, the problem becomes well posed. The regularization parameter α, chosen a priori, is a positive parameter chosen sufficiently small to obtain a solution close to the desired state, but not too small and also not too large as this leads to ill conditioning. This is similar to the familiar Tikhonov regularization. The variational (weak) formulation of (11) reads

$$(\nabla u, \nabla v) + (\mathbf{b} \cdot \nabla u, v) = (y, v) \quad \forall v \in H_0^1(\Omega). \tag{12}$$

The Lagrangian formulation associated with the optimization problem takes the form

$$\mathcal{L}(u,y,p) = J(u,y) + (\nabla u, \nabla p) + (\mathbf{b} \cdot \nabla u, p) - (y, p),$$

where $p \in H_0^1(\Omega)$ is the Lagrange multiplier corresponding to the constraint (12). The weak formulation of the corresponding first-order necessary conditions,

$$\left(\frac{\partial \mathcal{L}}{\partial u}, v\right) = 0 \quad \forall v \in H_0^1(\Omega)$$

$$\left(\frac{\partial \mathcal{L}}{\partial y}, z\right) = 0 \quad \forall z \in L^2(\Omega)$$

$$\left(\frac{\partial \mathcal{L}}{\partial p}, q\right) = 0 \quad \forall q \in H_0^1(\Omega)$$

gives now the system of optimality equations:

$$\begin{cases} (u,v) + (\nabla v, \nabla p) + (\mathbf{b}\cdot\nabla v, p) &= (u_d, v) \; \forall v \in H_0^1(\Omega) \\ \alpha(y,z) - (z,p) &= 0 \qquad \forall z \in L^2(\Omega) \\ (\nabla u, \nabla q) + (\mathbf{b}\cdot\nabla u, q) - (y,q) = 0 & \qquad\;\; \forall q \in H_0^1(\Omega) \end{cases},$$

which defines the solution $(u,y) \in H_0^1(\Omega) \times L^2(\Omega)$ of the optimal control problem with Lagrange multiplier $p \in H_0^1(\Omega)$. From the second equation, it follows that the control function y is related to the Lagrange multiplier as $y = \frac{1}{\alpha} p$. Eliminating y and applying the divergence theorem, this leads to the reduced system

$$\begin{aligned} (u,v) + (\nabla v, \nabla p) - (\mathbf{b}\cdot\nabla p, v) &= (u_d, v) \; \forall v \in H_0^1(\Omega) \\ (\nabla u, \nabla q) + (\mathbf{b}\cdot\nabla u, q) - \tfrac{1}{\alpha}(p,q) &= 0 \qquad \forall q \in H_0^1(\Omega). \end{aligned}$$

Since the problem is regularized, we may here use equal-order finite element approximations, for instance, piecewise linear basis functions on a triangular mesh (in 2D), for both the state variable u and the co-state variable p. This leads to a system of the form

$$\begin{bmatrix} M & K^T \\ K & -\alpha^{-1}M \end{bmatrix} \begin{bmatrix} u_h \\ p_h \end{bmatrix} = \begin{bmatrix} f_h \\ 0 \end{bmatrix},$$

where index h denotes the corresponding mesh parameter. Here M corresponds to a mass matrix and K, which has the same order as M, to the second-order elliptic operator with a first-order advection term.

By a change of sign of p_h, it can be put in the form

$$\begin{bmatrix} M & -K^T \\ K & \alpha^{-1}M \end{bmatrix} \begin{bmatrix} u_h \\ -p_h \end{bmatrix} = \begin{bmatrix} f_h \\ 0 \end{bmatrix}$$

and we can directly apply the preconditioner from Sects. 2 and 3, and the derived spectral condition number bounds. If

$$\int_{\Omega} \left(|\nabla u|^2 - \frac{1}{2}(\nabla \cdot \mathbf{b})u^2 \right) \geq 0,$$

i.e. if the operator is semi-coercive, then $K + K^T$ is positive semidefinite and it follows from Proposition 6 that the corresponding spectral condition number is bounded by 2, with eigenvalues in the interval $1/2 \leq \lambda \leq 1$.

Remark 6. In [7], a block-diagonal preconditioner,

$$\mathcal{D} = \begin{bmatrix} A + \alpha^{1/2}B & 0 \\ 0 & \alpha^{-1}A + \alpha^{-1/2}B \end{bmatrix},$$

is used for the saddle point matrix

$$\mathcal{A} = \begin{bmatrix} A & B \\ B & -\alpha^{-1}A \end{bmatrix},$$

where $B = B^T$ and A is symmetric and positive semidefinite, and $\ker(A) \cap \ker(B) = \{0\}$, so $A + \alpha^{1/2}B$ is symmetric and positive definite.

By assumptions made, from the generalized eigenvalue problem

$$A\mathbf{z} = \mu(A + \alpha^{1/2}B)\mathbf{z},$$

it follows that here $\mu \in [0,1]$ and it follows further readily that the preconditioned matrix $\mathcal{D}^{-1}\mathcal{A}$ has eigenvalues that satisfy

$$|\lambda| = \sqrt{\mu_i^2 + (1-\mu_i)^2} \quad \text{for some } \mu_i \in [0,1],$$

that is, $1/\sqrt{2} \leq |\lambda| \leq 1$. Hence, the eigenvalues are located in the double interval:

$$I = [-1, -1/\sqrt{2}] \cup [1/\sqrt{2}, 1].$$

For such eigenvalues in intervals on both sides of the origin, an iterative method of conjugate gradient type, such as MINRES, needs typically the double number of iterations, as for eigenvalues in a single interval on one (positive) side of the origin, to reach convergence; see e.g. [11]. This can be seen from the polynomial approximation problem

$$\min_{x \in I,\ P_k \in \pi_k^0} |P_k(x)| \leq \varepsilon$$

where π_k^0 denotes the set of polynomials of degree k, normalized at the origin, i.e. $P_k(0) = 1$.

Since the number of iterations increases as $O(\sqrt{\kappa})$, where $\kappa = |\lambda_{\max}| / |\lambda_{\min}|$ is the condition number, it follows that an indefinite interval condition number $\kappa = \sqrt{2}$ typically corresponds to a one-sided condition number of $4\sqrt{2}$.

The method proposed in the present paper has a condition number bounded by 2 and needs therefore a number of iterations about $\simeq \frac{\sqrt{2}}{2^{5/4}} = \frac{2^{1/4}}{2} \simeq 0.6$ times those for a corresponding block diagonal preconditioner. However, even if the block-diagonal preconditioning method requires more iterations, each iteration may be cheaper than in the method proposed in this paper. An actual comparison of the methods will appear.

4.2 Distributed Optimal Control of the Oseen Problem

In [7], Stokes equation is considered. Here, we extend the method to the Oseen equation and consider the velocity tracking problem for the stationary case, which reads as follows:

Find the velocity $\mathbf{u} \in H_0^1(\Omega)^d$; the pressure $p \in L_0^2(\Omega)$, where $L_0^2(\Omega) = \{q \in L^2(\Omega), \int_\Omega q\,dx = 1\}$; and the control function $\mathbf{f}$, which minimize the cost function

$$\mathcal{J}(\mathbf{u},\mathbf{f}) = \frac{1}{2}\|\mathbf{u}-\mathbf{u}_d\|^2 + \frac{1}{2}\alpha\|\mathbf{f}\|^2,$$

subject to state equation for an incompressible fluid velocity $\mathbf{u}$, such that

$$\begin{cases} -\Delta\mathbf{u} + (\mathbf{b}\cdot\nabla)\mathbf{u} + \nabla p = \mathbf{f} \text{ in } \Omega \\ \nabla\cdot\mathbf{u} = 0 \text{ in } \Omega \end{cases}$$

and boundary conditions $\mathbf{u} = \mathbf{0}$ on $\partial\Omega_1$, $\mathbf{u}\cdot\mathbf{n} = 0$ on $\partial\Omega_2 = \partial\Omega\backslash\partial\Omega_1$, where $\mathbf{n}$ denotes the outward normal vector to the boundary $\partial\Omega$.

Here $\mathbf{u}_d$ is the desired solution and $\alpha > 0$ is a regularization parameter, used to penalize too large values of the control function. Further, $\mathbf{b}$ is a given, smooth vector. For simplicity we assume that $\mathbf{b} = \mathbf{0}$ on $\partial\Omega_1$ and $\mathbf{b}\cdot\mathbf{n} = 0$ on $\partial\Omega_2$.

In a Navier–Stokes problem, solved by a Picard iteration using the frozen coefficient framework, $\mathbf{b}$ equals the previous iterative approximation of $\mathbf{u}$, in which case normally $\nabla\cdot\mathbf{u} = 0$ in Ω. For simplicity, we assume that this holds here also, that is, $\nabla\cdot\mathbf{b} = 0$.

The variational form of the state equation reads as follows:

$$\begin{cases} (\nabla\mathbf{u},\nabla\tilde{\mathbf{u}}) + (\mathbf{b}\cdot\nabla\mathbf{u},\tilde{\mathbf{u}}) - (\nabla\tilde{\mathbf{u}},p) = (\mathbf{f},\tilde{\mathbf{u}}) \ \forall\tilde{\mathbf{u}} \in H_0^1(\Omega) \\ (\nabla\cdot\mathbf{u},\tilde{p}) = 0 \quad \forall\tilde{p} \in L_0^2(\Omega) \end{cases}$$

The Lagrangian functional, corresponding to the optimization problem, is given by

$$\mathcal{L}(\mathbf{u},p,\mathbf{v},q,\mathbf{f}) = \mathcal{J}(\mathbf{u},\mathbf{f}) + (\nabla\mathbf{u},\nabla\mathbf{v}) + (\mathbf{b}\cdot\nabla\mathbf{u},\mathbf{v}) - (\nabla\cdot\mathbf{v},p) - (\nabla\cdot\mathbf{u},q) - (\mathbf{f},\mathbf{v})$$

where $\mathbf{v}$ is the Lagrange multiplier function for the state equation and q for its divergence constraint. Applying the divergence theorem, the divergence condition $\nabla \cdot \mathbf{b} = 0$ and the boundary conditions, we can write

$$\int_{\Omega} \mathbf{b} \cdot \nabla \tilde{\mathbf{u}} \cdot \mathbf{v} d\Omega = -\int_{\Omega} (\mathbf{b} \cdot \underline{\nabla} \mathbf{v}) \cdot \tilde{\mathbf{u}} d\Omega.$$

The five first-order necessary conditions for an optimal solution take then the form

$$\begin{aligned} (\mathbf{u},\tilde{\mathbf{u}}) + (\nabla \mathbf{v}, \nabla \tilde{\mathbf{u}}) - (\mathbf{b} \cdot \nabla \mathbf{v}, \tilde{\mathbf{u}}) - (\nabla \cdot \tilde{\mathbf{u}}, q) &= (\mathbf{u}_d, \tilde{\mathbf{u}}) \quad \forall \tilde{\mathbf{u}} \in H_0^1(\Omega)^d \\ (\nabla \cdot \mathbf{v}, \tilde{p}) &= 0 \qquad \forall \tilde{p} \in L_0^2(\Omega) \\ (\nabla \mathbf{u}, \nabla \tilde{\mathbf{v}}) + (\mathbf{b} \cdot \nabla \mathbf{u}, \tilde{\mathbf{v}}) - (\nabla \cdot \tilde{\mathbf{v}}, p) - (\mathbf{f}, \tilde{\mathbf{v}}) &= 0 \qquad \forall \tilde{\mathbf{v}} \in H_0^1(\Omega)^d \\ (\nabla \cdot \mathbf{u}, \tilde{q}) &= 0 \qquad \forall \tilde{q} \in L_0^2(\Omega) \\ \alpha(\mathbf{f}, \tilde{\mathbf{f}}) - (\tilde{\mathbf{f}}, \mathbf{v}) &= 0 \qquad \forall \tilde{\mathbf{f}} \in L^2(\Omega) \end{aligned} \tag{13}$$

Here $\mathbf{u}, p, \mathbf{f}$ are the solutions of the optimal control problem with $\mathbf{v}, q$ as Lagrange multipliers for the state equation, and $\tilde{\mathbf{u}}, \tilde{\mathbf{v}}, \tilde{p}, \tilde{q}, \tilde{\mathbf{f}}$ denote corresponding test functions.

As in the elliptic control problem, the control function $\mathbf{f}$ can be eliminated, $\mathbf{f} = \alpha^{-1}\mathbf{v}$, resulting in the reduced system,

$$\begin{aligned} (\mathbf{u},\tilde{\mathbf{u}}) + (\nabla \mathbf{v}, \nabla \tilde{\mathbf{u}}) - (\mathbf{b} \cdot \nabla \mathbf{v}, \tilde{\mathbf{u}}) - (\nabla \cdot \tilde{\mathbf{u}}, q) &= (\mathbf{u}_d, \tilde{\mathbf{u}}) \quad \forall \tilde{\mathbf{u}} \in H_0^1(\Omega)^d \\ (\nabla \mathbf{u}, \nabla \tilde{\mathbf{v}}) + (\mathbf{b} \cdot \nabla \mathbf{u}, \tilde{\mathbf{v}}) - (\nabla \cdot \tilde{\mathbf{v}}, p) - \alpha^{-1}(\mathbf{v}, \tilde{\mathbf{v}}) &= 0 \qquad \forall \tilde{\mathbf{v}} \in H_0^1(\Omega)^d \\ (\nabla \cdot \mathbf{v}, \tilde{p}) &= 0 \qquad \forall \tilde{p} \in L_0^2(\Omega) \\ (\nabla \cdot \mathbf{u}, \tilde{q}) &= 0 \qquad \forall \tilde{q} \in L_0^2(\Omega) \end{aligned}, \tag{14}$$

To discretize (14) we use an LBB-stable pair of finite element spaces for the pair $(\mathbf{u}, \mathbf{v})$ and (p, q). In [7] the Taylor–Hood pair with $\{Q2, Q2, Q1, Q1\}$ is used, namely, piecewise quadratic basis functions for $\mathbf{u}, \mathbf{v}$ and piecewise bilinear basis functions for p, q for a triangular mesh. The corresponding discrete system takes the form

$$\begin{bmatrix} M & -L+C & 0 & D^T \\ L+C & \alpha^{-1}M & D^T & 0 \\ 0 & D & 0 & 0 \\ D & 0 & 0 & 0 \end{bmatrix} \begin{bmatrix} \mathbf{u} \\ -\mathbf{v} \\ p \\ q \end{bmatrix} = \begin{bmatrix} M\mathbf{u}_d \\ \mathbf{0} \\ \mathbf{0} \\ \mathbf{0} \end{bmatrix}, \tag{15}$$

where we have changed the sign of $\mathbf{v}$. Here D comes from the divergence terms. Further, M is the mass matrix and $L+C$ is the discrete operator, corresponding to the convection–diffusion term $-\Delta\mathbf{u} + \mathbf{b} \cdot \nabla \mathbf{u}$ and $-L+C$ to $\Delta \mathbf{v} + \mathbf{b} \cdot \nabla \mathbf{v}$, respectively. Due to the use of an inf–sup (LBB)-stable pairs of finite element spaces, the divergence matrix D has full rank.

As for saddle point problems of similar type, one can use either a grad–div stabilization or a div–grad stabilization. In the first case we add the matrix

$D^TW^{-1}D$ to M and $\alpha^{-1}D^TW^{-1}D$ to $\alpha^{-1}M$, respectively, possibly multiplied with some constant factor, where W is a weight matrix. If W is taken as the discrete Laplacian matrix, then $D^TW^{-1}D$ becomes a projection operator onto the orthogonal complement of the solenoidal vectors.

The other type of stabilization consists of perturbing the zero block matrix in (15) by $\varepsilon \begin{bmatrix} \Delta & 0 \\ 0 & \Delta \end{bmatrix}$, where ε is a small parameter, typically $\varepsilon = O(h^2)$ with h being the space discretization parameter. In that case there is no need to use LBB-stable elements; see, e.g. [12] for more details. In the present paper, however, we use LBB-stable elements and there is no need to use any additional regularization at all but consider instead the solution of the system with the Schur complement matrix system:

$$\begin{bmatrix} 0 & D \\ D & 0 \end{bmatrix} \begin{bmatrix} M & -L+C \\ L+C & \alpha^{-1}M \end{bmatrix}^{-1} \left(\begin{bmatrix} 0 & D^T \\ D^T & 0 \end{bmatrix} \begin{bmatrix} p \\ q \end{bmatrix} - \begin{bmatrix} M\mathbf{u}_d \\ 0 \end{bmatrix} \right) = \begin{bmatrix} 0 \\ 0 \end{bmatrix} \tag{16}$$

This system can be solved by inner–outer iterations. To compute the residuals, we must then solve inner systems with the matrix $\begin{bmatrix} M & -L+C \\ L+C & \alpha^{-1}M \end{bmatrix}$, which takes place in the way discussed earlier in Sect. 2. To recall, only systems with $M + \sqrt{\alpha}(L+C)$ and $M + \sqrt{\alpha}(L-C)$ have to be solved. Further, as is seen from (16), the corresponding systems which actually arise have the form $D[M+\sqrt{\alpha}(L+C)]^{-1}D^T$ and $D[M+\sqrt{\alpha}(L-C)]^{-1}D^T$. At least for not too large convection terms, related to the diffusion term, these systems are well conditioned and can be preconditioned with a mass matrix or a mass matrix minus a small multiple times the Laplacian.

To avoid the need to solve inner systems and for stronger convections, it may be better to use a block-triangular factorization of the matrix in (15). For the arising inner systems with $M + \sqrt{\alpha}(L+C)$ and $M + \sqrt{\alpha}(L-C)$, it can be efficient to use some off-the-shelf software, such as some algebraic multigrid (AMG) method; see [13, 14]. In [15] and [13] numerical tests are reported, showing that AGMG [13], as one choice of an AMG method, performs much better than some other possible methods.

The perturbations due to the use of inner iterations with stopping criteria lead in general to complex eigenvalues. A generalized conjugate gradient method of GMRES [16] type can be used. Such methods go under different names and have been referred to as nonlinear conjugate gradient, variable preconditioned conjugate gradient [17] and flexible GMRES [18]. Since, due to the accurate preconditioning, there are few iterations, the additional cost for having a full length Krylov subspace, involving all previous search directions, is not much heavier than if a conjugate gradient method with vectors, orthogonal with respect to a proper inner product and, hence, short recursions, is used.

We remark, however, that such a method has been constructed for indefinite matrices in [19], based on inner products, defined by the matrix

$$\mathcal{D} = \begin{bmatrix} \hat{M} - \hat{M}_0 & 0 \\ 0 & S_0 \end{bmatrix},$$

where $\hat{M}_0$ is an approximation of $\hat{M}$, such that $\hat{M}_0 < \hat{M}$ and $S_0 < \hat{B}\hat{M}^{-1}\hat{B}^T$ is an spd approximation of the Schur complement matrix for the two-by-two block system $\begin{bmatrix} \hat{M} & \hat{B}^T \\ \hat{B} & 0 \end{bmatrix}$. This makes the matrix $\begin{bmatrix} \hat{M}_0 & 0 \\ \hat{B} & -S_0 \end{bmatrix}^{-1} \begin{bmatrix} \hat{M} & \hat{B}^T \\ \hat{B} & 0 \end{bmatrix}$ self-adjoint with respect to that inner product. The drawback of the method is the need to properly scale the approximation $\hat{M}_0$ to satisfy $\hat{M}_0 < \hat{M}$, and furthermore, $\hat{M}_0$ must be fixed, i.e. cannot be implicitly defined via variable inner iterations.

In our case, the corresponding preconditioning matrix defined in Sect. 2 satisfies $\hat{M}_0 > \hat{M}$, but there is no need to scale it. Furthermore, we may apply inner iterations for this preconditioner and also for the Schur complement matrix, hence the corresponding matrix $\hat{M}_0$ is in general not fixed so the above inner product method is not applicable.

The presentation of block-triangular factorization preconditioner and approximations of the arising Schur complement preconditioners with numerical tests will be devoted.

Acknowledgements This work was supported by the European Regional Development Fund in the IT4 Innovations Centre of Excellence project (CZ.1.05/1.1.00/02.0070).

Discussions with Maya Neytcheva on implementation aspects of the method are gratefully acknowledged.

References

1. van Rienen, U.: Numerical Methods in Computational Electrodynamics. Linear Systems in Practical Applications. Springer, Berlin (1999)
2. Axelsson, O., Kucherov, A.: Real valued iterative methods for solving complex symmetric linear systems. Numer. Lin. Algebra Appl. **7**, 197–218 (2000)
3. Benzi, M., Bertaccini, D.: Block preconditioning of real-valued iterative algorithms for complex linear systems. IMA J. Numer. Anal. **28**, 598–618 (2008)
4. Axelsson, O., Boyanova, P., Kronbichler, M., Neytcheva, M., Wu, X.: Numerical and computational efficiency of solvers for two-phase problems. Comput. Math. Appl. **65**, 301–314 (2012). http://dx.doi.org./10.1016/j.camva.2012.05.020
5. Boyanova, P.: On numerical solution methods for block-structured discrete systems. Doctoral thesis, Department of Information Technology, Uppsala University, Sweden (2012). http://urn.kb.se/resolve?urn=urn:nbn:se:uu:diva-173530
6. Lions, J.-L.: Optimal Control of Systems Governed by Partial Differential Equations. Springer, Berlin (1971)
7. Zulehner, W.: Nonstandard norms and robust estimates for saddle-point problems. SIAM J. Matrix Anal. Appl. **32**, 536–560 (2011)
8. Arridge, S.R.: Optical tomography in medical imaging. Inverse Probl. **15**, 41–93 (1999)
9. Egger, H., Engl, H.W.: Tikhonov regularization applied to the inverse problem of option pricing: convergence analysis and rates. Inverse Probl. **21**, 1027–1045 (2005)

10. Paige, C.C., Saunders, M.A.: Solution of sparse indefinite systems of linear equations. SIAM J. Numer. Anal. **12**, 617–629 (1975)
11. Axelsson, O., Barker, V.A.: Finite Element Solution of Boundary Value Problems. Theory and Computation. Academic, Orlando, FL (1984)
12. Brezzi, F., Fortin, M.: Mixed and Hybrid Finite Elements Methods. Springer, Berlin (1991)
13. Notay, Y.: The software package AGMG. http://homepages.ulb.ac.be/~ynotay/
14. Vassilevski, P.: Multilevel Block Factorization Preconditioners. Springer, New York (2008)
15. Notay, Y.: Aggregation-based algebraic multigrid for convection-diffusion equations. SIAM J. Sci. Comput. **34**, A2288–A2316 (2012)
16. Saad, Y., Schultz, M.H.: GMRES: a generalized minimal residual algorithm for solving nonsymmetric linear systems. SIAM J. Sci. Stat. Comput. **7**, 856–869 (1986)
17. Axelsson, O., Vassilevski, P.S.: A black box generalized conjugate gradient solver with inner iterations and variable-step preconditioning. SIAM J. Matrix Anal. Appl. **12**(4), 625–644 (1991)
18. Saad, Y.: A flexible inner-outer preconditioned GMRES algorithm. SIAM J. Sci. Comput. **14**, 461–469 (1993)
19. Bramble, J.H., Pasciak, J.E.: A preconditioning technique for indefinite systems resulting from mixed approximations of elliptic problems. Math. Comput. **50**, 1–17 (1988)

A Multigrid Algorithm for an Elliptic Problem with a Perturbed Boundary Condition

Andrea Bonito and Joseph E. Pasciak

Abstract We discuss the preconditioning of systems coupling elliptic operators in $\Omega \subset \mathbb{R}^d$, $d = 2,3$, with elliptic operators defined on hypersurfaces. These systems arise naturally when physical phenomena are affected by geometric boundary forces, such as the evolution of liquid drops subject to surface tension. The resulting operators are sums of interior and boundary terms weighted by parameters. We investigate the behavior of multigrid algorithms suited to this context and demonstrate numerical results which suggest uniform preconditioning bounds that are level and parameter independent.

Keywords Multigrid • Laplace-Beltrami • Surface Laplacian • Parameter dependent problems

Mathematics Subject Classification (2010): 65N30, 65N55

1 Introduction

There has been considerable interest in geometric differential equations in recent years as they play a crucial role in many applications. In this paper, we consider one aspect of developing efficient preconditioners for the systems of algebraic equations resulting from finite element approximation to these problems.

Let $\Omega \subset \mathbb{R}^d$, $d = 2,3$, be a bounded domain separated into two subdomains by an interface γ. We denote the subdomains by Ω_i, $i = 1,2$. This paper focuses on the

A. Bonito • J.E. Pasciak (✉)
Department of Mathematics, Texas A&M University, College Station, TX 77843-3368, USA
e-mail: bonito@math.tamu.edu; pasciak@math.tamu.edu

O.P. Iliev et al. (eds.), *Numerical Solution of Partial Differential Equations: Theory, Algorithms, and Their Applications*, Springer Proceedings in Mathematics & Statistics 45, DOI 10.1007/978-1-4614-7172-1_4,

study of an optimal multigrid algorithm for interactions between "bulk" perturbed elliptic operators and the surface Laplacian. Such variational problems involving interaction between diffusion operators on domains and surfaces appear in many different contexts, see, for example, [1,4,10,15,18,20,21].

As an illustration, we now describe an application involving capillary flow [1]. We consider the evolution of two different fluids inside a domain Ω separated by a moving interface $\gamma(t)$, $t > 0$. The interface γ is described as the deformation of a smooth reference domain $\hat{\gamma}$. We denote by $\mathbf{x}(t) : \hat{\gamma} \to \gamma(t)$ the mapping relating the two interfaces. Typically, $\mathbf{x}(t)$ is bi-Lipschitz, but we will require more smoothness on γ and therefore on $\mathbf{x}$. The fluids are assumed to be governed by the Stokes equations, i.e., the velocities $\mathbf{u}_i$ and the pressures p_i, $i = 1,2$, satisfy on each subdomain Ω_i:

$$\frac{\partial}{\partial t}\mathbf{u}_i - 2\mathrm{div}(D(\mathbf{u}_i)) + \nabla p_i = \mathbf{f}_i, \quad \mathrm{div}(\mathbf{u}_i) = 0, \qquad \text{on} \quad \Omega_i,$$

where $D(\mathbf{v}) := \frac{1}{2}((\nabla\mathbf{v}) + (\nabla\mathbf{v})^T)$ and $\{\mathbf{f}_i\}$ are given body forces. The surface tension effect appears together with the continuity of the velocity, i.e.,

$$\mathbf{u}_1 = \mathbf{u}_2, \quad \text{on} \quad \gamma,$$
$$(2D(\mathbf{u}_1) - p_1)\boldsymbol{\nu}_1 + (2D(\mathbf{u}_2) - p_2)\boldsymbol{\nu}_2 = \alpha\Delta_c\mathbf{x}, \quad \text{on} \quad \gamma,$$

where $\boldsymbol{\nu}_i$ are unit outward pointing normals, Δ_γ is the Laplace–Beltrami operator, and $\alpha > 0$ is the surface tension coefficient. The term $\Delta_\gamma\mathbf{x}$ is the total vector curvature (sum of principal curvatures in the normal direction) [13]. In addition, the system of equations is supplemented by the interface motion relation

$$\dot{\mathbf{x}} = \mathbf{u} \quad \text{on} \quad \gamma, \tag{1}$$

where $\mathbf{u} = \mathbf{u}_1 = \mathbf{u}_2$ is the fluid velocity at the interface γ.

Following an original idea of Dziuk [11] in the context of purely geometric flows (see also [12]), Bänsch [1] proposes a first-order scheme in time leading to a semi-implicit discretization of the curvature, thereby taking advantage of the stability property inherent to surface tension effects. It relies on an implicit Euler discretization of the interface motion (1)

$$\mathbf{x} \approx \mathbf{x}^{\mathrm{old}} + \tau\,\mathbf{u} \quad \text{on} \quad \gamma,$$

where τ is the time-stepping parameter. Injecting the above relation in the interface condition, we obtain an approximate interface relation

$$(D(\mathbf{u}_1) - p_1)\,\boldsymbol{\nu}_1 + (D(\mathbf{u}_2) - p_2)\,\boldsymbol{\nu}_2 \approx \tau\alpha\Delta_\gamma\mathbf{u} + \alpha\Delta_\gamma\mathbf{x}^{\mathrm{old}} \quad \text{on} \quad \gamma.$$

Hence, denoting by $\mathbf{u}$ the combined velocity, i.e., $\mathbf{u} = \mathbf{u}_i$ on Ω_i and by p the combined pressure, one arrives at a semi-discrete approximation in time: Given an interface position $\mathbf{x}^{\text{old}} \in W^1_\infty(\hat{\gamma})$ and a previous velocity $\mathbf{u}^{\text{old}} \in V^d$, seek $\mathbf{u} \in V^d$ and $p \in L_2(\Omega)$ such that for all $\mathbf{v} \in V^d$

$$\begin{aligned}\int_\Omega \mathbf{u}\cdot\mathbf{v} + 2\tau \int_\Omega D(\mathbf{u})\cdot D(\mathbf{v}) - \tau \int_\Omega p\cdot \operatorname{div}(\mathbf{v}) + \alpha\tau^2 \int_\gamma \nabla_\gamma \mathbf{u}\cdot\nabla_\gamma \mathbf{v} \\ = \int_\Omega \mathbf{u}^{\text{old}}\cdot\mathbf{v} + \tau\int_\Omega \mathbf{f}\cdot\mathbf{v} - \alpha\tau^2 \int_\gamma \nabla_\gamma \mathbf{x}^{\text{old}}\cdot\nabla_\gamma\mathbf{v}\end{aligned} \tag{2}$$

and for all $q \in L_2(\Omega)$

$$\int_\Omega \operatorname{div}(\mathbf{u})\, q = 0. \tag{3}$$

Here V denotes the set of functions in $H^1(\Omega)$ whose trace are in $H^1(\gamma)$. The above weak formulation assumes, in addition, that γ is closed and avoids additional terms involving $\partial\gamma$. A more general variational formulation taking into account the possible intersection of γ with the $\partial\Omega$ is considered by Bänsch [1].

There are a variety of well-known iterative methods for saddle-point problems whose efficiency depends on effective preconditioning of the velocity system and a Schur complement system [5,8,19]. This paper addresses the preconditioning of the velocity system. The efficient preconditioning of the Schur complement system is a topic of future research. Moreover, we report results for a simplified scalar system involving the form

$$A(u,v) := \alpha_0(u,v) + \alpha_1 D(u,v) + \alpha_2\, D_\gamma(u,v), u,v \in V, \tag{4}$$

where

$$(u,v) := \int_\Omega u\, v, \qquad D(u,v) := \int_\Omega \nabla u\cdot\nabla v$$

and

$$D_\gamma(u,v) := \int_\gamma \nabla_\gamma u \cdot \nabla_\gamma v.$$

Here α_i, $i = 0,1,2$ are nonnegative constants. From a preconditioning point of view, the problem of preconditioning the velocity system of (2) and that of (4) is more or less equivalent.

The goal of this paper is to investigate the behavior of multigrid algorithms applied to preconditioning the form $A(\cdot,\cdot)$. We shall demonstrate numerical results which suggest level- and parameter-independent convergence rates. In a subsequent manuscript [2], we shall provide theoretical results which guarantee such convergence in the case $\alpha_0 = 0$. Level- and parameter-independent convergence results for multigrid algorithms in the case when $\alpha_2 = 0$ have been considered before; see, e.g., [9]. The approach for the analysis in the case of $\alpha_0 = 0$ will also be described.

2 Preliminaries

For theoretical purposes, we take $\alpha_0 = 0$ and restrict our attention to the case where Ω is a polygonal or polyhedral domain in $\mathbb{R}^2$ or $\mathbb{R}^3$, respectively, which has been triangulated with an initial coarse mesh. Moreover, we consider the case when γ coincides with Γ, the boundary of Ω. Clearly, $\Gamma = \cup \bar{\Gamma}_j$ where $\{\Gamma_j\}$ denotes the set of polygonal faces of Γ.

We assume that we have a nested sequence of globally refined partitioning of Ω into triangles or tetrahedra, i.e., $\mathscr{T}_j$, $j = 1, 2, \ldots, J$. These are developed by uniform refinement of a coarse triangulation $\mathscr{T}_1$ of Ω and have a mesh size $h_j \approx \varepsilon^j$ for some $\varepsilon \in (0,1)$. In particular, we assume there are positive constants C and c satisfying

$$c\varepsilon^j \leq h_j \leq C\varepsilon^j.$$

The corresponding multilevel spaces of piecewise linear continuous functions are denote by W_j. The functions in W_j with zero mean value on Γ are denoted by V_j, and V_j restricted to Γ is denoted by M_j. Conceptually, we use $\theta_j^i := \varphi_j^i - |\Gamma|^{-1} \int_\Gamma \varphi_j^i$ as our computational basis for V_j, where $|\Gamma|$ denotes the measure of Γ, and φ_j^i are the nodal basis associated to the subdivision j. Technically, this means that our basis functions no longer have compact support. However, because the form $A(\cdot,\cdot)$ kills constants (for $\alpha_0 = 0$), the stiffness matrix is still sparse. The action of the smoother is more or less local as discussed in Remark 4.7 in [3].

There is one fundamental difference between the cases of $\alpha_1 = 0$ and $\alpha_1 \neq 0$. In the first case, the form $A(\cdot,\cdot)$ is indefinite and hence, for uniqueness, one computes in the subspace of reduced dimension, V_J. It is natural to develop the multigrid analysis on the sequence $\{V_j\}$. Keeping track of the mean value is mostly an implementation issue, see, e.g., [3]. Moreover, standard smoothing procedures work provided that the smoother on V_j is based on the natural decompositions in the larger space W_j. An alternative point of view for the indefiniteness issue in the multigrid context is taken in [16, 17].

The analysis of the multigrid algorithm involves the interaction between the quadratic form $A(\cdot,\cdot)$ and a base inner product. The analysis of [2] involves the use of a boundary extension operator $E^j : M_j \to V_j$. Let $\{x_j^i\}$ denote the grid points of the mesh $\mathscr{T}_j$. Given a function in M_j, we first define $E_j : M_j \to V_j$ by setting

$$(E_j u)(x_j^i) = \begin{cases} u(x_j^i): & \text{if } x_j^i \in \Gamma, \\ 0: & \text{otherwise.} \end{cases}$$

We then set

$$E^j u = \sum_{\ell=1}^{j} E_\ell((q_\ell - q_{\ell-1})u). \tag{5}$$

Here q_ℓ, for $\ell > 0$ denotes the $L^2(\Gamma)$ projection onto M_ℓ and $q_0 \equiv 0$. Note that even though E^j is based on the telescoping decomposition

$$u|_\Gamma = \sum_{\ell=1}^{j} ((q_\ell - q_{\ell-1})u)_\Gamma,$$

the sum in (5) does not telescope.

The critical property of this extension is given in the following proposition proven in [2].

Proposition 1. *For $s = 0, 1$, the extension $E^j : M_j \to V_j$ satisfies*

$$\|E^j u\|_{H^s(\Omega)} \le C\|u\|_{s-1/2,\Gamma}, \qquad \textit{for all } u \in M_j.$$

This extension operator was proposed by [14] for developing computable boundary extension operators for domain decomposition preconditioners. The $s = 1$ case of the above theorem was also given there.

3 The Multigrid Algorithm

The analysis of the multigrid algorithm requires the use of a base inner product. We note that even though the operators appearing in the multigrid algorithm below are defined in terms of the base inner product, the base inner product disappears in the implementation as long as the smoothers are defined by Jacobi or Gauss–Seidel iteration. We introduce the base norm (corresponding to $\alpha_0 = 0$):

$$|||u||| = \left[\alpha_1(\|u - E^J u\|^2_{L^2(\Omega)} + \|u\|^2_{-1/2,\Gamma}) + \alpha_2\|u\|^2_{L^2(\Gamma)}\right]^{1/2},$$

for $u \in V_J$. This is the diagonal of the inner product which we denote by $(((\cdot,\cdot)))$. This norm and inner product play a major role in the multigrid analysis in [2].

Following [7], we define the operators:

1. $A_j : V_j \to V_j$ is defined by

$$(((A_j v, \theta))) = A(v, \theta) \qquad \text{for all } v, \theta \in V_j.$$

2. $P_j : V \to V_j$ is defined by

$$A(P_j v, \theta) = A(v, \theta) \qquad \text{for all } v \in V, \theta \in V_j.$$

3. $\hat{Q}_j : V_J \to V_j$ is defined by

$$(((\hat{Q}_j v, \theta))) = (((v, \theta)))) \qquad \text{for all } v \in V_J, \theta \in V_j.$$

Along with these operators, we require a sequence of "smoothing" operators $R_j : V_j \to V_j$, $j = 2, 3, \ldots, J$. The smoothing iteration associated with R_j is the operator $S_j : V_j \times V_j \to V_J$ defined by $S_j(x, f) = x + R_j(f - A_j x)$. The adjoint of R_j with respect to the base inner product is denoted by R_j^t, and we set $S_j^*(w,f) = w + R_j^t(f - A_j w)$. The solution $w = A_j^{-1} f$ is a fixed point of the smoother iteration, and we find that for $x = A_j^{-1} f$, $(x - S_j(w,f)) = S_j(x-w,0) = (I - R_j A_j)(x-w) \equiv K_j(x-w)$. Thus, K_j relates the error before smoothing to that after. Similarly, we define $K_j^* = I - R_j^t A_j$ and note that K_j^* is the $A(\cdot,\cdot)$ adjoint of K_j, i.e.,

$$A(K_j x, y) = A(x, K_j^* y) \qquad \text{for all } x, y \in V_j.$$

The multigrid algorithms can be defined abstractly in terms of the above operators. We include this definition for completeness as it is certainly classical. For simplicity, we shall consider the V-cycle algorithm. The definitions of other variants such as the W-cycle or F-cycle algorithm are similar, and their analysis follows along the same lines. We define the multigrid operator as a map $Mg_j : V_j \times V_j \to V_j$ given as follows:

Multigrid Algorithm ($Mg_j : V_j \times V_j \to V_j$)

(a) If $j = 1$, set $Mg_1(V,F) = A_1^{-1} F$.
(b) Otherwise, for $j = 2, 3, \ldots, J$ define $Mg_j(W,F)$ from $Mg_{j-1}(\cdot,\cdot)$ by:

  (i) $V = S_j(W,F)$ (pre-smoothing).
  (ii) $U = V + Mg_{j-1}(0, \hat{Q}_{j-1}(F - A_j V))$ (correction).
  (iii) $Mg_j(W,F) = S_j^*(U,F)$ (post-smoothing).

4 Multigrid Analysis

The goal of the computational results of this paper and the analysis of [2] is the demonstration that the natural multigrid algorithm applied to our parameter-dependent problem converges uniformly independently of the parameters. We have developed a framework in [2] which allows the use of classical abstract multigrid theory to obtain parameter-independent convergence. The key to this is the introduction of the base norm and the analysis of the related projector:

$$\pi_j u = E_j u + Q_j(u - E_j u).$$

Here Q_j denotes the projection onto the subspace of V_j consisting of functions vanishing on Γ. The base inner product and above projector work as long as α_1 and α_0 are of the same magnitude. This framework fails to provide uniform convergence estimates when $\alpha_1 \ll \alpha_0$.

There are two fundamental ingredients in the algorithm of the previous section. We have already discussed the nested spaces $\{V_j\}$ and their natural embeddings. The other ingredient is the smoothing iterations. These are naturally defined in terms of a subspace decomposition of V_j, i.e.,

$$V_j = \cup_i V_j^i, \qquad i = 1, \ldots N_j.$$

The above decomposition may or may not be a direct sum. This gives rise to two distinct smoothing algorithms, specifically, block Jacobi and block Gauss–Seidel smoothing, see [6, 7]. Either of these gives rise to the operators R_j, S_j, K_j, R_j^t, S_j^* and K_j^*.

Remark 1 (Implementation). Even though the algorithm of the previous section is defined in terms of operators involving the base inner product $(((\cdot,\cdot)))$, this inner product never appears in the implementation. In fact, the implementation of the resulting multigrid algorithm only requires the sparse stiffness matrices on each of the levels, a solver for the stiffness matrices corresponding to $j = 1$ and the smoother subspaces, and a "prolongation" matrix which takes coefficients of the representation of a function $v_j \in V_j$ (in the basis for V_j) into the coefficients for v_j represented in the basis for V_{j+1}.

Our smoothers will be required to satisfy the following two conditions:

(C.1) For some $\omega \in (0,1]$ not depending on j,

$$A(K_j x, K_j x) \le A((I - \omega \lambda_j^{-1} A_j)x, x) \qquad \text{for all } x \in V_j, \tag{6}$$

where $\lambda_j := \sup_{u \in V_j} \frac{A(u,u)}{|||u|||^2}$.

(C.2) For some $\theta < 2$ not depending on j,

$$A(R_j v, R_j v) \le \theta(((R_j v, v))), \qquad \text{for all } v \in V_j. \tag{7}$$

These conditions are just (SM.1) and (SM.2) in [7].

To analyze the multigrid algorithm, we apply abstract results which can be found in [7]. Along with conditions (C.1) and (C.2) above, we introduce two additional conditions ((A.5) and (A.6) of [7]):

(C.3) There exists operators $\pi_j : V_J \to V_j$ (with $\pi_0 = \mathbf{0}$) satisfying

$$\sum_{j=1}^{J} \lambda_j |||(\pi_j - \pi_{j-1})v|||^2 \le C_a A(v,v), \qquad \text{for all } v \in V_J.$$

(C.4) There is an $\varepsilon_1 \in (0,1)$ and a positive constant C_{cs} such that for $v_j \in V_j$ and $w_\ell \in V_\ell$ with $\ell > j$,

$$A(v_j, w_\ell) \le C_{cs} \varepsilon_1^{\ell - j} A(v_j, v_j)^{1/2} (\lambda_\ell^{1/2} |||w_\ell|||).$$

The following theorem is Theorem 5.2 of [7].

Theorem 1. *Assume that conditions (C.1)–(C.4) hold. Then*

$$0 \le A(\mathscr{E}_J v, v) \le (1 - 1/C_M) A(v,v), \qquad \textit{for all } v \in V_J,$$

where

$$C_M = \left[\left(1+\frac{C_a}{\omega}\right)^{1/2} + \left(\frac{C_{cs}\varepsilon_1}{1-\varepsilon_1}\right)\left(\frac{C_a\theta}{2-\theta}\right)^{1/2}\right].$$

We use standard Jacobi or Gauss–Seidel smoothing but with subspaces associated with the nodal decomposition on W_j. The analysis of these smoothers is classical once one verifies the following proposition.

Proposition 2. *Assume that $\alpha_0 = 0$, $0 < \alpha_1$ and $0 < \alpha_2 \le \alpha_1$. Then there is a constant C not depending on j, such that*

$$\sum_i A(v_j^i, v_j^i) \le C\lambda_j |||v_j|||^2 \qquad \text{for all } v_J \in V_j.$$

Here $v_j = \sum_i v_j^i$ is the expansion of v_j into the finite element basis of V_j.

The above proposition implies [2] that (C.1) and (C.2) hold for the Gauss–Seidel smoother as well as a properly scaled Jacobi smoother. We first show [2] that

$$\left[\alpha_1(\|u - E^J u\|^2_{H^1(\Omega)} + \|u\|^2_{1/2,\Gamma}) + \alpha_2\|u\|^2_{H^1(\Gamma)}\right]^{1/2}$$

provides a norm that is equivalent to $A(u,u)$ for $u \in V_J$. This and Proposition 1 eventually lead to (C.3) and (C.4) (cf. [2]).

5 Numerical Results

To illustrate the theory suggested in the previous sections, we report the results of numerical computations. We consider two simple domains in $\mathbb{R}^2$, the first being the unit square and the second being the disk of radius one (centered at the origin).

The case of the unit square fits the theory discussed earlier. The square boundary can be mapped via a piecewise smooth map to the circle. Moving the square so that it is centered about the origin, the map takes the point (x,y) on the boundary to the point of unit absolute value in the same direction.

The multigrid algorithm is variational and so $\lambda = 1$ is always the largest eigenvalue of the preconditioned system. The coarsest mesh in the multigrid algorithm was $h = 1/4$, and we used one forward and one reverse sweep of the Gauss–Seidel iteration as a pre- and post-smoother (four sweeps per V-cycle iteration on each positive level). In Table 1, we report the condition number $K = 1/\lambda_0$[1] for the preconditioned multigrid algorithm when $\alpha_0 = 0$. We consider three cases corresponding to $(\alpha_1 = 1, \alpha_2 = 1)$, $(\alpha_1 = 1, \alpha_2 = 0.1)$, and $(\alpha_1 = 1, \alpha_2 = 0)$.

[1] λ_0 is the smallest eigenvalue of the preconditioned system.

Table 1 Condition numbers for the square, $\alpha_0 = 0$

h	$\alpha_1 = 1, \alpha_2 = 1$	$\alpha_1 = 1, \alpha_2 = 0.1$	$\alpha_1 = 1, \alpha_2 = 0$
$1/16$	1.162	1.167	1.176
$1/32$	1.180	1.194	1.200
$1/64$	1.207	1.208	1.211
$1/128$	1.214	1.214	1.216
$1/256$	1.217	1.217	1.218
$1/512$	1.219	1.219	1.219

Table 2 Condition numbers for the square, $\alpha_0 = 1$

h	$\alpha_1 = 1, \alpha_2 = 0$	$\alpha_1 = k, \alpha_2 = k^2, k = 0.1$	$\alpha_1 = k, \alpha_2 = k^2, k = 0.01$
$1/16$	1.173	1.144	1.078
$1/32$	1.198	1.180	1.133
$1/64$	1.209	1.200	1.172
$1/128$	1.215	1.210	1.195
$1/256$	1.218	1.215	1.208
$1/512$	1.219	1.219	1.214

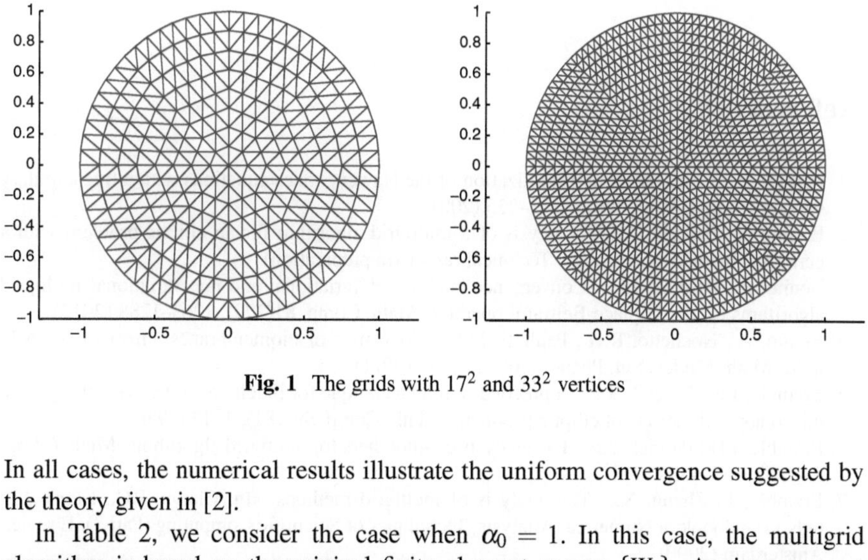

Fig. 1 The grids with 17^2 and 33^2 vertices

In all cases, the numerical results illustrate the uniform convergence suggested by the theory given in [2].

In Table 2, we consider the case when $\alpha_0 = 1$. In this case, the multigrid algorithm is based on the original finite element spaces $\{W_j\}$. Again we report the condition numbers for three cases. The first is $\alpha_1 = 1, \alpha_2 = 0$ and corresponds to a uniformly elliptic second-order problem. The second and third are singularly perturbed problems of the form $\alpha_1 = \tau$, $\alpha_2 = \tau^2$ with τ representing the time step size. Because of the lower-order term, this case does not fit into the theory of [2].

As a final example, we consider the case when Ω is the unit disk. We set up a sequence of triangulations providing successively better approximations to Ω. The coarsest grid contained 5^2 vertices or $h \approx 1/2$. The meshes with 17^2 and 33^2 vertices are given in Fig. 1. The resulting finite element spaces are no longer nested, and the multigrid algorithm is no longer variational. Non-variational multigrid algorithms for the surface Laplacian were investigated in [3]. We believe that the techniques in

Table 3 Condition numbers for the disk, $\alpha_0 = 0$

# Vertices	$\alpha_1 = 1, \alpha_2 = 1$	$\alpha_1 = 1, \alpha_2 = 0.1$	$\alpha_1 = 1, \alpha_2 = 0$
17^2	1.206	1.250	1.304
33^2	1.286	1.316	1.360
65^2	1.348	1.364	1.397
129^2	1.395	1.399	1.421
257^2	1.432	1.426	1.437
513^2	1.463	1.448	1.456

[2, 3] can be combined to give rise to uniform convergence for the non-variational V-cycle multigrid algorithm. The computational results reported in Table 3 clearly illustrate uniform parameter-independent convergence.

Acknowledgements This work was supported in part by award number KUS-C1-016-04 made by King Abdulla University of Science and Technology (KAUST). It was also supported in part by the National Science Foundation through Grant DMS-0914977 and DMS-1216551.

References

1. Bänsch, E.: Finite element discretization of the Navier-Stokes equations with a free capillary surface. Numer. Math. **88**(2), 203–235 (2001)
2. Bonito, A., Pasciak, J.E.: Analysis of a multigrid algorithm for an elliptic problem with a perturbed boundary condition. Technical report (in preparation)
3. Bonito, A., Pasciak, J.E.: Convergence analysis of variational and non-variational multigrid algorithms for the Laplace-Beltrami operator. Math. Comp. **81**(279), 1263–1288 (2012)
4. Bonito, A., Nochetto, R.H., Pauletti, M.S.: Dynamics of biomembranes: effect of the bulk fluid. Math. Model. Nat. Phenom. **6**(5), 25–43 (2011)
5. Bramble, J.H., Pasciak, J.E.: A preconditioning technique for indefinite systems resulting from mixed approximations of elliptic problems. Math. Comp. **50**(181), 1–17 (1988)
6. Bramble, J.H., Pasciak, J.E.: The analysis of smoothers for multigrid algorithms. Math. Comp. **58**(198), 467–488 (1992)
7. Bramble, J., Zhang, X.: The analysis of multigrid methods. In: Ciarlet, P.C., Lions, J.L. (eds.) Handbook of Numerical Analysis, Techniques of Scientific Computing (Part 3). Elsevier, Amsterdam (2000)
8. Bramble, J.H., Pasciak, J.E., Vassilev, A.T.: Analysis of the inexact Uzawa algorithm for saddle point problems. SIAM J. Numer. Anal. **34**(3), 1072–1092 (1997)
9. Bramble, J.H., Pasciak, J.E., Vassilevski, P.S.: Computational scales of Sobolev norms with application to preconditioning. Math. Comp. **69**(230), 463–480 (2000)
10. Du, Q., Liu, Ch., Ryham, R., Wang, X.: Energetic variational approaches in modeling vesicle and fluid interactions. Phys. D **238**(9–10), 923–930 (2009)
11. Dziuk, G.: An algorithm for evolutionary surfaces. Numer. Math. **58**(6), 603–611 (1991)
12. Dziuk, G., Elliott, C.M.: Finite elements on evolving surfaces. IMA J. Numer. Anal. **27**(2), 262–292 (2007)
13. Gilbarg, D., Trudinger, N.S.: Elliptic partial differential equations of second order. In: Classics in Mathematics. Springer, Berlin (2001). Reprint of the 1998 edition
14. Haase, G., Langer, U., Meyer, A., Nepomnyaschikh, S.V.: Hierarchical extension operators and local multigrid methods in domain decomposition preconditioners. East-West J. Numer. Math. **2**(3), 173–193 (1994)

15. Hysing, S.: A new implicit surface tension implementation for interfacial flows. Int. J. Numer. Methods Fluids **51**(6), 659–672 (2006)
16. Lee, Y.-Ju., Wu, J., Xu, J., Zikatanov, L.: Robust subspace correction methods for nearly singular systems. Math. Models Methods Appl. Sci. **17**(11), 1937–1963 (2007)
17. Lee, Y.-Ju., Wu, J., Xu, J., Zikatanov, L.: A sharp convergence estimate for the method of subspace corrections for singular systems of equations. Math. Comp. **77**(262), 831–850 (2008)
18. Pauletti, M.S.: Parametric AFEM for geometric evolution equation and coupled fluid-membrane interaction. ProQuest LLC, Ann Arbor, MI. Thesis (Ph.D.), University of Maryland, College Park (2008)
19. Rusten, T., Winther, R.: A preconditioned iterative method for saddlepoint problems. SIAM J. Matrix Anal. Appl. **13**(3), 887–904 (1992). Iterative Methods in Numerical Linear Algebra, Copper Mountain, CO (1990)
20. Sohn, J.S., Tseng, Y.-H., Li, S., Voigt, A., Lowengrub, J.S.: Dynamics of multicomponent vesicles in a viscous fluid. J. Comput. Phys. **229**(1), 119–144, (2010)
21. Walker, S.W., Bonito, A., Nochetto, R.H.: Mixed finite element method for electrowetting on dielectric with contact line pinning. Interfaces Free Bound. **12**(1), 85–119 (2010)

[illegible]

Parallel Unsmoothed Aggregation Algebraic Multigrid Algorithms on GPUs

James Brannick, Yao Chen, Xiaozhe Hu, and Ludmil Zikatanov

Abstract We design and implement a parallel algebraic multigrid method for isotropic graph Laplacian problems on multicore graphical processing units (GPUs). The proposed AMG method is based on the aggregation framework. The setup phase of the algorithm uses a parallel maximal independent set algorithm in forming aggregates, and the resulting coarse-level hierarchy is then used in a K-cycle iteration solve phase with a ℓ^1-Jacobi smoother. Numerical tests of a parallel implementation of the method for graphics processors are presented to demonstrate its effectiveness.

Keywords Multigrid methods • Unsmoothed aggregation • Adaptive aggregation

Mathematics Subject Classification (2010): 65N55, 65T08, 65F10

1 Introduction

We consider development of a multilevel iterative solver for large-scale sparse linear systems corresponding to graph Laplacian problems for graphs with balanced vertex degrees. A typical example is furnished by the matrices corresponding to the (finite

The first, third, and fourth author were supported in part by the grants NSF DMS-1217142, NSF OCI-0749202, and DoE DE-SC0006903.

J. Brannick • X. Hu • L. Zikatanov (✉)
Department of Mathematics, The Pennsylvania State University, University Park, PA 16802, USA
e-mail: brannick@psu.edu; hu_x@math.psu.edu; ludmil@psu.edu

Y. Chen
Microsoft Corporation, 1065 La Avenida, Mountain View, CA 94043, USA
e-mail: yaoc@microsoft.com

O.P. Iliev et al. (eds.), *Numerical Solution of Partial Differential Equations: Theory, Algorithms, and Their Applications*, Springer Proceedings in Mathematics & Statistics 45, DOI 10.1007/978-1-4614-7172-1_5,

difference)/(finite volume)/(finite element) discretizations of scalar elliptic equation with mildly varying coefficients on unstructured grids.

Multigrid (MG) methods have been shown to be very efficient iterative solvers for graph Laplacian problems, and numerous parallel MG solvers have been developed for such systems. Our aim here is to design an algebraic multigrid (AMG) method for solving the graph Laplacian system and discuss the implementation of such methods on multiprocessor parallel architectures, with an emphasis on implementation on graphical processing units (GPUs).

The programming environment which we use in this paper is the Compute Unified Device Architecture (CUDA) toolkit introduced in 2006 by NVIDIA which provides a framework for programming on GPUs. Using this framework in the last 5 years several variants of geometric multigrid (GMG) methods have been implemented on GPUs [6, 13, 14, 16–18], and a high level of parallel performance for the GMG algorithms on CUDA-enabled GPUs has been demonstrated in these works.

On the other hand, designing AMG methods for massively parallel heterogeneous computing platforms, e.g., for clusters of GPUs, is very challenging mainly due to the sequential nature of the coarsening processes (setup phase) used in AMG methods. In most AMG algorithms, coarse-grid points or basis are selected sequentially using graph theoretical tools (such as maximal independent sets and graph partitioning algorithms). Although extensive research has been devoted to improving the performance of parallel coarsening algorithms, leading to notable improvements on CPU architectures [8, 9, 11, 21, 22, 27, 28, 28], on a single GPU [4, 19, 26], and on multiple GPUs [12], the setup phase is still considered a bottleneck in parallel AMG methods. We mention the work in [4], where a smoothed aggregation setup is developed in CUDA for GPUs.

In this paper, we describe a parallel AMG method based on the unsmoothed aggregation AMG (UA-AMG) method. The setup algorithm we develop and implement has several notable design features. A key feature of our parallel aggregation algorithm (PAA) is that it first chooses coarse vertices using a parallel maximal independent set algorithm [11] and then forms aggregates by grouping coarse- level vertices with their neighboring fine-level vertices, which, in turn, avoids ambiguity in choosing fine-level vertices to form aggregates. Such a design eliminates both the memory write conflicts and conforms to the CUDA programming model. The triple matrix product needed to compute the coarse-level matrix (a main bottleneck in parallel AMG setup algorithms) simplifies significantly in the UA-AMG setting, reducing to summations of entries in the matrix on the finer level. The parallel reduction sums available in CUDA are quite an efficient tool for this task during the AMG setup phase. Additionally, the UA-AMG setup typically leads to low grid and operator complexities.

In the solve phase of the proposed algorithm, a K-cycle [1, 2, 29] is used to accelerate the convergence rate of the multilevel UA-AMG method. Such multilevel method optimizes the coarse-grid correction and results in an approximate two-level method. Two parallel relaxation schemes considered in our AMG implementation are a damped Jacobi smoother and a parameter-free ℓ^1-Jacobi smoother introduced in [25] and its weighted version in [7]. To further accelerate the convergence rate of the resulting K-cycle Method we apply it as a preconditioner to a nonlinear conjugate gradient method.

The remainder of the paper is organized as follows. In Sect. 2, we review the UA-AMG method. Then, in Sect. 3, a parallel graph aggregation method is introduced, which is our main contribution. The parallelization of the solve phase is discussed in Sect. 4. In Sect. 5, we present some numerical results to demonstrate the efficiency of the parallel UA-AMG method.

2 Unsmoothed Aggregation AMG

The linear system of interest has as coefficient matrix the graph Laplacian corresponding to an undirected connected graph $\mathcal{G} = (\mathcal{V}, \mathcal{E})$. Here, $\mathcal{V}$ denotes the set of vertices and $\mathcal{E}$ denotes the set of edges of $\mathcal{G}$. We set $n = |\mathcal{V}|$ (cardinality of $\mathcal{V}$). By $(\cdot,\cdot)$ we denote the inner product in $\ell^2(\mathbb{R}^n)$ and the superscript t denotes the adjoint with respect to this inner product. The *graph Laplacian* $A : \mathbb{R}^n \mapsto \mathbb{R}^n$ is then defined via the following bilinear form:

$$(Au,v) = \sum_{k=(i,j)\in\mathcal{E}} \omega_{ij}(u_i - u_j)(v_i - v_j) + \sum_{j\in\mathcal{S}} \omega_j^D u_j v_j, \quad \mathcal{S} \subset \mathcal{V}.$$

We assume that the weights ω_{ij} and ω_j^D are strictly positive for all i and j. The first summation is over the set of edges $\mathcal{E}$ (over $k \in \mathcal{E}$ connecting the vertices i and j), and u_i and u_j are the i-th and j-th coordinate of the vector $u \in \mathbb{R}^n$, respectively. We also assume that the subset of vertices $\mathcal{S}$ is such that the resulting matrix A is symmetric positive definite (SPD). If the graph is connected, $\mathcal{S}$ could contain only one vertex and A will be SPD. For matrices corresponding to the discretization scalar elliptic equation on unstructured grids, $\mathcal{S}$ is the set of vertices near (one edge away from) the boundary of the computational domain. The linear system of interest is then

$$Au = f. \tag{1}$$

With this system of equation, we associate a multilevel hierarchy which consists of spaces $V_0 \subset V_1 \subset \ldots \subset V_L = \mathbb{R}^n$; each of the spaces is defined as the range of interpolation/prolongation operator $P_{l-1}^l : \mathbb{R}^{n_{l-1}} \mapsto V_l$ with $\text{Range}(P_{l-1}^l) = V_{l-1}$.

Given the l-th level matrix $A_l \in \mathbb{R}^{n_l \times n_l}$, the aggregation-based prolongation matrix P_{l-1}^l is defined in terms of a non-overlapping partition of the n_l unknowns at level l into the n_{l-1} nonempty disjoint sets G_j^l, $j = 1,\ldots,n_{l-1}$, called aggregates. An algorithm for choosing such aggregates is presented in the next section. The prolongation P_{l-1}^l is the $n_l \times n_{l-1}$ matrix with columns defined by partitioning the constant vector, $\mathbb{1} = (1,\ldots,1)^t$, with respect to the aggregates:

$$(P_{l-1}^l)_{ij} = \begin{cases} 1 & \text{if } i \in G_j^l \\ 0 & \text{otherwise} \end{cases} \quad i = 1,\ldots,n_l, \quad j = 1,\ldots,n_{l-1}. \tag{2}$$

Algorithm 1: UA-AMG

Setup Phase:

Given n_0 (size of the coarsest level) and L (maximum levels)
$l \leftarrow L$,
while $N_l \geq n_0$ & $l > 0$ **do**
    Construct the aggregation $\mathcal{N}_l^i$, $i = 1, 2, \ldots, N_{l+1}$ based on A_l,
    Compute A_{l-1} by (4),
    $l \leftarrow l - 1$,
end while

Solve Phase:

if (On the coarsest level) **then**
    solve $A_l u_l = f_l$ exactly,
else
    Pre-smoothing: $u_l \leftarrow \texttt{smooth}(u_l, A_l, f_l)$,
    Restriction: compute $r_{l-1} = (P_{l-1}^l)^T (f_l - A_l u_l)$,
    Coarse grid correction: solve $A_{l-1} e_{l-1} = r_{l-1}$ approximately by recursively calling the AMG on coarser level $l-1$ and get e_{l-1},
    Prolongation: compute $u_l \leftarrow u_l + P_{l-1}^l e_{l-1}$,
    Post-smoothing: $u_l \leftarrow \texttt{smooth}(u_l, A_l, f_l)$.
end if

The resulting coarse-level matrix $A_{l-1} \in \mathbb{R}^{n_{l-1} \times n_{l-1}}$ is then defined by the so-called triple matrix product, namely,

$$A_{l-1} = (P_{l-1}^l)^t A_l (P_{l-1}^l). \tag{3}$$

Note that since we consider UA-AMG, the interpolation operators are Boolean matrices such that the entries in the coarse-grid matrix A_{l-1} can be obtained from a simple summation process:

$$(A_{l-1})_{ij} = \sum_{s \in G_i} \sum_{t \in G_j} (A_l)_{st}, \quad i, j = 1, 2, \cdots, n_{l-1}. \tag{4}$$

Thus, the triple matrix product, typically *the* costly procedure in an AMG setup, simplifies significantly for UA-AMG to reduction sums.

We now introduce a general UA-AMG method (see Algorithm 1), and in the subsequent sections we describe the implementation of each of the components of Algorithm 1 for GPUs.

3 The Setup Phase

Consider the system of linear equations (1) corresponding to an unweighted graph $\mathcal{G} = \{\mathcal{V}, \mathcal{E}\}$ partitioned into two subgraphs $\mathcal{G}_k = \{\mathcal{V}_k, \mathcal{E}_k\}, k = 1, 2$. Further assume that the two subgraphs are stored on separate computes. To implement a Jacobi or Gauss–Seidel smoother for the graph Laplacian equation with respect to $\mathcal{G}$, the communication between the two computers is proportional to the number of edge cuts of such a partitioning, given by

$$|\mathcal{E}\backslash(\mathcal{E}_1 \cup \mathcal{E}_2)|.$$

Therefore, a partition corresponding to the minimal edge cut in the graph results in the fastest implementation of such smoothers. This in turn gives a heuristic argument, as also suggested in [23, 24], that when partitioning the graph in subgraphs (aggregates), the subgraphs should have a similar number of vertices and have a small "perimeter." Such a partitioning can be constructed by choosing any vertex in the graph, naming it as a coarse vertex, and then aggregating it with its neighboring vertices. This heuristic motivates our aggregation method. The algorithm consists of a sequence of two subroutines: first, a parallel maximal independent set algorithm is applied to identify coarse vertices; then a parallel graph aggregation algorithm follows, so that subgraphs (aggregates) centered at the coarse vertices are formed.

In the algorithm, to reduce repeated global memory read access and write conflicts, we impose explicit manual scheduling on data caching and flow control in the implementations of both algorithms; the aim is to achieve the following goals:

1. (Read access coalescence): To store the data that a node uses frequently locally or on a fast connecting neighboring node.
2. (Write conflicting avoidance): To reduce or eliminate the situation that several nodes need to communicate with a center node simultaneously.

3.1 A Maximal Independent Set Algorithm

The idea behind such algorithm is to simplify the memory coalescence and design a random aggregation algorithm where there are as many as possible threads loading from a same memory location, while as few as possible threads writing to a same memory location. Therefore, it is natural to have one vertex per thread when choosing the coarse vertices. For vertices that are connected, the corresponding processing threads should be wrapped together in a group. By doing so, repeated memory loads from the global memory can be avoided.

However, we also need to ensure that no two coarse vertices compete for a fine-level point, because either atomic operations as well as inter-thread communication is costly on a GPU. Therefore, the coarse vertices are chosen in a way that any two of them are of distance 3 or more, which is the same as finding a maximal independent set of vertices for the graph corresponding to A^2, where A is the graph Laplacian of a given graph $\mathcal{G}$, so that each fine-level vertex can be determined independently which coarse vertex it associates with.

Given an undirected unweighted graph $\mathcal{G} = \{\mathcal{V}, \mathcal{E}\}$, we first find a set C of coarse vertices such that

$$d(i,j) \geq 3, \qquad \forall i,j \in C, i \neq j. \tag{5}$$

Here, $d(\cdot,\cdot)$ is the graph distance function defined recursively as

$$d(i,j) = \begin{cases} 0, & i = j; \\ \min\limits_{k:(i,k)\in\varepsilon} d(k,j) + 1, & i \neq j. \end{cases}$$

Assume we obtain such set C, or even a subset of C, we can then form aggregates, by picking up a vertex i in C and defining an aggregate as a set containing i and its neighbors. The condition (5) guarantees that two distinct vertices in C do not share any neighbors. The operation of marking the numberings of subgraphs on the fine-grid vertices is write conflict-free, and the restriction imposed by (5) ensures that aggregates can be formed independently and simultaneously.

The rationale of the independent set algorithm is as follows: First, a random vector v is generated, each component of which corresponds to a vertex in the graph. Then we define the set C as the following:

$$C = \{i \mid v_i > v_j, \forall j : 0 < d(i,j) < 3\}.$$

If C is not empty, then such construction results in a collection of vertices in C is of distance 3 or more. Indeed, assume that $d(i,j) < 3$ for $i, j \in C$; let $v_i > v_j$. From the definition of the set C, we immediately conclude that $i \notin C$. Of course, more caution is needed when C defined above is empty (a situation that may occur depending on the vector v). However, this can be remedied, by assuming that the vector v (with random entries) has a global maximum, which is also a local maximum. The C contains at least this vertex. The same algorithm can be applied then recursively to the remaining graph (after this vertex is removed). In practice, C does not contain one but more vertices.

3.2 *Parallel Graph Aggregation Algorithm*

We here give a description of the parallel aggregation algorithm (PAA, Algorithm 2), running the exact copies of the code on each thread.

Within each pass of the PAA, the following two steps are applied to each vertex i:

(A) Construct a set C which contains coarse vertices.
(B) Construct an aggregate for each vertex in C.

Note that these two subroutines can be executed in a parallel fashion. Indeed, step (A) does not need to be applied to the whole graph before starting step (B). Even if C is partially completed, any operation in step (B) will not interfere step (A), running on the neighboring vertices and completing the construction of C. A problem for this approach is that it usually cannot give a set of aggregates that cover the vertex set V after 1 pass of step (A) and step (B). We thus run several passes and the algorithm terminates when a complete cover is obtained. The number of passes is reduced if

Algorithm 2: Parallel Aggregation Algorithm (PAA)

(1) Generate a quasi-random number and store it in v_i, as

$$v_i \longleftarrow \text{quasi\_random}(i);$$

mark vertex i as "unprocessed"; wait until all threads complete these operations.

(2) (2a) Goto (2d) if i is marked "processed", otherwise continue to (2b).

(2b) Determine if the vertex i is a coarse vertex, by check if the following is true.

$$v_i > v_j, \qquad \forall j : (A^2)_{ij} \neq 0 \text{ and } j \text{ is unprocessed} .$$

If so, continue to (2c); if not, goto (2d).

(2c) Form an aggregate centered at i. Let S_i be a set of vertices defined as

$$S_i = \left\{ j \mid v_i \geq v_j, \forall j : A_{ij} \neq 0 \text{ and } j \text{ is unprocessed} \right\}.$$

Define a column vector w such that

$$w_k = \begin{cases} 1, & k \in S_i; \\ 0, & k \notin S_i. \end{cases}$$

Mark vertices $j \in S_i$ "processed" and request an atomic operation to update the prolongator P as

$$P \longleftarrow [P, w] .$$

(2d) Synchronize all threads (meaning: wait until all threads reach this step).

(2e) Stop if i is marked "processed", otherwise goto step (2a).

we make the set C as large as possible in each pass; therefore, the quasi-random vector v needs to have a lot of local maximums. Another heuristic argument is that C needs to be constructed in a way that every coarse vertex has a large number of neighboring vertices. Numerical experiments suggest that the following is a good way of generating the vector v with the desired properties:

$$v_i \longleftarrow \text{quasi\_random}(i) := d_i + \big((i \bmod 12) + \text{rand}()\big)/12, \tag{6}$$

where d_i is the degree of the vertex i and rand() generates a random number uniformly distributed on the interval $[0,1]$.

3.3 Aggregation Quality Improvements

To improve the quality of the aggregates, we can either impose some constraints during the aggregation procedure (which we call in-line optimization) or reshape an existing aggregation in order to improve it (which we call post-processing). One in-line strategy that we use to improve the quality of the aggregation is to limit the number of vertices in an aggregate during the aggregation procedure. However, such limitations may result in a small coarsening ratio. In such case, numerical results suggest that applying aggregation process twice, which is equivalent to skipping a level in a multilevel hierarchy, can compensate that. Our focus is on a post-processing strategy, which we name “rank one optimization.” It uses an a priori estimate to adjust the interface (boundary) of a pair of aggregates, so that the aggregation-based two-level method, with a fixed smoother, converges fast locally on those two aggregates.

We consider the connected graph formed by a union of aggregates (say, a pair of them, which will be the case of interest later), and let $\hat{n}$ be the dimension of the underlying vector space. Let $\hat{A} : \mathbb{R}^{\hat{n}} \mapsto \mathbb{R}^{\hat{n}}$ be a semidefinite weighted graph Laplacian (representing a local subproblem) and $\hat{R}$ be a given local smoother. As is usual for semidefinite graph Laplacians, we consider the subspace ℓ^2-orthogonal to the null space of $\hat{A}$ and we denote it by V. The ℓ^2 orthogonal projection on V is denoted here by Π_V. Let $\hat{S} = I - \hat{R}\hat{A}$ be the error propagation operator for the smoother $\hat{R}$. We consider the two-level method whose error propagation matrix is

$$E(V_c) = E(V_c;\hat{S}) = (I - Q_{\hat{A}}(V_c))(I - \hat{R}\hat{A}).$$

Here $V_c \subset V$ is a subspace and $Q_{\hat{A}}(V_c)$ is the $\hat{A}$-orthogonal projection of the elements of V onto the coarse space V_c. In what follows we use the notation $E(V_c;\hat{S})$ when we want to emphasize the dependence on $\hat{S}$. We note that $Q_{\hat{A}}(V_c)$ is well defined on V because $\hat{A}$ is SPD on V and hence it $(\hat{A}\cdot,\cdot)$ is an inner product on V. We also have that $Q_{\hat{A}}(V_c)$ self-adjoint on V and under the assumption $V_c \subset V$, we obtain $Q_{\hat{A}}(V_c) = \Pi_V Q_{\hat{A}}(V_c)$ and $\Pi_V Q_{\hat{A}}(V_c) = Q_{\hat{A}}(V_c)\Pi_V$. Also, $\hat{S}_V = \Pi_V \hat{S}$ is self-adjoint on V in the $(\hat{A}\cdot,\cdot)$ inner product iff $\hat{R}$ is self-adjoint in the ℓ^2-inner product on $\mathbb{R}^{\hat{n}}$.

We now introduce the operator $T(V_c)$ (recall that $V_c \subset V$)

$$T(V_c) = T(V_c;\hat{S}) = \hat{S} - E(V_c) = Q_{\hat{A}}(V_c)(I - \hat{R}\hat{A}) = Q_{\hat{A}}(V_c)\hat{S},$$

and from the definition of $Q_{\hat{A}}$ for all $v \in V$ we have

$$|E(V_c)v|_{\hat{A}}^2 = |\Pi_V E(V_c)v|_{\hat{A}}^2 = |\Pi_V \hat{S}v|_{\hat{A}}^2 - |\Pi_V T(V_c)v|_{\hat{A}}^2 = |\hat{S}_V v|_{\hat{A}}^2 - |T(V_c)v|_{\hat{A}}^2. \quad (7)$$

We note the following identities which follow directly from the definitions above and the assumption $V_c \subset V$:

$$|E(V_c;\hat{S})|_{\hat{A}} = |\Pi_V E(V_c;\hat{S}_V)|_{\hat{A}}, \quad |T(V_c;\hat{S})|_{\hat{A}} = |\Pi_V T(V_c;\hat{S}_V)|_{\hat{A}}. \quad (8)$$

The relation (7) suggests that, in order to minimize the seminorm $|E(V_c)v|_{\hat{A}}$ with respect to the coarse space V_c, we need to make $|T(V_c)|_{\hat{A}}$ maximal. The following lemma quantifies this observation and is instrumental in showing how to optimize locally the convergence rate when the subspaces V_c are one dimensional. In the statement of the lemma we use arg min to denote a subset of minimizers of a given, not necessarily linear, functional $F(x)$ on a space X. More precisely, we set

$$y \in \arg\min_{x \in X} F(x), \quad \text{if and only if,} \quad F(y) = \min_{x \in X} F(x).$$

We have similar definition (with obvious changes) for the set $\arg\max_{x \in X} F(x)$.

Lemma 3.1. *Let $\hat{S}_V = \Pi_V \hat{S}$ be the projection of the local smoother on V and $\mathcal{V}_c$ be the set of all one dimensional subspaces of V. Then we have the following:*

$$|\hat{S}_V|_{\hat{A}} = \max_{V_c \in \mathcal{V}_c} |T(V_c)|_{\hat{A}}, \tag{9}$$

$$\text{If } W_c \in \arg\max_{V_c \in \mathcal{V}_c} |T(V_c)|_{\hat{A}}, \text{ then } W_c \in \arg\min_{V_c \in \mathcal{V}_c} |E(V_c)|_{\hat{A}}, \tag{10}$$

where $E(V_c) = (I - Q_{\hat{A}}(V_c))\hat{S}$ and $T(V_c) = Q_{\hat{A}}(V_c)\hat{S}$.

Proof. From the identities (8) it follows that we can restrict our considerations on $V \subset \mathbb{R}^{\hat{n}}$ and that we only need to prove the Lemma with $E(V_c) = \Pi_V E(V_c; \hat{S}_V)$ and $T(V_c) = \Pi_V T(V_c; \hat{S}_V)$. In order to make the presentation more transparent, we denote $|\cdot| = |\cdot|_{\hat{A}}$, $\Pi = Q_{\hat{A}}$. Let us mention also that by orthogonality in this proof we mean orthogonality in the $(\hat{A}\cdot, \cdot)$ inner product on V. The proof then proceeds as follows.

Let $\varphi \in V$ be such that $|\hat{S}_V \varphi| = |\hat{S}_V||\varphi|$. We set $W_c = \text{span}\{\hat{S}_V \varphi\}$. Note that for such choice of W_c we have $\Pi(W_c)\hat{S}_V \varphi = \hat{S}_V \varphi$ and hence

$$|\hat{S}_V| = \frac{|\hat{S}_V \varphi|}{|\varphi|} = \frac{|T(W_c)\varphi|}{|\varphi|} \leq |T(W_c)|.$$

On the other hand, for all $V_c \in \mathcal{V}_c$ we have $|\Pi(V_c)| = 1$ and we then conclude that

$$|T(V_c)| = |\Pi(V_c)\hat{S}_V| \leq |\Pi(V_c)||\hat{S}_V| = |\hat{S}_V| \leq |T(W_c)|. \tag{11}$$

By taking a maximum on $\mathcal{V}_c$ in (11), we conclude the following thus prove (9):

$$|T(W_c)| \leq \max_{V_c \in \mathcal{V}_c} |T(V_c)| \leq |\hat{S}_V| \leq |T(W_c)|.$$

To prove (10), we observe that for any $W_c \in \arg\max_{V_c \in \mathcal{V}_c} |T(V_c)|$, the inequalities in (11) become equalities and hence

$$|\Pi(W_c)\hat{S}_V| = |\hat{S}_V| = |\hat{S}_V \Pi(W_c)|.$$

This implies that $|\hat{S}_V| = \max_{w \in W_c} \frac{|\hat{S}_V w|}{|w|}$. It is also clear that $|\hat{S}_V w| = |\hat{S}_V||w|$ for all $w \in W_c$, because W_c is one dimensional. In addition, since $\hat{S}_V$ is self-adjoint, it follows that W_c is the span of the eigenvector of $\hat{S}_V$ with eigenvalue of magnitude $|\hat{S}_V|$. Next, for any $V_c \in \mathcal{V}_c$ we have

$$|E(V_c)| = |(I - \Pi(V_c))\hat{S}_V| = |\hat{S}_V(I - \Pi(V_c))| = \max_{v \in V_c^{\perp}} \frac{|\hat{S}_V v|}{|v|}.$$

By the mini-max principle (see [10, pp. 31–35] or [20]) we have that $|E(V_c)| \geq \sigma_2$, where σ_2 is the second largest singular value of $\hat{S}_V$ and with equality holding iff $V_c = W_c$. This completes the proof. ■

We now move on to consider a pair of aggregates. Let $\hat{A}$ be the graph Laplacian of a connected positively weighted graph $\hat{\mathcal{G}}$ which is union of two aggregates $\mathcal{V}_1$ and $\mathcal{V}_2$. Furthermore, let $\mathbb{1}_{\mathcal{V}_1}$ be the characteristic vector for $\mathcal{V}_1$, namely, a vector with components equal to 1 at the vertices of $\mathcal{V}_1$ and equal to zero at the vertices of $\mathcal{V}_2$. Analogously we have a characteristic vector $\mathbb{1}_{\mathcal{V}_2}$ for $\mathcal{V}_2$. Finally, let $V_c(\mathcal{V}_1, \mathcal{V}_2)$ be the space of vectors that are linear combinations of $\mathbb{1}_{\mathcal{V}_1}$ and $\mathbb{1}_{\mathcal{V}_2}$. More specifically, the subspace V_c is defined as

$$V_c(\mathcal{V}_1, \mathcal{V}_2) = \text{span}\left\{ \left(\frac{1}{|\mathcal{V}_1|}\mathbb{1}_{\mathcal{V}_1} - \frac{1}{|\mathcal{V}_2|}\mathbb{1}_{\mathcal{V}_2} \right) \right\}.$$

Let $\mathcal{V}_c$ be the set of subspaces defined above for all possible pairs of $\mathcal{V}_1$ and $\mathcal{V}_2$, such that $\hat{\mathcal{G}} = \mathcal{V}_1 \cup \mathcal{V}_2$. Note that by the definition above, every pair $(\mathcal{V}_1, \mathcal{V}_2)$ gives us a space $V_c \in \mathcal{V}_c$ which is orthogonal to the null space of $\hat{A}$, i.e., orthogonal to $\mathbb{1} = \mathbb{1}_{\mathcal{V}_1} + \mathbb{1}_{\mathcal{V}_2}$.

We now apply the result of Lemma 3.1 and show how to improve locally the quality of the partition (the convergence rate $|E(V_c)|_{\hat{A}}$) by reducing the problem of minimizing the $\hat{A}$-norm of $E(V_c)$ to the problem of finding the maximum of the $\hat{A}$-norm of the rank one transformation $T(V_c)$. Under the assumption that the spaces V_c are orthogonal to the null space of $\hat{A}$ (which they satisfy by construction) from Lemma 3.1, we conclude that the spaces W_c which minimize $|E(V_c)|_{\hat{A}}$ also maximize $|T(V_c)|_{\hat{A}}$.

For the pair of aggregates, $|T(V_c)|_{\hat{A}}$ is the largest eigenvalue of $\hat{S}^T A Q_{\hat{A}}(V_c)\hat{S}\hat{A}^{\dagger}$, where $A^{\dagger}$ is the pseudo inverse of A. Clearly, the matrix $\hat{S}^T A Q_{\hat{A}}(V_c)\hat{S}\hat{A}^{\dagger}$ is also a rank one matrix and hence

$$|T(V_c)|_{\hat{A}} = \text{tr}(\hat{S}^T A Q_{\hat{A}}(V_c)\hat{S}\hat{A}^{\dagger}).$$

During optimization steps, we calculate the trace using the fact that for any rank one matrix W, we have

$$\text{tr}(W) = \frac{\widetilde{W}_k^T W_k}{W_{kk}} = \frac{\widetilde{W}_k^T W_k}{e_k^T W_k e_k}, \tag{12}$$

Algorithm 3: Subgraph Reshaping Algorithm

Input: Two set of vertices, $\mathcal{V}_1$ and $\mathcal{V}_2$, corresponding to a pair of neighboring subgraphs.
Output: Two sets of vertices, $\tilde{\mathcal{V}}_1$ and $\tilde{\mathcal{V}}_2$ satisfying that

$$\tilde{\mathcal{V}}_1 \cup \tilde{\mathcal{V}}_2 = \mathcal{V}_1 \cup \mathcal{V}_2, \qquad \text{and} \qquad \big||\tilde{\mathcal{V}}_1| - |\tilde{\mathcal{V}}_2|\big| \leq 1,$$

and the subgraphs corresponding to $\tilde{\mathcal{V}}_1$ and $\tilde{\mathcal{V}}_2$ are both connected.

(1) Let $n = |\tilde{\mathcal{V}}_1| + |\tilde{\mathcal{V}}_2|$, then compute $m = \lfloor n/2 \rfloor$.
(2) Run in parallel to generate all partitionings such that the vertices set

$$\tilde{\mathcal{V}}_1 \cup \tilde{\mathcal{V}}_2 = \mathcal{V}_1 \cup \mathcal{V}_2, \qquad |\tilde{\mathcal{V}}_1| = m,$$

and the subgraphs derived by $\tilde{\mathcal{V}}_1$ and $\tilde{\mathcal{V}}_2$ are connected.
(3) Run in parallel to compute the norm $|T(V_c)|_{\hat{A}}$ for all partitionings get from step (2), and return the partitioning that results in maximal $|T(V_c)|_{\hat{A}}$.

where W_{kk} is a nonzero diagonal entry (any nonzero diagonal entry), W_k is the k-th column of W and $\widetilde{W}_k^T$ is the k-th row of W. The formula (12) is straightforward to prove if we set $W = uv^T$ for two column vectors u and v and also suggests a numerical algorithm. We devise a loop computing $W_k = We_k$ and $W_{kk} = e_k^T W_k e_k$, for $k = 1, \ldots, m$, where m is the dimension of W. The loop is terminated whenever $W_{kk} \neq 0$, and we compute the trace via (12) for this k. In particular for the examples we have tested, $W = \hat{S}^T \hat{A} Q_{\hat{A}}(V_c) \hat{S} \hat{A}^\dagger$ is usually a full matrix and we observed that the loop almost always terminated when $k = 1$.

The algorithm which traverses all pairs of neighboring aggregates and optimizes their shape is as follows.

The subgraph reshaping algorithm fits well the programming model of a multi-core GPU. We demonstrate this algorithm on two example problems and later show its potential as a post-process for the PAA (Algorithm 2) outlined in the previous section. In the examples that follow next we use the rank one optimization and then measure the quality of the coarse space also by computing the energy norm of the $|Q|_{\hat{A}}$, where Q is the ℓ^2-orthogonal projection to the space W_c.

Example 3.2. Consider a graph Laplacian $\hat{A}$ corresponding to a graph which is a 4×4 square grid. The weights on the edges are all equal to 1. We start with an obviously non-optimal partitioning as shown on the left of Fig. 1, of which the resulting two-level method, consisting of ℓ^1-Jacobi pre- and post-smoothers and an exact coarse-level solver, has a convergence rate $|E|_{\hat{A}} = 0.84$ and $|Q|_{\hat{A}}^2 = 1.89$. After applying Algorithm 3, the refined aggregates have the shapes shown on the right of Fig. 1, of which the two-level method has the same convergence rate $|E|_{\hat{A}} = 0.84$ but the square of the energy seminorm is reduced to $|Q|_{\hat{A}}^2 = 1.50$.

Example 3.3. Consider a graph Laplacian $\hat{A}$ corresponding to a graph which is a 4×4 square grid, on which all horizontal edges are weighted 1 while all vertical edges are weighted 10. Such graph Laplacian represents anisotropic coefficient

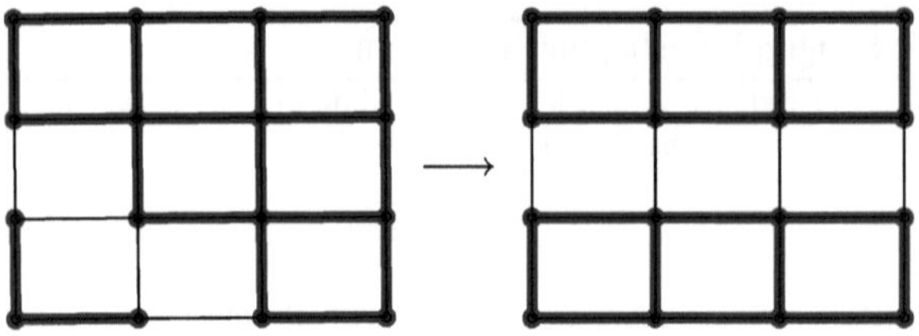

Fig. 1 Subgraph reshaping algorithm applied on a graph representing an isotropic coefficient elliptic PDE

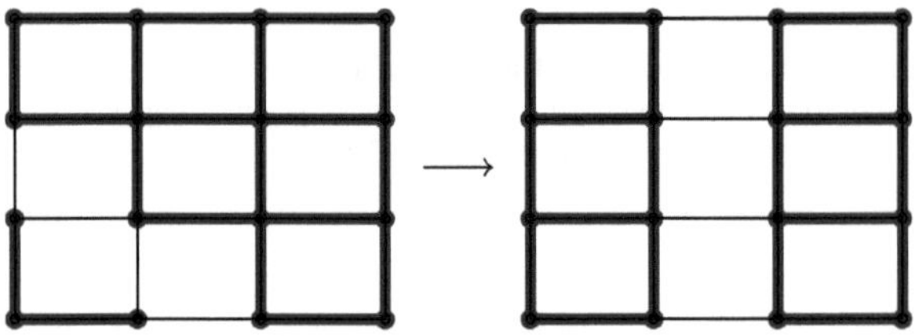

Fig. 2 Subgraph reshaping algorithm applied on a graph representing an anisotropic coefficient elliptic PDE

elliptic equations with Neumann boundary conditions. Start with a non-optimal partitioning as shown on the left of Fig. 2, of which the resulting two-level method has a convergence rate $|E|_{\hat{A}} = 0.96$ and $|Q|^2_{\hat{A}} = 4.88$. After applying Algorithm 3, the refined aggregates have the shapes shown on the right of Fig. 2, of which the two-level convergence rate is reduced to $|E|_{\hat{A}} = 0.90$ and the energy of the coarse-level projection is also reduced as $|Q|^2_{\hat{A}} = 1.50$.

4 Solve Phase

In this section, we discuss the parallelization of the solver phase on GPU. More precisely, we will focus on the parallel smoother, prolongation/restriction, MG cycle, and sparse matrix-vector multiplication.

4.1 Parallel Smoother

An efficient parallel smoother is crucial for the parallel AMG method. For the sequential AMG method, Gauss–Seidel relaxation is widely used and has been shown to have a good smoothing property. However the standard Gauss–Seidel is a sequential procedure that does not allow efficient parallel implementation. To improve the arithmetic intensity of the smoother and make it work better with

SIMD-based GPUs, we adopt the well-known Jacobi relaxation and introduce a damping factor to improve the performance of the Jacobi smoother. For a matrix $A \in \mathbb{R}^{n\times n}$ and its diagonals are denoted by $D = \mathrm{diag}(a_{11}, a_{22}, \cdots, a_{nn})$, the Jacobi smoother can be written in the following matrix form

$$x^{m+1} = x^m + \omega D^{-1} r^m, \quad \text{where } r^m = b - Ax^m,$$

or component-wise

$$x_i^{m+1} = x_i^m + \omega a_{ii}^{-1} r_i^m.$$

This procedure can be implemented efficiently on GPUs by assigning one thread to each component and update the corresponding components locally and simultaneously. We also consider the so-called ℓ^1-Jacobi smoother, which is parameter-free. Define

$$M = \mathrm{diag}(M_{11}, M_{22}, \cdots, M_{nn}),$$

where $M_{ii} = a_{ii} + d_{ii}$ with $d_{ii} = \sum_{j\neq i} |a_{ij}|$, and the ℓ^1-Jacobi has the following matrix form:

$$x^{m+1} = x^m + M^{-1} r^m, \quad \text{where } r^m = b - Ax^m,$$

or component-wise

$$x_i^{m+1} = x_i^m + M_{ii}^{-1} r_i^m.$$

In [7, 25] it has been show that if A is SPD, the smoother is always convergent and has multigrid smoothing properties comparable to full Gauss–Seidel smoother if $a_{ii} \geq \theta d_{ii}$ and θ is bounded away from zero. Moreover, because its formula is very similar to the Jacobi smoother, it can also be implemented efficiently on GPUs by assigning one thread to each component, and update the corresponding the component locally and simultaneously.

4.2 Prolongation and Restriction

For UA-AMG method, the prolongation and restriction matrices are piecewise constant and characterize the aggregates. Therefore, we can perform the prolongation and restriction efficiently in UA-AMG method. Here, the output array `aggregation` (column index of P), which contains the information of aggregates, plays an important rule.

- **Prolongation:** Let $v^{l-1} \in \mathbb{R}^{n_{l-1}}$, so that the action $v^l = P_{l-1}^l v^{l-1}$ can be written component-wise as follows:

$$(v^l)_i = (P_{l-1}^l v^{l-1})_i = (v^{l-1})_j, \quad j \in G_i^{l-1}$$

Assign each thread to one element of v^l, and the array `aggregation` can be used to obtain information about $j \in G_i^{l-1}$, i.e., $i = \texttt{aggregation}[j]$, so that prolongation can be efficiently implemented in parallel.

- **Restriction:** Let $v^l \in \mathbb{R}^{n_l}$, so that the action $(P_{l-1}^l)^T v^l$ can be written componentwise as follows:

$$(v^{l-1})_i = ((P_{l-1}^l)^T v^l)_i = \sum_{j \in G_i^{l-1}} (v^l)_j.$$

Therefore, each thread is assigned to an element of v^{l-1}, and the array `aggregation` can be used to obtain information about $j \in G_i^{j-1}$, i.e., to find all j such that $\texttt{aggregation}[j] = i$. By doing so, the action of restriction can also be implemented in parallel.

4.3 K-Cycle

Unfortunately, in general, UA-AMG with V-cycle is not an optimal algorithm in terms of convergence rate. But on the other hand, in many cases, UA-AMG using two-grid solver phase gives optimal convergence rate for graph Laplacian problems. This motivated us to use other cycles instead of V-cycle to mimic the two-grid algorithm. The idea is to invest more works on the coarse grid and make the method become closer to an exact two-level method; then hopefully, the resulting cycle will have optimal convergence rate.

The particular cycle we will discuss here is the so-called K-cycle (nonlinear AMLI-cycle), and we refer to [1,2,29] for details on its implementation in general.

4.4 Sparse Matrix-Vector Multiplication on GPUs

As the K-cycle will be used as a preconditioner for nonlinear preconditioned conjugate gradient (NPCG) method, the sparse matrix-vector multiplication (SpMV) has major contribution to the computational work involved. An efficient SpMV algorithm on GPU requires a suitable sparse matrix storage format. How different storage formats perform in SpMV is extensively studied in [3]. This study shows that the need for coalesce accessing of the memory makes ELLPACK (ELL) format one of the most efficient sparse matrix storage formats on GPUs when each row of the sparse matrix has roughly the same nonzeros. In our study, because our main focus is on the PAA and the performance of the UA-AMG method, we still use the compressed row storage (CSR) format, which has been widely used for the iterative linear solvers on CPU. Although this is not an ideal choice for GPU implementation, the numerical results in the next section already show the efficiency of our parallel AMG method.

5 Numerical Tests

In this section, we present numerical tests using the proposed parallel AMG methods. Whenever possible we compare the results with the CUSP libraries [15]. CUSP is an open source C++ library of generic parallel algorithms for sparse linear algebra and graph computations on CUDA-enabled GPUs. All CUSP's algorithms and implementations have been optimized for GPU by NVIDIA's research group. To the best of our knowledge, the parallel AMG method implemented in the CUSP package is the state-of-the-art AMG method on GPU. We use as test problems several discretizations of the Laplace equation.

5.1 Numerical Tests for PAA

Define Q, the ℓ^2 projection on the piecewise constant space Range(P), as the following:

$$Q = P(P^T P)^{-1} P^T.$$

We present several tests showing how the energy norm of this projection changes with respect to different parameters used in the PAA, since the convergence rate is an increasing function of $\|Q\|_A$.

The tests involving $\|Q\|_A$ further suggest two additional features necessary to get a multigrid hierarchy with predictable results. First, the sizes of aggregates need to be limited, and second, the columns of the prolongator P need to be ordered in a deterministic way, regardless of the order that aggregates are formed. The first requirement can be fulfilled simply by limiting the sizes of the aggregates in each pass of the PAA. We make the second requirement more specific. Let c_k to be the index of the coarse vertex of the k-th aggregate. We require that c_k should be an increasing sequence and then use the k-th column of P to record the aggregate with the coarse vertex numbered c_k. This can be done by using a generalized version of the prefix sum algorithm [5].

We first show in Table 1 the coarsening ratios (in the parenthesis in the table) and the energy norms $\|Q\|_A^2$ of a two-grid hierarchy, for a Laplace equation with Dirichlet boundary conditions on a structured grid containing n^2 vertices. The limit on the size of an aggregate is denoted by t, which suggests that any aggregate can include t vertices or less, which directly implies that the resulting coarsening ratio is less or equal to t.

For the same aggregations on the graphs that represent Laplace equations with Neumann boundary conditions, the corresponding coarsening ratio (in parenthesis) and $|Q|_A^2$ seminorms with respect to grid size n and limiting threshold t are shown in Table 2.

Table 1 (Coarsening ratios) and energy norm of a two-grid hierarchy of a Laplace equation on a uniform grid with Dirichlet boundary conditions

	$t = 2$	$t = 3$	$t = 4$	$t = 5$
$n = 128$	(1.99) 1.71	(2.03) 2.04	(2.41) 2.46	(2.97) 3.15
$n = 256$	(1.99) 1.72	(2.39) 2.57	(2.96) 2.59	(2.99) 3.20
$n = 512$	(2.00) 1.72	(2.01) 2.08	(2.40) 2.48	(2.99) 3.22

Table 2 (Coarsening ratios) and energy norm of a two-grid hierarchy of a Laplace equation on a uniform grid with Neumann boundary conditions

	$t = 2$	$t = 3$	$t = 4$	$t = 5$
$n = 128$	(1.99) 1.87	(2.03) 2.11	(2.41) 2.48	(2.97) 3.24
$n = 256$	(1.99) 1.74	(2.39) 2.59	(2.96) 2.62	(2.99) 3.24
$n = 512$	(2.00) 1.87	(2.01) 2.11	(2.40) 2.49	(2.99) 3.24

Table 3 (Coarsening ratios) and energy norm of a two-grid hierarchy of a Laplace equation on a uniform grid with Neumann boundary conditions

	$t = 2$	$t = 3$	$t = 4$	$t = 5$
$n = 126$	(2.00) 2.11	(2.00) 2.07	(2.36) 2.73	(2.36) 2.73
$n = 127$	(1.99) 1.86	(2.01) 1.98	(2.01) 2.49	(2.01) 2.34
$n = 128$	(1.99) 1.71	(2.03) 2.04	(2.41) 2.47	(2.97) 3.15
$n = 129$	(1.99) 1.84	(2.02) 2.04	(2.03) 2.31	(2.02) 2.42
$n = 130$	(1.99) 1.77	(2.40) 2.21	(2.92) 2.86	(2.94) 2.94
$n = 131$	(1.99) 2.61	(2.01) 2.41	(2.01) 2.45	(2.00) 2.49
$n = 132$	(1.98) 2.09	(2.21) 2.81	(2.33) 2.89	(2.26) 2.94

In Table 3 we present the computed bounds on the coarsening ratio and energy of a two-level hierarchy[2] when the fine level is an $n \times n$. Such results are valid for any structured grid with n^2 vertices (not just $n = 126, \ldots, 132$) This is seen as follows: (1) From (6), it follows that if we consider two grids of sizes $n_1 \times n_1$ and $n_2 \times n_2$, respectively, and such that

$$(n_1 - n_2) \equiv 0 \mod 12, \quad \text{or} \quad (n_1 + n_2) \equiv 0 \mod 12,$$

then our aggregation algorithm results in the same pattern of C points on these two grids; (2) as a consequence grids of size $n \times n$ for $n = 126 \equiv 6 \mod 12$ to $n = 132 \equiv 0 \mod 12$ give *all* possible coarsening patterns that can be obtained by our aggregation algorithm on *any* 2D tensor product grid. As a conclusion, the values

[2]By energy of a two-level hierarchy here, we mean the seminorm of the ℓ^2 projection on the coarse space.

Table 4 (Coarsening ratios) and energy norm of a two-grid hierarchy of a Laplace equation with Dirichlet boundary conditions discretized on an unstructured grid

	$t=2$	$t=3$	$t=4$	$t=5$	$t=\infty$
$n=128$	(1.80) 2.39	(2.53) 2.44	(3.17) 3.10	(3.73) 3.46	(4.91) 3.30
$n=256$	(1.79) 2.39	(2.52) 2.60	(3.15) 3.18	(3.69) 3.46	(4.91) 3.41
$n=512$	(1.80) 2.38	(2.55) 2.67	(3.19) 3.26	(3.72) 3.56	(4.93) 3.40

Table 5 (Coarsening ratios) and energy norm of a two-grid hierarchy of a Laplace equation with Neumann boundary conditions discretized on an unstructured grid

	$t=2$	$t=3$	$t=4$	$t=5$	$t=\infty$
$n=128$	(1.80) 2.39	(2.53) 2.54	(3.17) 3.18	(3.73) 3.48	(4.91) 3.33
$n=256$	(1.79) 2.49	(2.52) 2.65	(3.15) 3.20	(3.69) 3.48	(4.91) 3.41
$n=512$	(1.80) 2.47	(2.55) 2.80	(3.19) 3.27	(3.72) 3.57	(4.93) 3.53

of the coarsening ratios and the energy seminorm given in Table 3 are valid for any 2D structured grid.

We also apply this aggregation method on graphs corresponding to Laplace equations on two-dimensional unstructured grids with Dirichlet or Neumann boundary conditions. The unstructured grids are constructed by perturbing nodes in an $n \times n$ square lattice ($n = 128, 256, 512$), followed by triangulating the set of perturbed points using a Delaunay triangulation. The condition numbers of the Laplacians, derived using finite element discretization of the Laplace equations on the mentioned unstructured grids with Dirichlet boundary conditions, are about 1.2×10^4, 5.0×10^4, and 2.1×10^5, respectively. The coarsening ratios and $|Q|_A^2$ are listed in Tables 4 and 5. We remark here that we also apply the PAA without imposing limit on the size of an aggregate, and the corresponding numerical results are listed in columns named "$t = \infty$."

We note that in Tables 4 and 5, the coarsening ratios are not large enough to result in small operator complexity. We then estimate the energy norm $|Q|_A^2$ when Q is the ℓ^2 orthogonal projection from any level of the multigrid hierarchy to any succeeding sublevels. We start with a Laplace equation on a 128^2 structured square grid, set the limit of sizes of aggregates as $t = 5$ on each iteration of aggregation, and stop when the coarsest level is of less than 100 degrees of freedom. If we number the levels starting with the finest level (level 0), then, in this example, the coarsest level will be level 5. The coarsening ratios and energy norm $|Q|_A^2$ between any two levels are shown in Tables 6 and 7.

We observe that, on the diagonal of Tables 6 and 7, the energy norms are comparable to the coarsening ratios, until the last level where the grid becomes highly unstructured. This suggests that a linear or nonlinear AMLI solving cycle can give both a good convergence rate and a favorable complexity. It is also observed

Table 6 (Coarsening ratios) and energy norms squares of a multigrid hierarchy of a Laplace equation with Dirichlet boundary conditions discretized on a 128^2 grid

	0	1	2	3	4
1	(2.97) 3.15	–	–	–	–
2	(11.3) 7.34	(3.81) 3.09	–	–	–
3	(38.6) 15.3	(13.0) 6.74	(3.41) 2.68	–	–
4	(113.8) 31.9	(38.3) 13.9	(10.1) 5.25	(2.95) 3.91	–
5	(321.3) 54.5	(108.2) 22.5	(28.4) 9.09	(8.33) 4.53	(2.82) 4.77

Table 7 (Coarsening ratios) and energy norm of a multigrid hierarchy of a Laplace equation with Neumann boundary conditions discretized on a 128^2 grid

	0	1	2	3	4
1	(2.97) 3.23	–	–	–	–
2	(11.3) 7.68	(3.81) 3.28	–	–	–
3	(38.6) 16.9	(13.0) 7.99	(3.41) 2.94	–	–
4	(113.8) 42.5	(38.3) 20.6	(10.1) 7.07	(2.95) 4.23	–
5	(321.3) 98.6	(108.2) 48.3	(28.4) 17.0	(8.33) 7.11	(2.82) 5.75

that, on the lower triangular part of Tables 6 and 7, the energy norms are always smaller than the corresponding coarsening ratios, which suggests that flexible cycles that detect and skip unnecessary levels can be more efficient.

Another inspiring observation is that, in Table 1, even if we set a limit $t = 5$ for the maximal number of vertices in an aggregate, the resulting aggregates have an average number of vertices ranging from 2.97 to 2.99. We plot the aggregates of an unweighted graph corresponding to a 16×16 square grid on the left of Fig. 3 and observe that some aggregates contain five vertices and some contain only one. We then use the rank one optimization discussed in Sect. 3.3 and apply subgraph reshaping algorithm (Algorithm 3) as a post-process of the GPU PAA (Algorithm 2) and plot the resulting aggregates on the right of Fig. 3. Since the subgraph reshaping algorithm does not change the number of aggregates, the coarsening ratios on the left and right of Fig. 3 are identical and are equal to 2.72. The energy of the ℓ^2 projection is deceased from $|Q|_{\hat{A}}^2 = 2.51$ (left of Fig. 3) to $|Q|_{\hat{A}}^2 = 2.19$ (right of Fig. 3). However, two-level convergence rate increases from $|E|_{\hat{A}} = 0.67$ (left of Fig. 3) to $|E|_{\hat{A}} = 0.69$ (right of Fig. 3).

Some more comments on the reshaping algorithm are in order. For isotropic problems, the reshaping does not have significant impact of on the convergence rate because aggregation obtained by standard approach already results in a good convergence rate. However, for anisotropic problems, reshaping improves the convergence rate. In this case, starting with aggregates of arbitrary shape, the reshaping procedure results in aggregates aligned with the anisotropy and definitely improves the overall convergence rate.

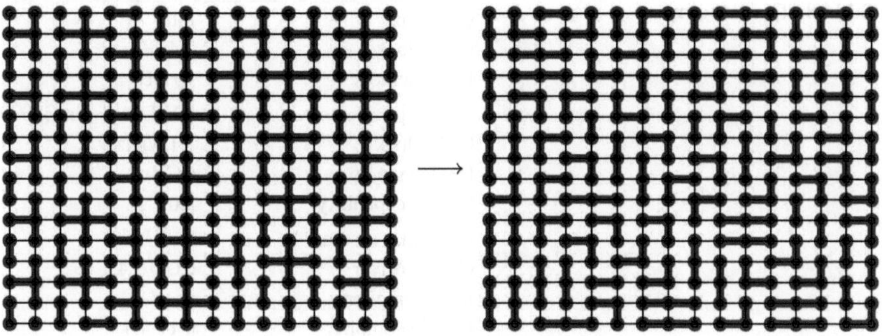

Fig. 3 Before (*left*) and after (*right*) the subgraph reshaping algorithm applied on partitioning given by PAA

In addition, even for isotropic case, the numerical results in the manuscript indicate that subgraph reshaping can be essential for a variety of cycling algorithms when aggressive coarsening is applied. As shown in Examples 3.2 and 3.3, the aggregation reshaping helps for some isotropic and anisotropic problems when coarsening ratio is 8. In Tables 6 and 7, we observe that such coarsening ratio can be achieved by skipping every other level in our current multilevel hierarchy.

Clearly, further investigation about the reshaping is needed for more general problems that have both anisotropic and isotropic regions. Analyzing such cases as well as testing how much improvement in the convergence can be achieved by subgraph reshaping for specific coarsening and cycling strategies is subject of an ongoing and future research.

5.2 *Numerical Tests for GPU Implementation*

In this section, we perform numerical experiments to demonstrate the efficiency of our proposed AMG method and discuss the specifics related to the use of GPUs as main platform for computations. We test the parallel algorithm on Laplace equation discretized on quasi-uniform grids in 2D. Our test and comparison platform is the NVIDIA Tesla C2070 together with a Dell computing workstation. Details in regard to the machine are given in Table 8.

Because our aim is to demonstrate the improvement of our algorithm on GPUs, we concentrate on comparing the method we describe here with the parallel smoothed aggregation AMG method implemented in the CUSP package [15].

We consider the standard linear finite element method for the Laplace equation on unstructured meshes. The results are shown in Table 9. Here, CUSP uses smoothed aggregation AMG method with V-cycles, and our method is UA-AMG with K-cycles. The stopping criterion is that the ℓ^2 norm of the relative residual is less than 10^{-6}. According to the results, we can see that our parallel UA-AMG

Table 8 Test platform

CPU type	Intel
CPU clock	2.4 GHz
Host memory	16 GB
GPU type	NVIDIA Tesla C2070
GPU clock	1.15GHz
Device memory	6 GB
CUDA capability	2.0
Operating system	RedHat
CUDA driver	CUDA 4.1
Host complier	gcc 4.1
Device complier	nvcc 4.1
CUSP	v0.3.0

Table 9 Comparison between the parallel AMG method in CUSP package (smoothed aggregation AMG with V-cycles) and our new parallel AMG method (UA-AMG with K-cycles)

	#DoF = 1 million				#DoF = 4 million			
	# Iter.	Setup	Solve	Total	# Iter.	Setup	Solve	Total
CUSP	36	0.63	0.35	0.98	41	2.38	1.60	3.98
New	19	0.13	0.47	0.60	19	0.62	2.01	2.63

method converges uniformly with respect to the problem size. This is due the improved aggregation algorithm constructed by our parallel aggregation method and the K-cycle used in the solver phase. We can see that our method is about 3 to 4 times faster in setup phase, which demonstrate the efficiency of our PAA. In the solver phase, due to the factor that we use K-cycle, which does much more work on the coarse grids, our solver phase is a little bit slower than the solver phase implemented in CUSP. However, the use of a K-cycle yields a uniformly convergent UA-AMG method, which is an essential property for designing scalable solvers. When the size of the problem gets larger, we expect the computational time of our AMG method to scale linearly, whereas the AMG method in CUSP seems to grows faster than linear and will be slower than our solver phase eventually. Overall, our new AMG solver is about 1.5 times faster than the smoothed aggregation AMG method in CUSP in terms of total computational time, and the numerical tests suggest that it converges uniformly for the Poisson problem.

References

1. Axelsson, O., Vassilevski, P.S.: Algebraic multilevel preconditioning methods, I. Numer. Math. **56**(2–3), 157–177 (1989)
2. Axelsson, O., Vassilevski, P.S.: Algebraic multilevel preconditioning methods, II. SIAM J. Numer. Anal. **27**(6), 1569–1590 (1990)

3. Bell, N., Garland, M.: Efficient sparse matrix-vector multiplication on CUDA. Technical Report NVR-2008-004, NVIDIA Corporation, 2008
4. Bell, N., Dalton, S., Olson, L.: Exposing fine-grained parallelism in algebraic multigrid methods. SIAM J. Sci. Comput. **34**(4), C123–C152 (2012)
5. Blelloch, G.E.: Prefix sums and their applications. In: Synthesis of Parallel Algorithms. Morgan Kaufmann, Los Altos (1993)
6. Bolz, J., Farmer, I., Grinspun, E., Schröder, P.: Sparse matrix solvers on the gpu: conjugate gradients and multigrid. ACM Trans. Graph. **22**(3), 917–924 (2003)
7. Brezina, M., Vassilevski, P.S.: Smoothed aggregation spectral element agglomeration AMG: SA-ρAMGe. In: Proceedings of the 8th International Conference on Large-Scale Scientific Computing. Lecture Notes in Computer Science, vol. 7116, pp. 3–15. Springer, Berlin (2012)
8. Chow, E., Falgout, R., Hu, J., Tuminaro, R., Yang, U.: A survey of parallelization techniques for multigrid solvers. In: Parallel Processing for Scientific Computing, vol. 20, pp. 179–201. SIAM, Philadelphia (2006)
9. Cleary, A.J., Falgout, R.D., Henson, V.E., Jones, J.E.: Coarse-grid selection for parallel algebraic multigrid. In: Solving Irregularly Structured Problems in Parallel. Lecture Notes in Computer Science, vol. 1457, pp. 104–115. Springer, Berlin (1998)
10. Courant, R., Hilbert, D.: Methods of Mathematical Physics, vol. I. Wiley Classics Library. Wiley, New York (1989)
11. De Sterck, H., Yang, U.M., Heys, J.J.: Reducing complexity in parallel algebraic multigrid preconditioners. SIAM. J. Matrix Anal. Appl. **27**(4), 1019–1039 (2006)
12. Emans, M., Liebmann, M., Basara, B.: Steps towards GPU accelerated aggregation AMG. In: 2012 11th International Symposium on Parallel and Distributed Computing (ISPDC), pp. 79–86 (2012)
13. Feng, Z., Zeng, Z.: Parallel multigrid preconditioning on graphics processing units (gpus) for robust power grid analysis. In: Proceedings of the 47th Design Automation Conference, DAC '10, pp. 661–666. ACM, New York (2010)
14. Feng, C., Shu, S., Xu, J., Zhang, C.-S.: Numerical Study of Geometric Multi- grid Methods on CPU-GPU Heterogeneous Computers. ArXiv e-prints (2012), arXiv:1208.4247. http://arxiv.org/abs/1208.4247
15. Garland, M., Bell, N.: CUSP: Generic parallel algorithms for sparse matrix and graph computations. CUSP Software Library. (2010). http://code.google.com/p/cusp-library
16. Goddeke, D., Strzodka, R., Mohd-Yusof, J., McCormick, P., Wobker, H., Becker, C., Turek, S.: Using GPUs to improve multigrid solver performance on a cluster. Int. J. Comput. Sci. Eng. **4**(1), 36–55 (2008)
17. Goodnight, N., Woolley, C., Lewin, G., Luebke, D., Humphreys, G.: A multigrid solver for boundary value problems using programmable graphics hardware. ACM SIGGRAPH 2005 Courses on SIGGRAPH 05, p. 193 (2003)
18. Grossauer, H., Thoman, P.: GPU-based multigrid: Real-time performance in high resolution nonlinear image processing. Int. J. Comput. Vis. **5008**, 141–150 (2008)
19. Haase, G., Liebmann, M., Douglas, C., Plank, G.: A parallel algebraic multigrid solver on graphics processing units. In: High Performance Computing and Applications, Springer, Berlin, Heidelberg, New York, **5938**, 38–47 (2010)
20. Halmos, P.R.: Finite-dimensional vector spaces. In: Undergraduate Texts in Mathematics, 2nd edn. Springer, New York (1974)
21. Henson, V.E., Yang, U.M.: BoomerAMG: A parallel algebraic multigrid solver and preconditioner. Appl. Numer. Math. **41**(1), 155–177 (2002)
22. Joubert, W., Cullum, J.: Scalable algebraic multigrid on 3500 processors. Electron. Trans. Numer. Anal. **23**, 105–128 (2006)
23. Karypis, G., Kumar, V.: A parallel algorithm for multilevel graph partitioning and sparse matrix ordering. J. Parallel Distrib. Comput. **48**, 71–85 (1998)
24. Karypis, G., Kumar, V.: Parallel multilevel k-way partitioning scheme for irregular graphs. SIAM Rev. **41**(2), 278–300 (1999)

25. Kolev, T.V., Vassilevski, P.S.: Parallel auxiliary space AMG for $H(\text{curl})$ problems. J. Comput. Math. **27**(5), 604–623 (2009)
26. Kraus, J., Förster, M.: Efficient AMG on heterogeneous systems. In: Facing the Multicore - Challenge II. Lecture Notes in Computer Science, vol. 7174, pp. 133–146. Springer, Berlin (2012)
27. Krechel, A., Stüben, K.: Parallel algebraic multigrid based on subdomain blocking. Parallel Comput. **27**(8), 1009–1031 (2001)
28. Tuminaro, R.S.: Parallel smoothed aggregation multigrid: aggregation strategies on massively parallel machines. In: Proceedings of the 2000 ACM/IEEE Conference on Supercomputing (CDROM). IEEE, New York (2000)
29. Vassilevski, P.S.: Multilevel Block Factorization Preconditioners. Springer, New York (2008)

Aspects of Guaranteed Error Control in CPDEs

C. Carstensen, C. Merdon, and J. Neumann

Abstract Whenever numerical algorithms are employed for a reliable computational forecast, they need to allow for an error control in the final quantity of interest. The discretization error control is of some particular importance in computational PDEs (CPDEs) where guaranteed upper error bound (GUB) are of vital relevance. After a quick overview over energy norm error control in second-order elliptic PDEs, this paper focuses on three particular aspects: first, the variational crimes from a nonconforming finite element discretization and guaranteed error bounds in the discrete norm with improved postprocessing of the GUB; second, the reliable approximation of the discretization error on curved boundaries; and finally, the reliable bounds of the error with respect to some goal functional, namely, the error in the approximation of the directional derivative at a given point.

Keywords Guaranteed error control • Equilibration error estimators • Poisson model problem • Conforming finite element methods • Crouzeix–Raviart nonconforming finite element methods • Curved boundaries • Guaranteed goal-oriented error control

Mathematics Subject Classification (2010): 65N30, 65N15

C. Carstensen (✉) • C. Merdon
Humboldt-Universität zu Berlin, Unter den Linden 6,
10099 Berlin, Germany

Department of Computational Science and Engineering,
Yonsei University, 120-749 Seoul, Korea
e-mail: cc@mathematik.hu-berlin.de; merdon@mathematik.hu-berlin.de

J. Neumann
Weierstraß-Institut, Mohrenstr. 39, 10117 Berlin, Germany
e-mail: Johannes.Neumann@wias-berlin.de

O.P. Iliev et al. (eds.), *Numerical Solution of Partial Differential Equations: Theory, Algorithms, and Their Applications*, Springer Proceedings in Mathematics & Statistics 45, DOI 10.1007/978-1-4614-7172-1_6,

1 Introduction

A posteriori finite element error control of second-order elliptic boundary value problems usually involves residuals of the prototype

$$\operatorname{Res}(v) = \int_\Omega (fv - \sigma_h \cdot \nabla v)\,\mathrm{d}x \quad \text{for } v \in V := H_0^1(\Omega) \tag{1}$$

with some given Lebesgue integrable function f and the discrete flux σ_h [10, 16]. Its dual norm with respect to some energy norm $|||\cdot|||$ reads

$$|||\operatorname{Res}|||_\star := \sup_{v\in V} \operatorname{Res}(v)/\,|||v|||.$$

For instance, the Poisson model problem seeks $u \in V$ with $f + \Delta u = 0$ and leads to the variational formulation

$$\int_\Omega \nabla u \cdot \nabla v\,\mathrm{d}x = \int_\Omega fv\,\mathrm{d}x \quad \text{for all } v \in V.$$

In this example, the energy norm reads $|||\cdot||| := \|\nabla\cdot\|_{L^2(\Omega)}$, and $\sigma_h = \nabla u_h$ might be the gradient of the piecewise affine conforming finite element solution u_h.

Section 2 summarizes techniques and recent advances from the ongoing computational surveys [4, 11, 13] to compute guaranteed upper bounds for $|||\operatorname{Res}|||_\star$, or error majorants in the sense of Repin [22], via the design of some $q \in H(\operatorname{div},\Omega)$ such that, by a triangle inequality,

$$|||\operatorname{Res}|||_\star \le |||f + \operatorname{div} q|||_\star + |||\operatorname{div}(q - \sigma_h)|||_\star.$$

While $|||f + \operatorname{div} q|||_\star$ may lead to oscillations or other higher-order terms, the second term is often estimated suboptimally as $|||\operatorname{div}(q-\sigma_h)|||_\star \le \|q - \sigma_h\|_{L^2(\Omega)}$. A new generation of equilibration error estimators is based on

$$|||\operatorname{div}(q - \sigma_h)|||_\star = \min_{v\in H^1(\Omega)} \|q - \sigma_h - \operatorname{Curl} v\|_{L^2(\Omega)}$$

and the novel postprocessing from [13] improves the efficiency at almost no extra costs. Section 2 reports on the superiority of those error estimates with an application to the conforming P_1 finite element method for the Poisson model problem.

Section 3 examines the nonconforming Crouzeix–Raviart approximations u_{CR} and its discrete flux $\sigma_h = \nabla_{\mathrm{NC}} u_{\mathrm{CR}}$ for the Poisson model problem. The Helmholtz decomposition allows a split of the broken energy error norm into

$$|||u - u_{\mathrm{CR}}|||_{\mathrm{NC}}^2 = |||\operatorname{Res}|||_\star^2 + |||\operatorname{Res}_{\mathrm{NC}}|||_\star^2.$$

The two residuals Res and $\operatorname{Res}_{\mathrm{NC}}$ allow an estimation via all known a posteriori error estimators. Furthermore, the special structure of the nonconforming residual $\operatorname{Res}_{\mathrm{NC}}$ allows an alternative analysis by the design of conforming companions of u_{CR} [12].

In this paper, we also apply the postprocessed equilibration error estimators to the first residual for even sharper error control beyond [12].

Section 4 extends guaranteed error control to domains with curved boundaries and exemplifies the modifications for some sector domain.

Section 5 establishes guaranteed goal-oriented error estimation where the error $u - u_h$ between the exact and the discrete P_1-FEM solution is *not* measured in the energy norm but with respect to some goal functional $Q \in H^1(\Omega)$. Its Riesz representation solves some dual problem [3,5] that links the error $Q(e)$ to the energy norms of two perturbed Poisson problems [21]. Lower and upper bounds for those quantities lead to guaranteed bounds for $Q(u - u_h)$.

2 Review of Guaranteed Energy Norm Error Control

This section deals with guaranteed upper bounds for dual norms of residuals by equilibration error estimators. An application to the P_1 conforming finite element method for the Poisson model problem concludes the section.

2.1 Notation

Consider a regular triangulation $\mathscr{T}$ of the simply connected, polygonal, and bounded Lipschitz domain $\Omega \subset \mathbb{R}^2$ into triangles with edges $\mathscr{E}$, nodes $\mathscr{N}$, boundary nodes $\mathscr{N}(\partial\Omega)$, and free nodes $\mathscr{N}(\Omega) := \mathscr{N} \setminus \mathscr{N}(\partial\Omega)$. The midpoints of all edges are denoted by $\mathrm{mid}(\mathscr{E}) := \{\mathrm{mid}(E) \,|\, E \in \mathscr{E}\}$, and the boundary edges along $\partial\Omega$ are denoted by $\mathscr{E}(\partial\Omega) := \{E \in \mathscr{E} \,|\, E \subseteq \partial\Omega\}$, while $\mathscr{E}(\Omega) := \mathscr{E} \setminus \mathscr{E}(\partial\Omega)$ denotes the set of interior edges. The set $\mathscr{T}(E) := \{T \in \mathscr{T} \,|\, E \subset \partial T\}$ contains the neighboring triangles of the edge $E \in \mathscr{E}$. The open set $\omega_z := \{x \in \Omega \,|\, \varphi_z(x) > 0\}$ for the nodal basis function φ_z is the interior of $\bigcup \mathscr{T}(z)$ for the subtriangulation $\mathscr{T}(z) := \{T \in \mathscr{T} \,|\, z \in \mathscr{N}(T)\}$. The diameter $\mathrm{diam}(T)$ of a triangle T is denoted by h_T. The red refinement $\mathrm{red}(\mathscr{T})$ of $\mathscr{T}$ is a regular triangulation that refines each triangle $T \in \mathscr{T}$ into four congruent subtriangles by straight lines through the midpoints of the three edges. With the set $P_k(\mathscr{T})$ of elementwise polynomials of total degree $\leq k$, the Raviart–Thomas finite element space of order m reads

$$\mathrm{RT}_m(\mathscr{T}) := \Big\{ q \in H(\mathrm{div}, \Omega) \,\big|\, \forall T \in \mathscr{T} \, \exists a_T, b_T, c_T \in P_m(T) \\ \forall x \in T,\, q(x) = a_T x + \big(b_T, c_T\big) \Big\}.$$

The set $C_0(\Omega)$ contains continuous functions with zero boundary conditions along $\partial\Omega$.

2.2 Equilibration Error Estimators

Consider some residual of the form (1) with source function $f \in L^2(\Omega)$ and discrete flux $\sigma_h \in P_0(\mathscr{T};\mathbb{R}^2)$ such that $\operatorname{Res}(\varphi_z) = 0$ for all $z \in \mathscr{N}(\Omega)$. Equilibration error estimators design some quantity $q \in H(\operatorname{div},\Omega)$ such that $|||f + \operatorname{div} q|||_\star$ is of higher order and

$$|||\operatorname{Res}|||_\star \le |||f + \operatorname{div} q|||_\star + |||\operatorname{div}(\sigma_h - q)|||_\star.$$

Two examples for such a design are given through the Braess equilibration error estimator [6, 8] and the Luce–Wohlmuth error estimator [13, 18] which solve at most one-dimensional linear systems of equations around each node $z \in \mathscr{N}$ and design some Raviart–Thomas function $q_{\mathrm{B}} \in \mathrm{RT}_0(\mathscr{T})$ or $q_{\mathrm{LW}} \in \mathrm{RT}_0(\mathscr{T}^\star)$ on the dual triangulation $\mathscr{T}^\star$.

The dual mesh $\mathscr{T}^\star$ divides every triangle $T \in \mathscr{T}$ into six subtriangles of same area by connection of the center $\operatorname{mid}(T)$ with the three vertices and the three edge midpoints of T. This results in the two guaranteed upper bounds:

$$\eta_{\mathrm{B}} := \|h_{\mathscr{T}}(f - f_{\mathscr{T}})\|_{L^2(\Omega)}/j_{1,1} + \|\sigma_h - q_{\mathrm{B}}\|_{L^2(\Omega)}, \tag{2}$$

$$\eta_{\mathrm{LW}} := \|h_{\mathscr{T}}(f - f^\star)\|_{L^2(\Omega)}/j_{1,1} + \|\sigma_h - q_{\mathrm{LW}}\|_{L^2(\Omega)} \tag{3}$$

for the piecewise integral mean $f_{\mathscr{T}} \in P_0(\mathscr{T})$, i.e., $f_{\mathscr{T}}|_T := \fint_T f\,dx$ for $T \in \mathscr{T}$ and $f^\star \in P_0(\mathscr{T}^\star)$ with $f^\star|_{T^\star} := 3\fint_T f\varphi_z\,dx$ on the two subtriangles $T^\star \in \mathscr{T}^\star(z)$ of $T \in \mathscr{T}(z)$. The function $f^\star$ is our preferred approximation of f in the Luce–Wohlmuth design [13, 15] that allows this very easy estimation of $|||f - f^\star|||_\star$. The number $j_{1,1}$ is the first positive root of the Bessel function J_1 from the Poincaré constant [17].

Definitions (2)–(3) employ the estimate $|||\operatorname{div}(\sigma_h - q)|||_\star \le \|\sigma_h - q\|_{L^2(\Omega)}$, which is suboptimal, because of

$$|||\operatorname{div}(q - \sigma_h)|||_\star = \min_{\gamma \in H^1(\Omega)} \|q - \sigma_h - \operatorname{Curl}\gamma\|_{L^2(\Omega)}.$$

The novel postprocessing from [13] designs some piecewise affine γ_h that is cheap to compute and leads to sharper estimates. The computation runs some simple PCG scheme with k iterations on a refined triangulation $\operatorname{red}(\mathscr{T})$ or $\operatorname{red}^2(\mathscr{T})$ for η_{B} and $\mathscr{T}^\star$ for η_{LW}. In the numerical examples below, the number of cg iterations of the postprocessing is added to the label in brackets. Every additional "r" in front of this number is related to one red refinement. For example, the error estimator $\eta_{\mathrm{Brr}(3)}$ is the postprocessed η_{B} on two red refinements with 3 cg iterations. The case $k = \infty$ means an exact solve and leads to the best possible γ; further details on the algorithm are included in [13].

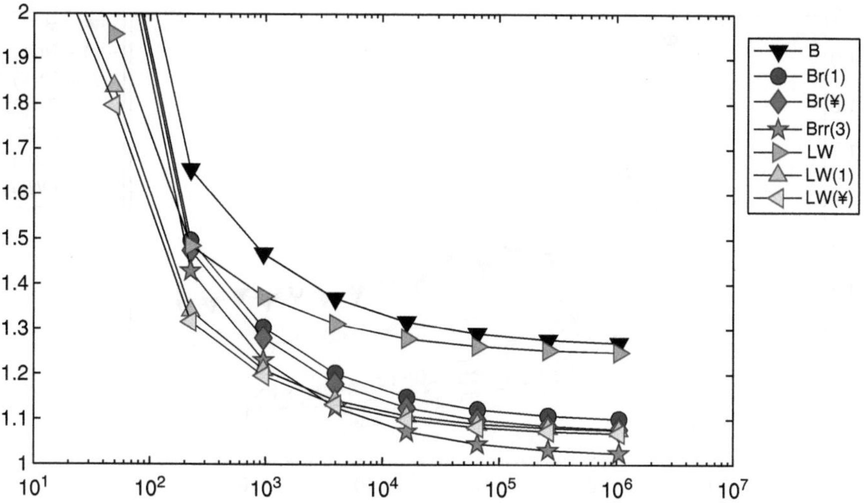

Fig. 1 History of efficiency indices $\eta_{xyz}/|||e|||$ of the standard and postprocessed Braess and Luce–Wohlmuth error estimators η_{xyz} labeled *xyz* as functions of ndof on uniform meshes in Sect. 2.3

2.3 Poisson Model Problem with Big Oscillations

The Poisson model problem seeks $u \in H_0^1(\Omega)$ with $f + \Delta u = 0$ for some source function $f \in L^2(\Omega)$ on the unit square $\Omega := (0,1)^2$. The conforming FEM seeks $u_h \in V_C := P_1(\mathscr{T}) \cap C_0(\Omega)$ with

$$\int_\Omega \nabla u_h \cdot \nabla v_h \, dx = \int_\Omega f v_h \, dx \quad \text{for all } v_h \in V_C.$$

This leads to the residual (1) with $\sigma_h = \nabla u_h$ and $V_C \subseteq \ker \mathrm{Res}$. Elementary calculations, e.g., in [9], reveal that $|||\mathrm{Res}|||_\star = |||u - u_h||| := \|\nabla(u - u_h)\|_{L^2(\Omega)}$.

The remaining parts of this section concern the benchmark problem with an oscillating source term $f := -\Delta u$ that matches the exact solution:

$$u(x,y) = x(x-1)y(y-1)\exp(-100(x-1/2)^2 - 100(y-117/1000)^2) \in H_0^1(\Omega).$$

Figures 1 and 2 show the efficiency indices $\eta_{xyz}/|||u - u_h|||$ for various GUB η_{xyz} after Braess and Luce–Wohlmuth for uniform and adaptive mesh refinement. The Dörfler marking drives the adaptive mesh refinement with the refinement indicators:

$$\eta(T)^2 := |T| \, \|f\|_{L^2(T)}^2 + |T|^{1/2} \sum_{E \in \mathscr{E}(T)} \|[\sigma_h]_E \cdot \nu_E\|_{L^2(E)}^2. \tag{4}$$

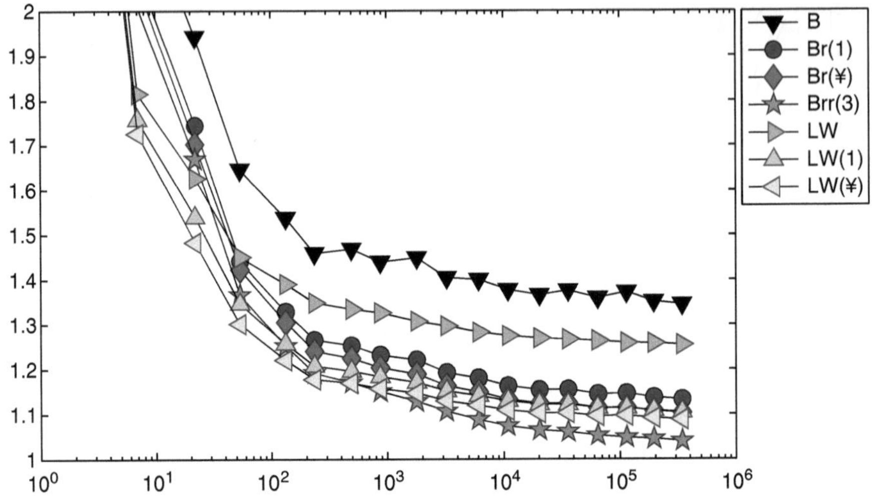

Fig. 2 History of efficiency indices $\eta_{\rm xyz}/|||e|||$ of the standard and postprocessed Braess and Luce–Wohlmuth error estimators $\eta_{\rm xyz}$ labeled xyz in the figure as functions of ndof on adaptive meshes in Sect. 2.3

On coarse triangulations, the oscillations dominate the guaranteed upper bounds, and the postprocessing is almost effectless. However, as the number of degrees of freedom grows and the oscillations decrease, the efficiency improves and the postprocessing unfolds its full effectivity.

The postprocessing $\eta_{\rm Br(1)}$ of $\eta_{\rm B}$ based on red$(\mathscr{T})$ and the postprocessing $\eta_{\rm LW(1)}$ of $\eta_{\rm LW}$ based on $\mathscr{T}^\star$ reduce the efficiency indices about 20 % to values between 1.1 and 1.15, respectively. The optimal postprocessing with $k = \infty$ shows only very little further improvement over the postprocessing with $k = 1$. The postprocessing $\eta_{\rm Brr(3)}$ of $\eta_{\rm B}$ based on two red refinements red$^2(\mathscr{T})$ and $k = 3$ iterations even leads to striking efficiency indices of about 1.05.

Similar treatment is possible for conforming obstacle problems [14].

3 Guaranteed Error Control for CR-NCFEM

This section develops sharp guaranteed upper bounds for the broken energy norm

$$|||u - u_{\rm CR}|||_{\rm NC}^2 := \sum_{T\in\mathscr{T}} \|\nabla(u - u_{\rm CR})\|_{L^2(T)}^2$$

for the error between the exact solution u and the Crouzeix–Raviart nonconforming FEM (CR-NCFEM) solution $u_{\rm CR}$.

3.1 Main Result

The CR-NCFEM employs the Crouzeix–Raviart functions:

$$\mathrm{CR}^1(\mathscr{T}) := \{v \in P_1(\mathscr{T}) \,|\, v \text{ is continuous at } \mathrm{mid}(\mathscr{E})\},$$
$$\mathrm{CR}^1_0(\mathscr{T}) := \{v \in \mathrm{CR}^1(\mathscr{T}) \,|\, \forall E \in \mathscr{E}(\partial\Omega),\ v(\mathrm{mid}(E)) = 0\}.$$

The nonconforming finite element approximation $u_{\mathrm{CR}} \in \mathrm{CR}^1_0(\mathscr{T})$ for the Poisson model problem with its piecewise gradient $\nabla_{\mathrm{NC}} u_{\mathrm{CR}}$ satisfies

$$\int_\Omega \nabla_{\mathrm{NC}} u_{\mathrm{CR}} \nabla_{\mathrm{NC}} v_{\mathrm{CR}} dx = \int_\Omega f v_{\mathrm{CR}}\, \mathrm{d}x \quad \text{for all } v_{\mathrm{CR}} \in \mathrm{CR}^1_0(\mathscr{T}).$$

The main result from [12] for the 2D case with a simply connected domain Ω and homogeneous Dirichlet boundary conditions requires the Helmholtz decomposition of $\nabla_{\mathrm{NC}}(u - u_{\mathrm{CR}}) = \nabla\alpha + \mathrm{curl}\,\beta$ for $\alpha \in H^1_0(\Omega)$ and $\beta \in H^1(\Omega)$. It follows

$$|||u - u_{\mathrm{CR}}|||^2_{\mathrm{NC}} = |||\alpha|||^2 + \|\mathrm{curl}\,\beta\|^2_{L^2(\Omega)} = |||\mathrm{Res}|||^2_\star + |||\mathrm{Res}_{\mathrm{NC}}|||^2_\star$$

with the residuals

$$\mathrm{Res}(v) := \int_\Omega f v\, \mathrm{d}x - \int_\Omega \nabla_{\mathrm{NC}} u_{\mathrm{CR}} \cdot \nabla v\, \mathrm{d}x \qquad \text{for } v \in H^1_0(\Omega),$$
$$\mathrm{Res}_{\mathrm{NC}}(v) := -\int_\Omega \mathrm{curl}_{\mathrm{NC}}\, u_{\mathrm{CR}} \cdot \nabla v\, \mathrm{d}x \qquad \text{for } v \in H^1(\Omega).$$

The dual norm of the second residual allows the alternative characterization

$$|||\mathrm{Res}_{\mathrm{NC}}|||_\star = \min_{v \in V} |||u_{\mathrm{CR}} - v|||_{\mathrm{NC}} \leq |||u - u_{\mathrm{CR}}|||_{\mathrm{NC}}. \tag{5}$$

3.2 Guaranteed Upper Bounds for $|||\mathrm{Res}|||_\star$

The dual norm of the first residual is controlled [1, 12] by the explicit bound

$$|||\mathrm{Res}|||^2_\star \leq \eta^2 := \sum_{T \in \mathscr{T}} \left(\frac{h_T}{j_{1,1}} \|f - f_{\mathscr{T}}\|_{L^2(T)} + \frac{f_T}{2} \|\bullet - \mathrm{mid}(T)\|_{L^2(T)} \right)^2. \tag{6}$$

Here, $\mathrm{mid}(T)$ denotes the triangle center of $T \in \mathscr{T}$, and the quantity $\mathrm{osc}(f, \mathscr{T}) := \|h_{\mathscr{T}}(f - f_{\mathscr{T}})\|_{L^2(\Omega)}$ denotes the oscillations of f. Since $V_C \subseteq \ker \mathrm{Res}$, $|||\mathrm{Res}|||_\star$ can also be estimated by any other guaranteed error estimator [10], e.g., the equilibration error estimators from Sect. 2.

Table 1 Guaranteed upper bounds for $|||\mathrm{Res}|||_\star$ by η and the equilibration error estimators η_{B}, η_{LW}, and some of their postprocessings for uniform mesh refinements in the example of Sect. 3.2

ndof	8	40	176	736	3008	12160	48896
$|||u-u_{\mathrm{CR}}|||_{\mathrm{NC}}$	0.0583	0.0527	0.0287	0.0198	0.0103	0.00517	0.00259
$\mathrm{osc}(f)$	0.223	0.0952	0.0391	0.00938	0.00243	0.000613	0.000154
η	0.233	0.112	0.0521	0.0190	0.00769	0.00336	0.00156
B	0.253	0.140	0.0672	0.0219	0.00835	0.00352	0.00160
LW	0.230	0.116	0.0490	0.0178	0.00737	0.00328	0.00154
Br(1)	0.249	0.133	0.0657	0.0210	0.00796	0.00333	0.00151
Br(∞)	0.248	0.131	0.0654	0.0210	0.00795	0.00333	0.00151
LW(1)	0.229	0.113	0.0477	0.0172	0.00705	0.00312	0.00146
LW(∞)	0.228	0.112	0.0474	0.0172	0.00704	0.00312	0.00146
Brr(3)	0.247	0.128	0.0645	0.0206	0.00782	0.00327	0.00148

The oscillations $\mathrm{osc}(f)$ are displayed to show its declining influence to η

The numerical example from Sect. 2.3 allows for a comparison of the performance of η with that of the Braess and the Luce–Wohlmuth error estimator from Sect. 2 for the estimation of $|||\mathrm{Res}|||_\star$. Table 1 shows that there is only small improvement of up to 8 % possible compared to η by $\eta_{\mathrm{LW}(1)}$; the estimator η_{B} is even worse than η. This led to the decision in [12] to employ only η for the estimation of $|||\mathrm{Res}|||_\star$ in the error control for the nonconforming FEM for the Poisson problem. It seems more favorable to spend effort in the sharp estimation of $|||\mathrm{Res}_{\mathrm{NC}}|||_\star$.

3.3 Guaranteed Upper Bounds for $|||\mathrm{Res}_{\mathrm{NC}}|||_\star$

Since $\mathrm{Res}_{\mathrm{NC}}(\varphi_z) = 0$ for all $z \in \mathscr{N}$, any equilibration error estimator from Sect. 2 is applicable (with $\sigma_h = \mathrm{curl}\, u_{\mathrm{CR}}$ and $f \equiv 0$ in (1)) and leads, e.g., via $q_{\mathrm{xyz}} = q_{\mathrm{B}}$ or q_{LW}, to the upper bounds

$$|||\mathrm{Res}_{\mathrm{NC}}|||_\star \leq \|\mathrm{curl}\, u_{\mathrm{CR}} - q_{\mathrm{xyz}}\|_{L^2(\Omega)} =: \mu_{\mathrm{xyz}}.$$

The second characterization (5) of $|||\mathrm{Res}_{\mathrm{NC}}|||_\star$ allows an upper bound for $|||\mathrm{Res}_{\mathrm{NC}}|||_\star$ by the design of conforming functions $v_{\mathrm{xyz}} \in V$ such that

$$|||\mathrm{Res}_{\mathrm{NC}}|||_\star \leq |||u_{\mathrm{CR}} - v_{\mathrm{xyz}}|||_{\mathrm{NC}} =: \mu_{\mathrm{xyz}}.$$

Since $q_{\mathrm{xyz}} := \mathrm{curl}\, v_{\mathrm{xyz}} \in H(\mathrm{div}, \Omega)$, those can also be seen as equilibration error estimators and allow the postprocessing of Sect. 2.2. Three designs for some v_{xyz} from [1, 12] are repeated in the sequel.

Ainsworth [1] designs some piecewise linear $v_{\mathrm{A}} \in P_1(\mathscr{T}) \cap C_0(\Omega)$ by averaging on node patches $\mathscr{T}(z) := \{T \in \mathscr{T} \,|\, z \in T\}$,

$$v_{\mathrm{A}}(z) := \begin{cases} 0 & \text{if } z \in \mathcal{N} \setminus \mathcal{N}(\Omega), \\ \left(\sum_{T \in \mathcal{T}(z)} u_{\mathrm{CR}}|_T(z)\right) / \|\mathcal{T}(z)\| & \text{if } z \in \mathcal{N}(\Omega). \end{cases}$$

The averaging of the auxiliary function from [2,7,23]

$$v^0 := u_{\mathrm{CR}} - f_{\mathcal{T}} \psi/2 \in P_2(\mathcal{T}),$$

where $\psi(x) := \|x - \mathrm{mid}(T)\|^2/2 - \fint_T \|y - \mathrm{mid}(T)\|^2 \mathrm{d}y$ for $x \in T \in \mathcal{T}$, leads to $v_{\mathrm{AP2}} \in P_2(\mathcal{T}) \cap C_0(\Omega)$ via

$$v_{\mathrm{AP2}}(z) := \begin{cases} 0 & \text{if } z \in \mathcal{N}(\partial\Omega) \cup \mathrm{mid}(\mathcal{E}(\partial\Omega)), \\ \left(\sum_{T \in \mathcal{T}(z)} v^0|_T(z)\right) / \|\mathcal{T}(z)\| & \text{if } z \in \mathcal{N}(\Omega), \\ \left(\sum_{T \in \mathcal{T}(E)} v^0|_T(z)\right) / \|\mathcal{T}(E)\| & \text{if } z = \mathrm{mid}(E), E \in \mathcal{E}(\Omega). \end{cases}$$

The novel design from [12] employs the red-refined triangulation and defines $v_{\mathrm{RED}}(z) \in P_1(\mathrm{red}(\mathcal{T})) \cap C_0(\Omega)$ via

$$v_{\mathrm{RED}}(z) := \begin{cases} u_{\mathrm{CR}}(z) & \text{for } z \in \mathrm{mid}(\mathcal{E}(\Omega)), \\ 0 & \text{for } z \in \mathcal{N}(\partial\Omega) \cup \mathrm{mid}(\mathcal{E}(\partial\Omega)), \\ v_z & \text{for } z \in \mathcal{N}(\Omega). \end{cases}$$

The values v_z for $z \in \mathcal{N}(\Omega)$ may be chosen by an averaging as above or by patchwise minimization as in [12]; this leads to the two averagings v_{ARED} and v_{PMRED}.

3.4 Numerical Experiment with Big Oscillations

This section concludes with the revisit of the example of Sect. 2.3 for the CR-NCFEM. Figures 3 and 4 display the efficiency indices $\eta_{\mathrm{xyz}}/|||e|||$ for all error estimators of Sect. 3.3. Under the label B and LW, both residuals were estimated with the same error estimator, i.e., $|||u - u_{\mathrm{CR}}|||_{\mathrm{NC}}$ is bound by $\eta_{\mathrm{B}} + \mu_{\mathrm{B}}$ and $\eta_{\mathrm{LW}} + \mu_{\mathrm{LW}}$, respectively. The error estimators based on conforming interpolations xyz $\in \{\mathrm{A}, \mathrm{AP2}, \mathrm{ARED}, \mathrm{PMRED}\}$ involve $|||\mathrm{Res}|||_\star \leq \eta$ and hence bound $|||u - u_{\mathrm{CR}}|||_{\mathrm{NC}}$ by $\eta + \mu_{\mathrm{xyz}}$. The same holds for their postprocessings. Notice that r(3) applied to ARED or PMRED means altogether two red refinements.

The energy error is estimated very effectively with efficiency indices between 1.5 for unpostprocessed estimators like η_{B} and η_{A} and about 1.05 for the postprocessed estimators $\eta_{\mathrm{Brr(3)}}$ or $\eta_{\mathrm{Arr(3)}}$.

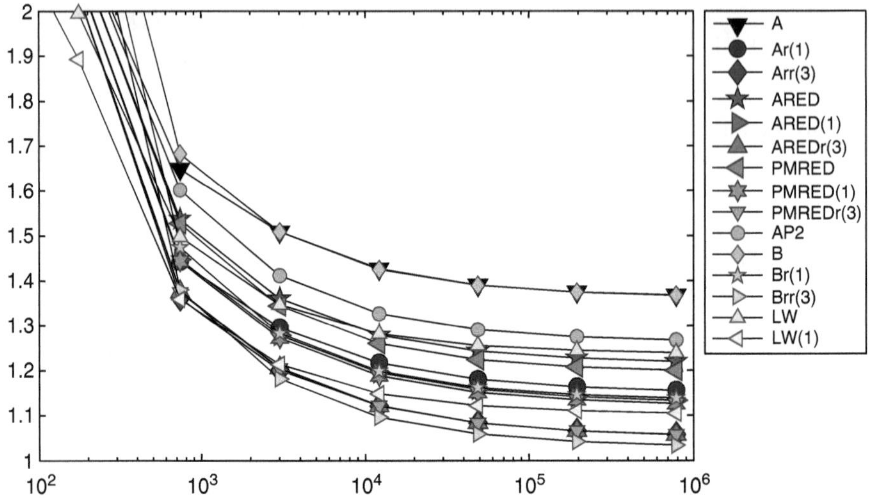

Fig. 3 History of efficiency indices $\eta_{xyz}/|||e|||$ of various error estimators η_{xyz} labeled *xyz* as functions of ndof on uniform meshes in Sect. 3

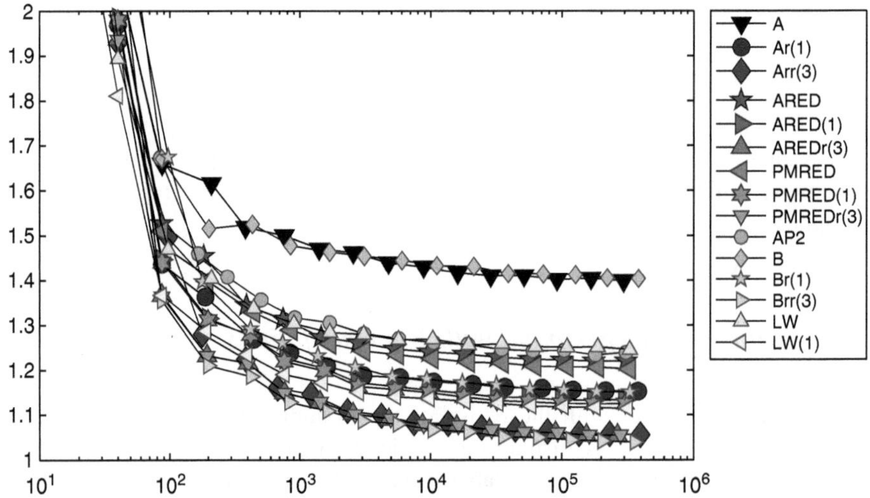

Fig. 4 History of efficiency indices $\eta_{xyz}/|||e|||$ of various error estimators η_{xyz} labeled *xyz* in the figure as functions of ndof on adaptive meshes in Sect. 3

4 Guaranteed Error Control for Curved Boundaries

Particular attention requires the inexact approximation of the geometry by the polygonal boundary of a triangulation into triangles. This section is devoted to an example for a convex boundary where there is no real need of curved finite elements. The benchmark problem on the sector domain

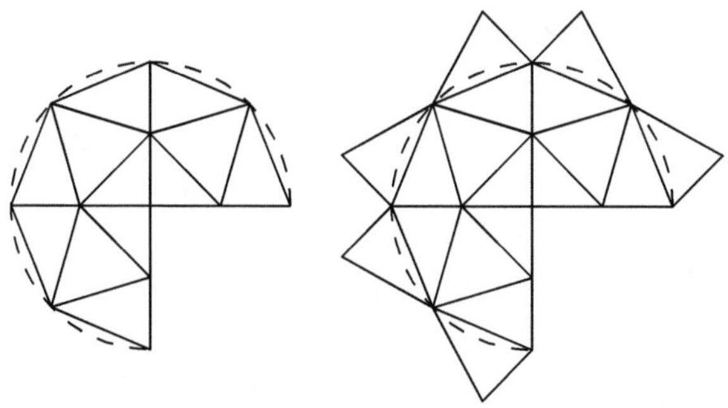

Fig. 5 Triangulation $\mathscr{T}$ (*left, solid lines*) and extended triangulation $\hat{\mathscr{T}}$ (*right, solid lines*) with $\bigcup\mathscr{T} \subseteq \Omega \subseteq \bigcup\hat{\mathscr{T}}$ for the sector domain Ω (*dashed lines*) from Sect. 4

$$\Omega = \left\{x = (r\cos\varphi, r\sin\varphi) \,\middle|\, 0 < \varphi < 3\pi/2,\ 0 < r < 1\right\}$$

from [1] employs the exact solution $u(r,\varphi) = (r^{2/3} - r^2)\sin(2\varphi/3)$ with a typical corner singularity at the reentrant corner.

Since the domain is not matched exactly, $\bigcup\mathscr{T} \subset \overline{\Omega}$ requires extra considerations for u_h extended by zero outside of $\bigcup\mathscr{T}$ such that $u_h \in H_0^1(\Omega)$. The reflection of boundary triangles of Fig. 5 yields an extended triangulation $\hat{\mathscr{T}}$ with $\Omega \subset \bigcup\hat{\mathscr{T}}$ where the extended source function $f(\varphi) = 32\sin(2\varphi/3)/9$ is well defined. The new triangles involve only Dirichlet nodes and allow the Braess or Luce–Wohlmuth design of an equilibration q_{B} or q_{LW} from Sect. 2.2 on the extended triangulation, possibly with a postprocessing $\gamma_h \in H^1(\bigcup\hat{\mathscr{T}})$. This results in

$$|||\mathrm{Res}|||_\star \le \left\|h_{\hat{\mathscr{T}}}(f + \operatorname{div}\hat{q})\right\|_{L^2(\bigcup\hat{\mathscr{T}})}/j_{1,1} + \|\hat{q} - \sigma_h - \operatorname{Curl}\gamma_h\|_{L^2(\Omega)}.$$

The integration of $\hat{q} - \sigma_h - \operatorname{Curl}\gamma_h$ over the non-polygonal domain Ω separates into an exact integration over triangles in $\mathscr{T}$ and an integration over intersections $T \cap \Omega$ of triangles $T \in \hat{\mathscr{T}} \setminus \mathscr{T}$. The latter integration employs polar coordinates and Gauss quadrature with at least 100 quadrature points.

To consider also the domain approximation error in the adaptive refinement, the refinement indicators (4) are replaced by

$$\eta(T)^2 + 2\,\mathrm{width}(\hat{T} \cap \Omega)/\pi\, \|f\|_{L^2(\hat{T}\cap\Omega)} \quad \text{for } T \in \mathscr{T} \text{ with a reflection } \hat{T} \in \hat{\mathscr{T}} \setminus \mathscr{T}.$$

Additionally, modified refinement routines shift the midpoints of all bisected edges along the curved boundary onto the unit circle. For simplicity, the postprocessing of Sect. 2.2 is only applied to vertices $z \in \mathscr{N}$ with $\hat{\omega}_z \subseteq \Omega$ where $\hat{\omega}_z$ is the patch with respect to the extended triangulation $\hat{\mathscr{T}}$. Undocumented experiments show to us that otherwise the efficiency becomes worse.

The oscillations in this example are not as large as in the square example from Sect. 2.3, but the conclusions appear similar. Figure 6 displays the efficiency indices

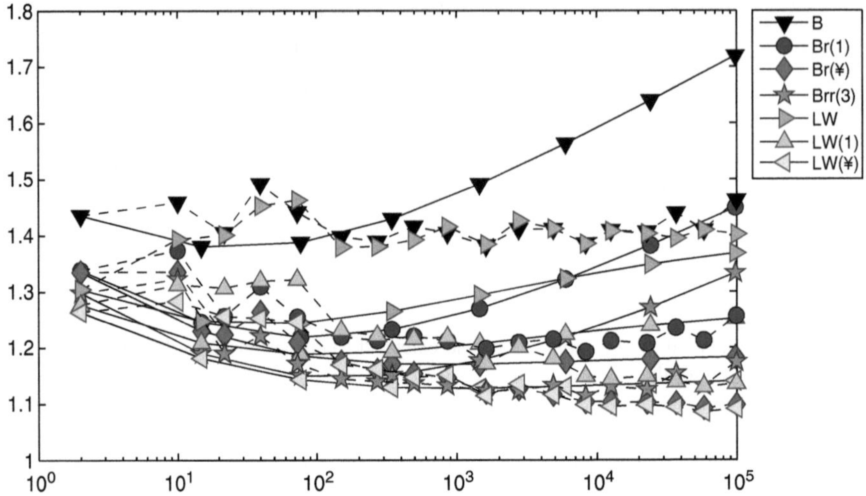

Fig. 6 History of efficiency indices $\eta_{\mathrm{xyz}}/|||e|||$ of the standard and postprocessed Braess and Luce–Wohlmuth error estimators η_{xyz} labeled *xyz* in the figure as functions of the number of unknowns on uniform (*solid lines*) and adaptive (*dashed lines*) meshes for the sector example of Sect. 4

of the two error estimators η_{LW} and η_{B}. The postprocessed equilibration error estimator $\eta_{\mathrm{LW}(1)}$ or $\eta_{\mathrm{Brr}(3)}$ permits efficiency indices around 1.2, while $\eta_{\mathrm{Br}(1)}$ leads to 1.3 for adaptive mesh refinement. Due to the simple extension of the solution from $\mathcal{T}$ to $\hat{\mathcal{T}}$, there is a large refinement along the circular boundary edges, but the efficiency is almost as good as in the other examples. As a result, even for curved boundaries, reliable error control is possible and accurate.

For the nonconforming solution u_{CR}, a similar treatment is possible (cf. [12] for details).

5 Guaranteed Goal-Oriented Error Estimation

This section is devoted to guaranteed error control with respect to some functional like the derivative $-\partial/\partial x_1 \delta_{x_0}$ evaluated at a point $x_0 = (\pi/7, 49/100)$. Section 5.1 describes a way to recast that problem into a computable term plus a linear and bounded goal functional $Q \in H^{-1}(\Omega)$ which in Sect. 5.2 is controlled via the parallelogram identity in terms of energy error estimates. Figure 7 displays the numerical results for a benchmark with an overestimation by a guaranteed bound by just one order of magnitude.

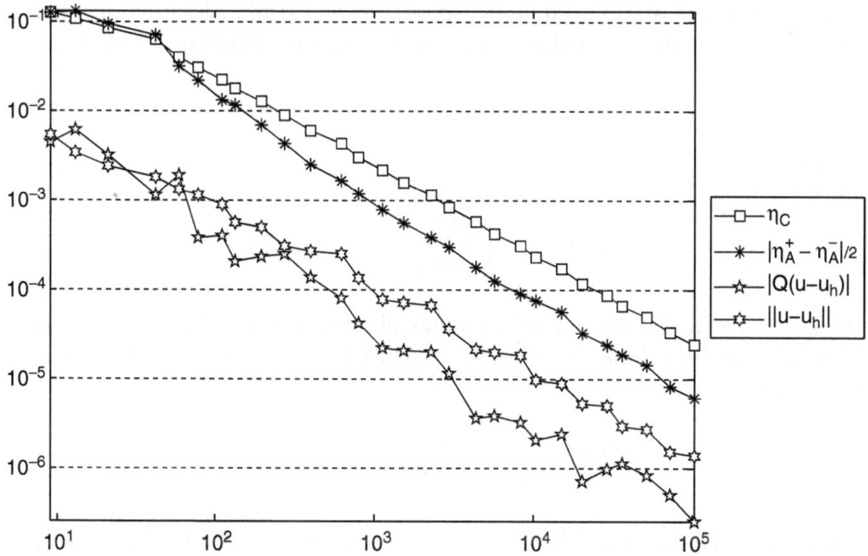

Fig. 7 Convergence history of the error $\|Q(u-u_h)\|$, $\|\eta_A^+ - \eta_A^-\|/2$, η_C, and $\|u-u_h\|_{L^2(\Omega)}$

5.1 Reduction to L^2 Functionals

Given some fixed point x_0 in the domain $\Omega = (0,1)^2$, this section aims at guaranteed error bounds of the x_1 derivative $\partial u(x_0)/\partial x_1$. This point value $-\partial\delta_{x_0}/\partial x_1$ is not well defined for any Sobolev function. This subsection discusses a split of

$$\partial \delta u(x_0)/\partial x_1 = Q(u) + M(f)$$

in a bounded functional $Q(u)$ and an unbounded functional $M(f)$ independent of u [19] that can be computed beforehand. The fundamental solution of the Laplace operator Δ in 2D is $\log r/2\pi$ in polar coordinates (r,ϕ) at x_0, in symbolic notation $2\pi\partial\delta_{x_0} = \Delta \log r$. The derivative $-2\pi\partial\delta_{x_0}/\partial x_1 = \Delta \cos\phi/r$ leads to the formula (recall $x = x_0 + r(\cos\phi, \sin\phi)$)

$$2\pi \frac{\partial v(x_0)}{\partial x_1} = \int_\Omega (\cos\phi/r)\Delta v(x)\,\mathrm{d}x \quad \text{for all } v \in \mathscr{D}(\Omega). \tag{7}$$

This identity is the clue to cast the point derivative of the solution of the Laplace equation as a function of the right-hand side $f \in L^2(\Omega) \cap L^p(U)$ for some neighborhood U of x_0 and some $p > 2$. By local elliptic regularity, u is C^1 in a neighborhood of x_0 and $\Delta v = -f$ allows for $f/r \in L^1(U)$. Hence, formula (7) makes sense for the exact solution u. The boundary conditions, however, do not allow to utilize the formula directly for $v = u$ in (7) and so involve some cutoff function χ, which is identically one in some neighborhood of x_0 and vanishes outside U.

In the example of this section, the spline function η of order 6 on the interval $(0.1, 0.45)$ with natural boundary conditions has been evaluated with MATLAB by

```
spapi(6,[0.1*ones(1,5),0.275,0.45*ones(1,5)],[zeros(1,5),1,zeros(1,5)])
```

to define

$$1-\chi(r,\phi) := \frac{\int_0^r \eta(s)\,\mathrm{d}s}{\int_0^1 \eta(s)\,\mathrm{d}s} \quad \text{for } 0<r<1.$$

With $v := \chi u$ in (7), $\Delta u = f$ in some neighborhood of x_0 where $r=0$ is some singularity in the volume integral. The product rule $\Delta v = \chi f + 2\nabla\chi\cdot\nabla u + u\Delta\chi$ shows that

$$\frac{\partial u(x_0)}{\partial x_1} = \int_\Omega \frac{\cos\phi}{2\pi r}\chi(x) f(x)\,\mathrm{d}x + Q(u). \tag{8}$$

The point is that the linear functional $Q(u)$ involves smooth functions like $\nabla\chi/r$ (which vanishes near x_0) as well as u and its derivative ∇u and hence is linear, bounded, and $Q \in H^{-1}(\Omega)$. Indeed, some further integration by parts reveals that

$$Q(u) = \int_\Omega g(x)u(x)\,\mathrm{d}x \quad \text{for } g(x) := -\nabla\chi(x)\cdot\nabla\left(\frac{\cos\phi}{\pi r}\right) - \frac{\cos\phi}{2\pi r}\Delta\chi. \tag{9}$$

Recall that $\chi\equiv 1$ in a neighborhood of $r=0$, and so $g\in L^2(\Omega)$ is smooth. Since the first integral on the right-hand side of (8) is known and computable, the computation of the unbounded functional $-\partial\delta_{x_0}/\partial x_1$ is reduced to that of the bounded functional Q of the following subsection.

5.2 Guaranteed Bounds for Goal Functionals

Given some L^2 function g and the goal functional Q from (9), the estimation of $Q(u-u_h)$ is driven by $g\in L^2(\Omega)$ as the right-hand side, the exact solution z, and the discrete solution z_h of the adjoint problem [3,5]. Then, the parallelogram identity for any $\alpha\neq 0$ yields

$$Q(u-u_h) = \frac{1}{4}\left|\left|\left|\alpha(u-u_h)+\frac{z-z_h}{\alpha}\right|\right|\right|^2 - \frac{1}{4}\left|\left|\left|\alpha(u-u_h)-\frac{z-z_h}{\alpha}\right|\right|\right|^2. \tag{10}$$

As in [21], upper and lower bounds for the energy norm terms imply corresponding bounds for the error $Q(u-u_h)$. Note that lower bounds can be designed from upper bounds and vice versa with the hyper circle identity

$$\|p - p_{\mathrm{RT}}\|^2_{L^2(\Omega)} + \|p - \nabla u_h\|^2_{L^2(\Omega)} = \|p_{\mathrm{RT}} - \nabla u_h\|^2_{L^2(\Omega)} + 2\,(u - u_h, f - f_{\mathscr{T}})$$

for the Raviart–Thomas solution $p_{\mathrm{RT}} \in \mathrm{RT}_0(\mathscr{T})$ [6, 20]. The upper bound

$$\|p - p_{\mathrm{RT}}\|^2_{L^2(\Omega)} \le \frac{\mathrm{osc}^2(f,\mathscr{T})}{j_{1,1}^2} + \mathrm{dist}^2\big(p_{\mathrm{RT}}, \nabla H^1_0(\Omega)\big)$$

employs the Helmholtz decomposition $p - p_{\mathrm{RT}} = \nabla\alpha + \mathrm{Curl}\,\beta$ with $\nabla\alpha \perp \mathrm{Curl}\,\beta$ and the Poincaré constant from Sect. 2.2. Any $v \in H^1_0(\Omega)$ satisfies

$$\begin{aligned}
\|p - p_{\mathrm{RT}}\|^2_{L^2(\Omega)} &= |||\alpha|||^2 + |||\beta|||^2 = (\nabla\alpha, p - p_{\mathrm{RT}}) + (\mathrm{Curl}\,\beta, p - p_{\mathrm{RT}})\\
&= -\,(\alpha, \operatorname{div} p - \operatorname{div} p_{\mathrm{RT}}) + (\mathrm{Curl}\,\beta, \nabla v - p_{\mathrm{RT}})\\
&= (\alpha - \alpha_{\mathscr{T}}, f - f_{\mathscr{T}}) + (\mathrm{Curl}\,\beta, \nabla v - p_{\mathrm{RT}})\\
&\le |||\alpha|||\frac{\mathrm{osc}(f,\mathscr{T})}{j_{1,1}} + |||\beta|||\,\mathrm{dist}\big(p_{\mathrm{RT}}, \nabla H^1_0(\Omega)\big)\\
&\le \left(\frac{\mathrm{osc}^2(f,\mathscr{T})}{j_{1,1}} + \mathrm{dist}^2\big(p_{\mathrm{RT}}, \nabla H^1_0(\Omega)\big)\right)^{1/2} \big(|||\alpha|||^2 + |||\beta|||^2\big)^{1/2}.
\end{aligned}$$

The upper bound $|||u - u_h||| \le \mathrm{osc}(f,\mathscr{T})\,/\,j_{1,1} + |||u_M - u_h|||$ incorporates a function u_M similar to v^0 from Sect. 3.3, but here u_{CR} is the *CR* solution for the right-hand side $f_{\mathscr{T}}$ to ensure $\nabla_{NC} u_M = p_{\mathrm{RT}}$ [23]. This leads to

$$\begin{aligned}
|||u - u_h||| &= \sup_{|||v|||=1} \big(F(v) - a(u_h, v)\big) = \sup_{|||v|||=1} \big((f - \operatorname{div} p_{\mathrm{RT}}, v) + (p_{\mathrm{RT}} - \nabla u_h, \nabla v)\big)\\
&\le \frac{\mathrm{osc}(f,\mathscr{T})}{j_{1,1}} + \sup_{|||v|||=1} (\nabla v, \nabla_{NC} u_M - \nabla u_h).
\end{aligned}$$

With the convention scheme $u^+ = \alpha u + z/\alpha$, $u^- = \alpha u - z/\alpha$, $f^+ = \alpha f + g/\alpha$, and $f^- = \alpha f - g/\alpha$, those bounds imply guaranteed upper and lower bounds for (10). As in Sect. 3.3, an averaging of u_M results in a continuous $P_2(\mathscr{T})$ function u_A which gives an upper bound for $\mathrm{dist}\big(p_{\mathrm{RT}}, \nabla H^1_0(\Omega)\big)$. Altogether, this leads to guaranteed upper and lower bounds for $Q(u - u_h)$:

$$\begin{aligned}
\eta_A^+ = \frac{1}{4}\Bigg(&\left(\frac{\mathrm{osc}(f^+,\mathscr{T})}{j_{1,1}} + |||u_M^+ - u_h^+|||\right)^2 - \|p_{\mathrm{RT}}^- - \nabla u_h^-\|^2_{L^2(\Omega)} + \frac{3\,\mathrm{osc}^2(f^-,\mathscr{T})}{2 j_{1,1}^2}\\
&+ \|p_{\mathrm{RT}}^- - \nabla u_A\|_{L^2(\Omega)} + 2\,|||u_M^- - u_h^-|||\frac{\mathrm{osc}(f^-,\mathscr{T})}{j_{1,1}}\Bigg),
\end{aligned}$$

$$\eta_A^- = \frac{1}{4}\left(\|p_{\mathrm{RT}}^+ - \nabla u_h^+\|_{L^2(\Omega)}^2 - \frac{3\,\mathrm{osc}^2(f^+,\mathscr{T})}{2j_{1,1}^2} - \|p_{\mathrm{RT}}^+ - \nabla u_A\|_{L^2(\Omega)} \right.$$
$$\left. -2\,|||u_M^+ - u_h^+|||\frac{\mathrm{osc}(f^+,\mathscr{T})}{j_{1,1}} - \left(\frac{\mathrm{osc}(f^-,\mathscr{T})}{j_{1,1}} + |||u_M^- - u_h^-|||\right)^2\right).$$

Elementary calculations show that $\alpha_A := (|||z_M - z_h|||/\,|||u_M - u_h|||)^{1/2}$ is the optimal choice for the parameter α. The same bounds yield an upper bound η_C for the Cauchy inequality $\|Q(u-u_h)\| \leq |||u-u_h|||\;|||z-z_h||| \leq \eta_C$.

5.3 Benchmark Example

The function $f{=}2x{-}2x^2{+}2y{-}2y^2$ with the analytical solution $u{=}x(1{-}x)y(1{-}y)$ and the reduction from Sect. 5.1 leads to some smooth known function g. Standard quadrature resolves the unbounded functional, and adaptive goal-oriented FEM handles the bounded functional Q. The adaptive mesh-refinement algorithm employs the refinement rules from [19]. They employ Dörfler marking separately for the primal and the dual problem and choose the smaller set of marked edges for the final mesh refinement.

Figure 7 displays the error $\|Q(u-u_h)\|$, η_C, the guaranteed error bound $\|\eta_A^+ - \eta_A^-\|/2$ for $\|Q(u-u_h) - (\eta_A^+ + \eta_A^-)/2\|$, and the L^2 norm of the error $u-u_h$ in the primal problem. The a posteriori error control of the L^2 error $\|u-u_h\|_{L^2(\Omega)}$ in the primal problem is possible in this example on a convex domain but significantly harder for nonconvex polygons. In the general case, the duality argument requires the precise values for the reduced elliptic regularity to deduce guaranteed error bounds.

Acknowledgements This work was written while the first author enjoyed the kind hospitality of the Oxford PDE Centre.

References

1. Ainsworth, M.: Robust a posteriori error estimation for nonconforming finite element approximation. SIAM J. Numer. Anal. **42**(6), 2320–2341 (2004)
2. Ainsworth, M.: A posteriori error estimation for lowest order Raviart-Thomas mixed finite elements. SIAM J. Sci. Comput. **30**(1), 189–204 (2007/2008)
3. Bangerth, W., Rannacher, R.: Adaptive finite element methods for differential equations. In: Lectures in Mathematics ETH Zürich. Birkhäuser, Basel (2003)
4. Bartels, S., Carstensen, C., Klose, R.: An experimental survey of a posteriori Courant finite element error control for the Poisson equation. Adv. Comput. Math. **15**(1–4), 79–106 (2001)
5. Becker, R., Rannacher, R.: An optimal control approach to a posteriori error estimation in finite element methods. Acta Numer. **10**, 1–102 (2001)

6. Braess, D.: Finite elements. In: Theory, Fast Solvers, and Applications in Elasticity Theory, 3rd edn. Cambridge University Press, Cambridge (2007)
7. Braess, D.: An a posteriori error estimate and a comparison theorem for the nonconforming P_1 element. Calcolo **46**(2), 149–155 (2009)
8. Braess, D., Schöberl, J.: Equilibrated residual error estimator for edge elements. Math. Comp. **77**(262), 651–672 (2008)
9. Brenner, S.C., Carstensen, C.: Finite Element Methods, Encyclopedia of Computational Mechanics (Chap. 4). Wiley, New York (2004)
10. Carstensen, C.: A unifying theory of a posteriori finite element error control. Numer. Math. **100**(4), 617–637 (2005)
11. Carstensen, C., Merdon, C.: Estimator competition for Poisson problems. J. Comp. Math. **28**(3), 309–330 (electronic) (2010)
12. Carstensen, C., Merdon, C.: Computational survey on a posteriori error estimators for nonconforming finite element methods for Poisson problems. J. Comput. Appl. Math. **249**, 74–94 (2013). http://dx.doi.org/10.1016/j.cam.2012.12.021. DOI: 10.1016/j.cam.2012.12.021
13. Carstensen, C., Merdon, C.: Effective postprocessing for equilibration a posteriori error estimators. Numer. Math., **123**(3), 425–459 (2013). http://dx.doi.org/10.1007/s00211-012-0494-4. DOI: 10.1007/s00211-012-0494-4
14. Carstensen, C., Merdon, C.: A posteriori error estimator competition for conforming obstacle problems. Numer. Methods Partial Differential Eq. **29**, 667–692 (2013). doi: 10.1002/num.21728
15. Carstensen, C., Merdon, C.: Refined fully explicit a posteriori residual-based error control (2013+) (submitted)
16. Carstensen, C., Eigel, M., Hoppe, R.H.W., Loebhard, C.: Numerical mathematics: Theory, methods and applications. Numer. Math. Theory Methods Appl. **5**(4), 509–558 (2012)
17. Laugesen, R.S., Siudeja, B.A.: Minimizing Neumann fundamental tones of triangles: an optimal Poincaré inequality. J. Differ. Equ. **249**(1), 118–135 (2010)
18. Luce, R., Wohlmuth, B.I.: A local a posteriori error estimator based on equilibrated fluxes. SIAM J. Numer. Anal. **42**(4), 1394–1414 (2004)
19. Mommer, M.S., Stevenson, R.: A goal-oriented adaptive finite element method with convergence rates. SIAM J. Numer. Anal. **47**(2), 861–886 (2009)
20. Prager, W., Synge, J.L.: Approximations in elasticity based on the concept of function space. Q. Appl. Math. **5**, 241–269 (1947)
21. Prudhomme, S., Oden, J.T.: On goal-oriented error estimation for elliptic problems: application to the control of pointwise errors. Comput. Methods Appl. Mech. Eng. **176**(1–4), 313–331 (1999). New Advances in Computational Methods (Cachan, 1997)
22. Repin, S.: A posteriori estimates for partial differential equations. In: Radon Series on Computational and Applied Mathematics, vol. 4. Walter de Gruyter, Berlin (2008)
23. Vohralík, M.: A posteriori error estimates for lowest-order mixed finite element discretizations of convection-diffusion-reaction equations. SIAM J. Numer. Anal. **45**(4), 1570–1599 (2007)

A Finite Volume Element Method for a Nonlinear Parabolic Problem

P. Chatzipantelidis and V. Ginting

Abstract We study a finite volume element discretization of a nonlinear parabolic equation in a convex polygonal domain. We show the existence of the discrete solution and derive error estimates in L_2- and H^1-norms. We also consider a linearized method and provide numerical results to illustrate our theoretical findings.

Keywords Nonlinear parabolic problem • Finite volume element method • Error estimates

Mathematics Subject Classification (2010): 65M60, 65M15

1 Introduction

We consider the nonlinear parabolic problem for $t \in [0,T]$, $T > 0$,

$$u_t - \nabla \cdot (A(u)\nabla u) = f, \text{ in } \Omega, \quad u = 0, \text{ on } \partial\Omega, \quad \text{with } u(0) = u^0, \text{ in } \Omega, \tag{1}$$

where Ω is a bounded convex polygonal domain in $\mathbb{R}^2$ and $A(v) = \mathrm{diag}\,(a_1(v), a_2(v))$, a strictly positive definite and bounded real-valued matrix function, such that there exists $\beta > 0$.

P. Chatzipantelidis (✉)
Department of Mathematics, University of Crete, Heraklion, GR–71409, Greece
e-mail: chatzipa@math.uoc.gr

V. Ginting
Department of Mathematics, University of Wyoming, Laramie, WY 82071, USA
e-mail: vginting@uwyo.edu

O.P. Iliev et al. (eds.), *Numerical Solution of Partial Differential Equations: Theory, Algorithms, and Their Applications*, Springer Proceedings in Mathematics & Statistics 45,
DOI 10.1007/978-1-4614-7172-1_7,

$$|x^\top A'(y)x| \leq \beta x^\top x, \quad \forall y \in \mathbb{R},\ \forall x \in \mathbb{R}^2. \tag{2}$$

Further, we assume that A' is Lipschitz continuous, i.e., $\exists L > 0$

$$|a_i'(y) - a_i'(\tilde{y})| \leq L|y - \tilde{y}|, \quad \forall y, \tilde{y} \in \mathbb{R},\ i = 1,2, \tag{3}$$

and that there exists a sufficiently smooth unique solution u of (1).

Questions about the existence and regularity of solutions for (1) have been intensively investigated, for example, in [7, Chap. 5]. Nonlinear parabolic problems such as (1) occur in many applied fields. To name a few, in the chemotaxis model, see Keller and Segel [6]; in groundwater hydrology, see L.A. Richards [10]; and in modeling and simulation of oil recovery techniques in the presence of capillary pressure, see [3].

We shall study fully discrete approximations of (1) by the finite volume element method (FVEM). The FVEM, which is also called finite volume method or covolume method in some literatures, is a class of important numerical methods for solving differential equations, especially those arising from conservation laws including mass, momentum, and energy, because this method possesses local conservation property, which is crucial in many applications. It is popular in computational fluid mechanics, groundwater hydrology, reservoir simulations, and others. Many researchers have studied this method for linear and nonlinear problems. We refer to the monographs [5, 9] for the general presentation of this method and references therein for details.

The approximate solution will be sought in the space of piecewise linear functions

$$\mathcal{X}_h = \{\chi \in \mathcal{C} : \chi|_K \text{ linear}, \forall K \in \mathcal{T}_h;\ \chi|_{\partial\Omega} = 0\},$$

where $\mathcal{T}_h$ is a family of quasiuniform triangulations $T_h = \{K\}$ of Ω, with h denoting the maximum diameter of the triangles $K \in \mathcal{T}_h$ and $\mathcal{C} = \mathcal{C}(\Omega)$ the space of continuous functions on $\bar{\Omega}$.

The FVEM is based on a local conservation property associated with the differential equation. Namely, integrating (1) over any region $V \subset \Omega$ and using Green's formula we obtain for $t \in [0, T]$

$$\int_V u_t\, dx - \int_{\partial V} (A(u)\nabla u) \cdot n\, d\sigma = \int_V f\, dx, \tag{4}$$

where n denotes the unit exterior normal vector to ∂V. The semidiscrete FVEM approximation $u_h(t) \in \mathcal{X}_h$ will satisfy (4) for V in a finite collection of subregions of Ω called control volumes, the number of which will be equal to the dimension of the finite element space $\mathcal{X}_h$. These control volumes are constructed in the following way. Let z_K be the barycenter of $K \in \mathcal{T}_h$. We connect z_K with line segments to the midpoints of the edges of K, thus partitioning K into three quadrilaterals K_z, $z \in Z_h(K)$, where $Z_h(K)$ are the vertices of K. Then with each

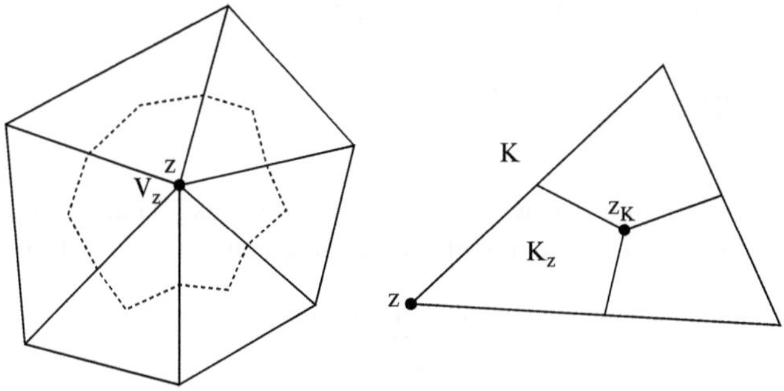

Fig. 1 *Left:* a union of triangles that have a common vertex z; the *dotted line* shows the boundary of the corresponding control volume V_z. *Right:* a triangle K partitioned into the three subregions K_z

vertex $z \in Z_h = \cup_{K \in \mathcal{T}_h} Z_h(K)$ we associate a control volume V_z, which consists of the union of the subregions K_z, sharing the vertex z (see Fig. 1). We denote the set of interior vertices of Z_h by Z_h^0. The semidiscrete FVEM for (1) is then to find $u_h(t) \in \mathcal{X}_h$, for $t \in [0,T]$, such that

$$\int_{V_z} u_{h,t}\, dx - \int_{\partial V_z} (A(u_h)\nabla u_h) \cdot n\, ds = \int_{V_z} f\, dx, \quad \forall z \in Z_h^0, \tag{5}$$

with $u_h(0) = u_h^0$, where $u_h^0 \in \mathcal{X}_h$ is a given approximation of u^0. Note that different choices for z_K, e.g., the circumcenter of K, lead to other methods than the one considered here; see [8, 12].

In our analysis of the FVEM we use existing results associated with the finite element method approximation $\tilde{u}_h(t) \in \mathcal{X}_h$ of $u(t)$, defined by

$$(\tilde{u}_{h,t}, \chi) + a(\tilde{u}_h; \tilde{u}_h, \chi) = (f, \chi), \quad \forall \chi \in \mathcal{X}_h, \quad \text{for } t > 0, \tag{6}$$

with $(f,g) = \int_\Omega fg\, dx$, $a(w;v,g) = (A(w)\nabla v, \nabla g)$ and $\|w\| = (w,w)^{1/2}$ the norm in $L_2 = L_2(\Omega)$. Further let $H_0^1 = H_0^1(\Omega)$ be the standard Sobolev space with zero boundary conditions. Thus, in order to rewrite (5) in a weak formulation, we introduce the finite dimensional space of piecewise constant functions

$$\mathcal{Y}_h = \{\eta \in L_2 :\ \eta|_{V_z} = \text{constant},\ \forall z \in Z_h^0;\ \eta|_{V_z} = 0,\ \forall z \in Z_h \setminus Z_h^0\}.$$

We now multiply (5) by $\eta(z)$ for an arbitrary $\eta \in \mathcal{Y}_h$ and sum over all $z \in Z_h^0$ to obtain the Petrov–Galerkin formulation for $t \in [0,T]$

$$(u_{h,t}, \eta) + a_h(u_h; u_h, \eta) = (f, \eta), \quad \forall \eta \in \mathcal{Y}_h, \quad \text{with } u_h(0) = u_h^0, \tag{7}$$

where $a_h(\cdot;\cdot,\cdot) : \mathcal{X}_h \times \mathcal{X}_h \times \mathcal{Y}_h \to \mathbb{R}$ is defined by

$$a_h(w;v,\eta) = -\sum_{z\in Z_h^0} \eta(z) \int_{\partial V_z} (A(w)\nabla v)\cdot n\, d\sigma, \quad \forall v,w \in \mathcal{X}_h,\ \eta \in \mathcal{Y}_h. \tag{8}$$

We shall now rewrite the Petrov–Galerkin method (7) as a Galerkin method in $\mathcal{X}_h$. For this purpose, we introduce the interpolation operator $J_h : \mathcal{C} \mapsto \mathcal{Y}_h$ by

$$J_h w = \sum_{z\in Z_h^0} w(z)\Psi_z,$$

where Ψ_z is the characteristic function of the control volume V_z. It is known that J_h is self-adjoint and positive definite (see [4]), and hence the following defines an inner product $\langle\cdot,\cdot\rangle$ on $\mathcal{X}_h$:

$$\langle \chi,\psi\rangle = (\chi, J_h\psi), \quad \forall \chi,\psi \in \mathcal{X}_h. \tag{9}$$

Further, in [4] it is shown that the corresponding norm is equivalent to the L_2norm, uniformly in h, i.e., with $C \geq c > 0$,

$$c\|\chi\| \leq |||\chi||| \leq C\|\chi\|, \quad \forall \chi \in \mathcal{X}_h, \quad \text{where } |||\chi||| \equiv \langle\chi,\chi\rangle^{1/2}.$$

With this notation, (7) may equivalently be written in Galerkin form as

$$\langle u_{h,t},\chi\rangle + a_h(u_h;u_h,J_h\chi) = (f,J_h\chi), \quad \forall \chi \in \mathcal{X}_h, \quad \text{for } t \geq 0. \tag{10}$$

Then let $N \in \mathbb{N}$, $N \geq 1$, $k = T/N$, and $t^n = nk$, $n = 0,\dots,N$. Discretizing in time (10), with the backward Euler method, we approximate $u(t^n)$ by $U^n \in \mathcal{X}_h$, for $n = 1,\dots,N$, such that

$$\langle \bar{\partial}U^n,\chi\rangle + a_h(U^n;U^n,J_h\chi) = (f^n,J_h\chi), \quad \forall \chi \in \mathcal{X}_h, \quad \text{with } U^0 = u_h^0, \tag{11}$$

where $\bar{\partial}U^n = (U^n - U^{n-1})/k$ and $f^n = f(t^n)$.

To show the existence of the semidiscrete solution $\tilde{u}_h$ of the finite element method (6), one can employ Brouwer's fixed point theorem and the coercivity property of $a(\cdot;\cdot,\cdot)$:

$$a(w;\chi,\chi) \geq \alpha\|\nabla\chi\|^2, \quad \forall \chi \in \mathcal{X}_h,\ \forall w \in L_2 \tag{12}$$

(see [11]). However, the corresponding coercivity property for $a_h(\cdot;\cdot,\cdot)$,

$$a_h(w;\chi,J_h\chi) \geq \tilde{\alpha}\|\nabla\chi\|^2, \quad \forall \chi \in \mathcal{X}_h, \tag{13}$$

holds for $\|\nabla w\|_{L_\infty}$ in a bounded ball, where $\|w\|_{L_\infty} = \sup_{x\in\Omega}|w(x)|$. For this reason, we will employ a different argument than the one in [11] to show the existence of U^n. It is known that for fixed w, in general, the bilinear form $a_h(w;\psi,J_h\chi)$ is nonsymmetric on S_h, but (for a linear problem) it is not far from being symmetric, or $|a_h(\chi,J_h\psi)-a_h(\psi,J_h\chi)|\le Ch\|\nabla\chi\|\,\|\nabla\psi\|$, cf. [4]. Note that if z_K is the circumcenter of K, it is shown in [8] that (13) is satisfied for $w\in L_2$, and thus, one may show the existence of the solution of the finite volume method analogously to the one for the finite element method. We show the existence and uniqueness of the solution U^n of (11) and derive error estimates in L_2- and H^1-norms; see Theorems 3.1 and 4.1. Recently in [12], a two-grid FVEM was considered, for circumcenter-based control volumes, with suboptimal estimates in L_2- and H^1-norms.

Our analysis follows the corresponding one for the FVEM nonlinear elliptic and linear parabolic problems in [1,2]. This is based in bounds for the error functionals $\varepsilon_h(\cdot,\cdot)$ defined by

$$\varepsilon_h(f,\chi)=(f,J_h\chi)-(f,\chi),\quad \forall f\in L_2,\ \chi\in\mathcal{X}_h, \tag{14}$$

and $\varepsilon_a(\cdot;\cdot,\cdot)$ defined by

$$\varepsilon_a(w;v_h,\chi)=a_h(w;v_h,J_h\chi)-a(w;v_h,\chi)\quad \forall v_h,\chi\in\mathcal{X}_h,\ w\in L_2. \tag{15}$$

Following [11], we introduce the projection $R_h:H_0^1\to\mathcal{X}_h$ defined by

$$a(v;R_hv,\chi)=a(v;v,\chi),\quad \forall\chi\in\mathcal{X}_h. \tag{16}$$

In [11] optimal order error estimates in L_2- and H^1-norms were established for the difference $R_hu(t)-u(t)$. Here we combine these error estimates with bounds for the difference $\vartheta^n=U^n-R_hu^n$, which satisfies

$$\langle\bar\partial\vartheta^n,\chi\rangle+a_h(U^n;\vartheta^n,J_h\chi)=\delta(t^n;U^n,\chi),\quad \text{for }\chi\in\mathcal{X}_h, \tag{17}$$

with

$$\begin{aligned}\delta(t^n;v,\chi)\equiv&-(\omega^n,J_h\chi)-\varepsilon_h(f^n-u_t^n,\chi)+\varepsilon_a(v;R_hu^n,\chi)\\&+((A(u^n)-A(v))\nabla R_hu^n,\nabla\chi)\equiv\sum_{j=1}^4 I_j,\end{aligned} \tag{18}$$

and $\omega^n=(R_h-I)\bar\partial u^n+(\bar\partial u^n-u_t^n)$. Further we analyze a linearized fully discrete scheme and provide numerical examples to illustrate our results.

The rest of the paper is organized as follows. In Sect. 2 we recall known results and derive error bounds for the error functional δ. In Sect. 3 we derive error estimates and in Sect. 4 existence of the nonlinear fully discrete method. In Sect. 5 we consider a linearized version of the backward Euler scheme, and finally in Sect. 6 we present our numerical examples.

2 Preliminaries

In this section we recall known results about the projection R_h defined by (16) and the error functionals ε_h and ε_a introduced in (14) and (15). We also derive bounds for the error functional δ defined in (18).

We consider quasiuniform triangulations $\mathcal{T}_h$ for which the following inverse inequalities hold (see, e.g., [11]):

$$\|\nabla\chi\| \le Ch^{-1}\|\chi\|, \quad \text{and} \quad \|\nabla\chi\|_{L_\infty} \le Ch^{-1}\|\nabla\chi\|, \quad \text{for } \chi \in \mathcal{X}_h. \tag{19}$$

In such meshes, it is shown in [11, Lemma 13.2] that there exists $M_0 > 0$, independent of h, such that

$$\|\nabla u(t)\|_{L_\infty} + \|\nabla R_h u(t)\|_{L_\infty} \le M_0, \quad \text{for } t \le T, \tag{20}$$

and the following error estimates for $R_h u - u$.

Lemma 2.1. *With R_h defined by* (16) *and $\rho = R_h u - u$, we have under the appropriate regularity assumptions on u, with $C_u > 0$ independent of t,*

$$\|\nabla^s D_t^\ell \rho(t)\| \le C_u h^{2-s}, \quad 0 < t \le T, \quad \textit{and} \quad s,\ell = 0,1, \quad \textit{where } D_t = \partial/\partial t.$$

Our analysis is based on error estimates for the difference $\vartheta^n = U^n - R_h u^n$. Thus, in view of the error equation (17) for ϑ^n, we recall necessary bounds for the error functionals ε_h and ε_a derived in [1,2].

Lemma 2.2. *For the error functional ε_h, defined by* (14)*, we have*

$$|\varepsilon_h(f,\chi)| \le Ch^2\|\nabla f\|\,\|\nabla\chi\|, \quad \forall f \in H^1,\ \chi \in \mathcal{X}_h.$$

To this end, for $M = \max(2M_0, 1)$, we consider

$$\mathcal{B}_M = \{\chi \in \mathcal{X}_h : \|\nabla\chi\|_{L_\infty} \le M\}.$$

Lemma 2.3. *For the error functional ε_a, defined in* (15)*, we have*

$$|\varepsilon_a(w_h; v_h, \chi)| \le Ch\|\nabla w_h \cdot \nabla v_h\|\,\|\nabla\chi\|, \ \forall w_h, v_h, \chi \in \mathcal{X}_h. \tag{21}$$

Further, if u is the solution of (1)*, then for $v \in \mathcal{B}_M$,*

$$|\varepsilon_a(v; R_h u(t), \chi)| \le Ch^2\|\nabla\chi\|. \tag{22}$$

Proof. The first bound is shown in [1, Lemma 2.3]. The second bound is a direct result of Lemma 2.1, [1, Lemma 2.4], and the fact that $v \in \mathcal{B}_M$. □

Then, in view of Lemma 2.3 there exists a constant $c > 0$ such that for h sufficiently small, the coercivity property (13) for a_h holds for $w \in \mathcal{B}_M$. Further, in [1] we showed the following "Lipschitz"-type estimation for ε_a.

Lemma 2.4. *For the error functional ε_a, defined in* (15)*, there exists a constant C, independent of h, such that for $\chi, \psi \in \mathcal{X}_h$*

$$|\varepsilon_a(v;\psi,\chi) - \varepsilon_a(w;\psi,\chi)| \le Ch\|\nabla\psi\|_{L_\infty}(1+\|\nabla w\|_{L_\infty})\|\nabla(v-w)\|\,\|\nabla\chi\|.$$

Finally, we show appropriate bounds for the functional δ, defined by (18).

Lemma 2.5. *For δ defined by* (18)*, we have for $\chi \in \mathcal{X}_h$ and $v \in \mathcal{B}_M$*

$$|\delta(t^n;v,\chi)| \le C(k+h^2)\|\chi\| + Ch^2\|\nabla\chi\| + \begin{cases} C\|v-R_h u^n\|\,\|\nabla\chi\| \\ C\|\nabla(v-R_h u^n)\|\,\|\chi\|. \end{cases}$$

Proof. Using the splitting in (18) we bound each of the terms I_j, $j = 1,\dots,4$. Recall that $\omega^n = (R_h - I)\bar{\partial}u^n + (\bar{\partial}u^n - u_t^n)$; then in view of Lemma 2.1, we have

$$\|\omega^n\| \le Ck^{-1}\int_{t^{n-1}}^{t^n}\|\rho_t\|\,ds + C\int_{t^{n-1}}^{t^n}\|u_{tt}\|\,ds \le C(k+h^2), \tag{23}$$

and hence

$$|I_1| \le C(k+h^2)\|\chi\|. \tag{24}$$

To bound $I_2 + I_3$, we use Lemma 2.2 and (22) to get

$$|I_2 + I_3| \le Ch^2\|\nabla\chi\|. \tag{25}$$

Finally, employing (2) and (20) and adding and subtracting $R_h u^n$ and using Lemma 2.1, we get

$$\begin{aligned} |I_4| &= |((A(u^n) - A(v))\nabla R_h u^n, \nabla\chi)| \le C\|v-u^n\|\,\|\nabla\chi\| \\ &\le Ch^2\|\nabla\chi\| + C\|v-R_h u^n\|\|\nabla\chi\|. \end{aligned} \tag{26}$$

Combining now (24)–(26) we get the first one of the desired bounds. To show the second estimate of this lemma, we bound I_4 differently. Using integration by parts, we rewrite I_4 as

$$\begin{aligned} I_4 &= ((A(u^n) - A(R_h u^n))\nabla R_h u^n, \nabla\chi) + ((A(R_h u^n) - A(v))\nabla R_h u^n, \nabla\chi) \\ &= ((A(u^n) - A(R_h u^n))\nabla R_h u^n, \nabla\chi) + (\operatorname{div}[(A(R_h u^n) - A(v))\nabla R_h u^n], \chi) \\ &= I_4^i + I_4^{ii}. \end{aligned}$$

Then, in view of (2), Lemma 2.1, and (20), we have

$$|I_4^i| \le Ch^2 \|\nabla \chi\|. \tag{27}$$

Further, employing (2), (3), and (20), we obtain

$$\begin{aligned}|I_4^{ii}| &\le C(\|(A'(R_h u^n) - A'(v))\nabla R_h u^n\| + \|A'(v)\nabla(R_h u^n - v)\|)\|\chi\| \\ &\le C(\|v - R_h u^n\| + \|\nabla(v - R_h u^n)\|)\|\chi\|.\end{aligned} \tag{28}$$

Therefore combining (27) and (28), we have

$$|I_4| \le C\|\nabla(v - R_h u^n)\|\|\chi\| + Ch^2\|\nabla\chi\|. \tag{29}$$

Thus, combining (24), (25), (29), and (26), we obtain the second of the desired estimates of the lemma. □

3 Error Estimates for the Backward Euler Method

In this section we derive error estimates for the FVEM (11) in L_2- and H^1-norms, under the assumption that $U^j \in \mathcal{B}_M$, for $j = 0, \dots, n$. In Sect. 4 we will show the existence of $U^n \in \mathcal{B}_M$.

Theorem 3.1. *Let U^n and u be the solutions of* (11) *and* (1), *with $U^0 = R_h u^0$. If $U^j \in \mathcal{B}_M$, for $j = 0, \dots, n$, $n \ge 1$, and k, h be sufficiently small, then there exist $C > 0$, independent of k and h, such that*

$$\|\nabla^s(U^n - u^n)\| \le C(k + k^{-s/2}h^{2-s}), \quad \textit{for } s = 0, 1. \tag{30}$$

Proof. Using the error splitting $U^n - u^n = (U^n - R_h u^n) + (R_h u^n - u^n) = \vartheta^n + \rho^n$ and Lemma 2.1, it suffices to show

$$\|\nabla^s \vartheta^n\| \le C_s(k + k^{-s/2}h^{2-s}), \quad \text{for } s = 0, 1. \tag{31}$$

We start with the estimation of $\|\vartheta^n\|$. Due to the symmetry of $\langle \chi, \psi \rangle$, we have the following identity:

$$\langle \bar\partial \vartheta^n, \vartheta^n \rangle = \frac{1}{2k}(|||\vartheta^n|||^2 - |||\vartheta^{n-1}|||^2) + \frac{1}{2k}|||\vartheta^n - \vartheta^{n-1}|||^2. \tag{32}$$

Choosing $\chi = \vartheta^n$ in (17) and using the fact that $U^n \in \mathcal{B}_M$, (13), and (32), we get after eliminating $|||\vartheta^n - \vartheta^{n-1}|||$

$$\frac{1}{2k}(|||\vartheta^n|||^2 - |||\vartheta^{n-1}|||^2) + \tilde\alpha\|\nabla\vartheta^n\|^2 \le \delta(t^n; U^n, \vartheta^n). \tag{33}$$

Employing now the first estimate of Lemma 2.5, with $v = U^n$ and $\chi = \vartheta^n$, to bound the right-hand side of (33), we obtain

$$\frac{1}{2k}(|||\vartheta^n|||^2 - |||\vartheta^{n-1}|||^2) + \tilde{\alpha}\|\nabla\vartheta^n\|^2 \le C(k+h^2)\|\vartheta^n\| + C(k\|\vartheta^n\| + h^2)\|\nabla\vartheta^n\|.$$

Then, after eliminating $\|\nabla\vartheta^n\|^2$ and moving $|||\vartheta^n|||^2$ to the left, we have for k sufficiently small

$$|||\vartheta^n|||^2 \le (1+Ck)|||\vartheta^{n-1}|||^2 + CkE, \quad \text{with } E = O(k^2+h^4).$$

Hence, using the fact that $\vartheta^0 = 0$, we obtain

$$|||\vartheta^n|||^2 \le CkE\sum_{\ell=0}^{n}(1+Ck)^{n-\ell+1} \le C(k^2+h^4).$$

Thus, there exists $C_0 > 0$, such that $|||\vartheta^n||| \le C_0(k+h^2)$. Since $|||\cdot|||$ and $\|\cdot\|$ are equivalent norms, the first part of the proof is complete.

Next we turn to the estimation of $\|\nabla\vartheta^n\|$. Choosing this time $\chi = \bar{\partial}\vartheta^n$ in (17), we obtain

$$|||\bar{\partial}\vartheta^n|||^2 + a(U^n;\vartheta^n,\bar{\partial}\vartheta^n) = \delta(t^n;U^n,\bar{\partial}\vartheta^n) + \varepsilon_a(U^n;\vartheta^n,\bar{\partial}\vartheta^n). \tag{34}$$

Note now that since $a(\cdot;\cdot,\cdot)$ is symmetric, we have the identity

$$2ka(U^n;\vartheta^n,\bar{\partial}\vartheta^n) = a(U^n;\vartheta^n,\vartheta^n) - a(U^n;\vartheta^{n-1},\vartheta^{n-1}) + k^2a(U^n;\bar{\partial}\vartheta^n,\bar{\partial}\vartheta^n).$$

Using now this and (12) in (34), we get, after subtracting $a(U^{n-1};\vartheta^{n-1},\vartheta^{n-1})$ from both parts of (34),

$$\begin{aligned}
&2k|||\bar{\partial}\vartheta^n|||^2 + a(U^n;\vartheta^n,\vartheta^n) - a(U^{n-1};\vartheta^{n-1},\vartheta^{n-1}) + \alpha k^2\|\nabla\bar{\partial}\vartheta^n\|^2 \\
&\le 2k\delta(t^n;U^n,\bar{\partial}\vartheta^n) + 2k\varepsilon_a(U^n;\vartheta^n,\bar{\partial}\vartheta^n) \\
&+ \{a(U^n;\vartheta^{n-1},\vartheta^{n-1}) - a(U^{n-1};\vartheta^{n-1},\vartheta^{n-1})\} = I + II + III.
\end{aligned} \tag{35}$$

Employing the second bound of Lemma 2.5, with $v = U^n$ and $\chi = \bar{\partial}\vartheta^n$, we have

$$\begin{aligned}
|I| &\le Ck(k+h^2)\|\bar{\partial}\vartheta^n\| + Ckh^2\|\nabla\bar{\partial}\vartheta^n\| + Ck\|\nabla\vartheta^n\|\,\|\bar{\partial}\vartheta^n\| \\
&\le k|||\bar{\partial}\vartheta^n|||^2 + Ck\|\nabla\vartheta^n\|^2 + \frac{\alpha k^2}{2}\|\nabla\bar{\partial}\vartheta^n\|^2 + CkE,
\end{aligned} \tag{36}$$

with $E = O(k^2 + k^{-1}h^4)$. Next, using Lemma 2.3 and the fact that $U^n \in \mathcal{B}_M$, we obtain

$$|II| \leq Ckh\|\nabla U^n\|_{L_\infty} \|\nabla \vartheta^n\| \, \|\nabla \bar{\partial} \vartheta^n\| \leq Ch^2\|\nabla \vartheta^n\|^2 + \frac{\alpha k^2}{2}\|\nabla \bar{\partial} \vartheta^n\|^2. \tag{37}$$

Finally, using again (2), the fact that $\vartheta^{n-1} \in \mathcal{B}_{2M}$, and (23), we have

$$\begin{aligned} |III| &\leq Ck\| \, |\nabla \vartheta^{n-1}| \, |\bar{\partial} U^n| \, \| \, \|\nabla \vartheta^{n-1}\| \\ &\leq Ck(\| \, |\nabla \vartheta^{n-1}| \, |\bar{\partial} \vartheta^n| \, \| + \| \, |\nabla \vartheta^{n-1}| \, |R_h \bar{\partial} u^n| \, \|) \, \|\nabla \vartheta^{n-1}\| \\ &\leq k|||\bar{\partial} \vartheta^n|||^2 + Ck\|\nabla \vartheta^{n-1}\|^2. \end{aligned} \tag{38}$$

Therefore applying (36)–(38), in (35), eliminating $|||\bar{\partial}\vartheta^n|||$ and $\|\nabla \bar{\partial} \vartheta^n\|$ and using (12), we obtain for k and h sufficiently small,

$$a(U^n; \vartheta^n, \vartheta^n) \leq (1 + Ck) a(U^{n-1}; \vartheta^{n-1}, \vartheta^{n-1}) + CkE.$$

Thus, using the fact that $\vartheta^0 = 0$ and A is strictly positive definite, we get

$$c\|\nabla \vartheta^n\|^2 \leq a(U^n; \vartheta^n, \vartheta^n) \leq CkE \sum_{\ell=0}^{n} (1 + Ck)^{n-\ell+1} \leq C(k^2 + k^{-1}h^4).$$

Thus, there exists $C_1 > 0$, such that

$$\|\nabla \vartheta^n\| \leq C_1(k + k^{-1/2}h^2), \tag{39}$$

which completes the second part of the proof. □

4 Existence of the Backward Euler Approximation

Here we show the existence of the solution of the nonlinear fully discrete scheme (11), if $U^0 = R_h u^0$ and the discretization parameters k and h are sufficiently small and satisfy $k = O(h^{1+\varepsilon})$, with $0 < \varepsilon < 1$.

Let $G_n : \mathcal{X}_h \to \mathcal{X}_h$, be defined by

$$\langle G_n v - U^{n-1}, \chi \rangle + k a_h(v; G_n v, J_h \chi) = k(f^n, J_h \chi), \quad \forall \chi \in \mathcal{X}_h. \tag{40}$$

Obviously, if G_n has a fixed point v, then $U^n = v$ is the solution of (11).

In view of (39), recall that if $U^{n-1} \in \mathcal{B}_M$, then

$$\|\nabla(U^{n-1} - R_h u^{n-1})\| \leq C_1(k + k^{-1/2}h^2). \tag{41}$$

Then the following two lemmas hold:

Lemma 4.6. *Let $U^{n-1} \in \mathcal{B}_M$ such that (41) holds. Then for $k = O(h^{1+\varepsilon})$ with $0 < \varepsilon < 1$, there exists a constant $C_2 > 0$, independent of h, sufficiently large such that $U^{n-1} \in \tilde{\mathcal{B}}$, where*

$$\tilde{\mathcal{B}}_n = \{w \in \mathcal{X}_h : \|\nabla(w - R_h u^n)\| \le C_2 h^{1+\tilde{\varepsilon}}\}, \quad \text{with } \tilde{\varepsilon} = \min(\varepsilon, \frac{1-\varepsilon}{2}). \tag{42}$$

Proof. Using the stability property of R_h and the fact that $k = O(h^{1+\varepsilon})$, we have

$$\begin{aligned}\|\nabla(U^{n-1} - R_h u^n)\| &\le \|\nabla(U^{n-1} - R_h u^{n-1})\| + k\|\nabla R_h \bar{\partial} u^n\| \\ &\le C_1(k + k^{-1/2}h^2) + k\|\nabla \bar{\partial} u^n\| \le C_2 h^{1+\tilde{\varepsilon}}. \quad \square\end{aligned}$$

Lemma 4.7. *Let $U^{n-1}, v \in \mathcal{B}_M$ such that (41) holds and $v \in \tilde{\mathcal{B}}_n$, with $\tilde{\mathcal{B}}_n$ defined by (42). Then for $k = \mathcal{O}(h^{1+\varepsilon})$, with $0 < \varepsilon < 1$, $G_n v \in \tilde{\mathcal{B}}_n$.*

Proof. Let us now denote by $\xi^n = G_n v - R_h u^n$ and $\xi^{n-1} = U^{n-1} - R_h u^{n-1}$. Then, using (40), (1), and (16), ξ^n satisfies a similar equation to (17), with ξ^n and v instead of ϑ^n and U^n; hence,

$$\langle \bar{\partial}\xi^n, \chi \rangle + a_h(v; \xi^n, J_h\chi) = \delta(t^n; v, \chi), \quad \text{for } \chi \in \mathcal{X}_h. \tag{43}$$

Choosing $\chi = \bar{\partial}\xi^n$ in (43) and following the proof of Theorem 3.1, we obtain the corresponding inequality to (35), without the last term III, with ξ^n and v in the place of ϑ^n and U^n:

$$\begin{aligned}&2k|||\bar{\partial}\xi^n|||^2 + a(v; \xi^n, \xi^n) - a(v; \xi^{n-1}, \xi^{n-1}) + \alpha k^2 \|\nabla \bar{\partial}\xi^n\|^2 \\ &\le 2k\delta(t^n; v, \bar{\partial}\xi^n) + 2k\varepsilon_a(v; \xi^n, \bar{\partial}\xi^n) = I + II.\end{aligned} \tag{44}$$

Similarly as before we obtain the corresponding estimates to (36) and (37), with ξ^n and v in the place of ϑ^n and U^n. Thus,

$$|I| \le 2k|||\bar{\partial}\xi^n|||^2 + \frac{\alpha k^2}{2}\|\nabla \bar{\partial}\xi^n\|^2 + Ck\|\nabla(v - R_h u^n)\|^2 + CkE, \tag{45}$$

with $E = O(k^2 + k^{-1}h^4)$ and

$$|II| \le Ch^2 a(v; \xi^n, \xi^n) + \frac{\alpha k^2}{2}\|\nabla \bar{\partial}\xi^n\|^2. \tag{46}$$

Then using (45) and (46) in (44) and eliminating $|||\bar{\partial}\xi^n|||^2$ and $\|\nabla \bar{\partial}\xi^n\|^2$, we get for h sufficiently small

$$a(v; \xi^n, \xi^n) \le (1 + Ck)a(v; \xi^{n-1}, \xi^{n-1}) + Ck\|\nabla(v - R_h u^n)\|^2 + CkE.$$

Finally, using in this inequality, (41), the facts that $v \in \tilde{\mathcal{B}}_n$ and $\varepsilon < 1$ and (13), we obtain the desired bound for k sufficiently small. □

Theorem 4.1. *Let $\mathcal{T}_h$ satisfy the inverse assumption* (19) *and $U^{n-1}, v \in \mathcal{B}_M$ such that* (41) *holds. Then for h sufficiently small and $k = \mathcal{O}(h^{1+\varepsilon})$, with $0 < \varepsilon < 1$, there exists $U^n \in \mathcal{B}_M$ satisfying* (11).

Proof. Obviously, in view of Lemmas 4.6 and 4.7, starting with $v_0 = U^{n-1}$, through G_n, we obtain a sequence of elements $v_{j+1} = G_n v_j \in \tilde{\mathcal{B}}_n$, $j \geq 0$. Thus, combining this with (20) and the facts that $M > M_0$ and $\tilde{\varepsilon} > 0$, we get $G_n v_j \in \mathcal{B}_M$ for h sufficiently small, i.e.,

$$\|\nabla G_n v_j\|_{L_\infty} \leq \|\nabla R_h u^n\|_{L_\infty} + Ch^{-1}\|\nabla(G_n v_j - R_h u^n)\| \leq M, \ j \geq 0.$$

To show now the existence of $U^n \in \mathcal{B}_M$, it suffices that

$$|||G_n v - G_n w||| < L|||v - w|||, \quad \forall v, w \in \mathcal{B}_M, \quad \text{with } 0 < L < 1.$$

Employing (40) for $v, w \in \mathcal{B}_M$ and $\chi \in \mathcal{X}_h$, we obtain

$$\langle G_n v - G_n w, \chi \rangle + k a_h(v; G_n v, J_h \chi) - k a_h(w; G_n w, J_h \chi) = 0.$$

Hence, for $\chi = G_n v - G_n w$, this gives

$$\begin{aligned} |||\chi|||^2 + k a_h(w; \chi, J_h \chi) &= k(a_h(w; G_n v, J_h \chi) - a_h(v; G_n v, J_h \chi)) \\ &= k(a(w; G_n v, \chi) - a(v; G_n v, \chi)) \\ &\quad + k(\varepsilon_a(v; G_n v, \chi) - \varepsilon_a(w; G_n v, \chi)) = I + II. \end{aligned} \tag{47}$$

To bound I we use (2) and the fact that $G_n v \in \mathcal{B}_M$ to get

$$|I| \leq Ck\,\|\nabla G_n v\|_{L_\infty}\,\|v - w\|\,\|\nabla \chi\| \leq Ck\,\|v - w\|\,\|\nabla \chi\|. \tag{48}$$

For II, we use Lemma 2.4, the inverse inequality (19), and the fact that $v, G_n v \in \mathcal{B}_M$ to obtain

$$|II| \leq Ckh\|\nabla(v - w)\|\,\|\nabla \chi\| \leq Ck\|v - w\|\,\|\nabla \chi\|. \tag{49}$$

Employing now (13), (48), and (49) into (47), we have

$$|||\chi|||^2 + k\tilde{\alpha}\,\|\nabla \chi\|^2 \leq Ck\,\|v - w\|\,\|\nabla \chi\| \leq Ck\|v - w\|^2 + k\tilde{\alpha}\,\|\nabla \chi\|^2,$$

which in view of the fact that $\|\cdot\|$ and $|||\cdot|||$ are equivalent norms gives for sufficiently small k the desired bound. □

5 A Linearized Fully Discrete Scheme

In this section we analyze a linearized backward Euler (LBE) scheme for the approximation of (1). This time for $U^0 = R_h u^0$, we define the nodal approximations $U^n \in \mathcal{X}_h$ to u^n, $n = 1, \ldots, N$, by

$$\langle \bar{\partial} U^n, \chi \rangle + a_h(U^{n-1}; U^n, J_h\chi) = (f^n, J_h\chi), \quad \forall \chi \in \mathcal{X}_h,\ n \geq 1. \tag{50}$$

Theorem 5.2. *Let U^n and u be the solutions of (50) and (1), with $U^0 = R_h u^0$. Then, for $U^{n-1} \in \mathcal{B}_M$, h sufficiently small and $k = O(h^{1+\varepsilon})$, with $0 < \varepsilon < 1$, we have $U^n \in \mathcal{B}_M$ and*

$$\|\nabla^s (U^n - u(t^n))\| \leq C(k + k^{-s/2} h^{2-s}), \quad \text{with } s = 0, 1.$$

Proof. Since the discrete scheme (50) is linear, the existence of $U^n \in \mathcal{X}_h$ is obvious. The proof is analogous to that for Theorem 3.1; thus, it suffices to bound $\|\nabla^s \vartheta^n\|$, $s = 0, 1$. This time ϑ^n satisfies a similar equation to (17) with U^{n-1} in the place of U^n:

$$\langle \bar{\partial} \vartheta^n, \chi \rangle + a_h(U^{n-1}; \vartheta^n, J_h\chi) = \delta(t^n; U^{n-1}, \chi), \quad \forall \chi \in \mathcal{X}_h.$$

We start with the estimation for $\|\vartheta^n\|$. In an analogous way to (33), we obtain the following inequality:

$$\frac{1}{2k}(|||\vartheta^n|||^2 - |||\vartheta^{n-1}|||^2) + \tilde{\alpha}\|\nabla \vartheta^n\|^2 \leq \delta(t^n; U^{n-1}, \vartheta^n).$$

To bound now the right-hand side of this inequality we employ the first estimate of Lemma 2.5, with $v = U^{n-1}$ and $\chi = \vartheta^n$, using the fact that $U^{n-1} - R_h u^n = \vartheta^{n-1} - kR_h \bar{\partial} u^n$ and the stability of R_h, to get

$$\begin{aligned}
&\frac{1}{2k}(|||\vartheta^n|||^2 - |||\vartheta^{n-1}|||^2) + \tilde{\alpha}\|\nabla \vartheta^n\|^2 \\
&\leq C(k + h^2)\|\vartheta^n\| + C(k\|U^{n-1} - R_h u^n\| + h^2)\|\nabla \vartheta^n\| \\
&\leq C|||\vartheta^n|||^2 + \tilde{\alpha}\|\nabla \vartheta^n\|^2 + Ck|||\vartheta^{n-1}|||^2 + CE, \quad \text{with } E = O(k^2 + h^4).
\end{aligned}$$

Next, after eliminating $\|\nabla \vartheta^n\|$, we get for k sufficiently small

$$|||\vartheta^n|||^2 \leq (1 + Ck)|||\vartheta^{n-1}|||^2 + CkE.$$

Hence, since $\vartheta^0 = 0$, we have by repeated application $|||\vartheta^n||| \leq C(k + h^2)$, which, in view of the fact that $|||\cdot|||$ and $\|\cdot\|$ are equivalent norms, completes the first part of the proof. Next we turn to the bound for $\|\nabla \vartheta^n\|$. In an analogous way to (34), we get

$$|||\bar{\partial} \vartheta^n|||^2 + a(U^{n-1}; \vartheta^n, \bar{\partial} \vartheta^n) = \delta(t^n; U^{n-1}, \bar{\partial} \vartheta^n) + \varepsilon_a(U^{n-1}; \vartheta^n, \bar{\partial} \vartheta^n).$$

Hence, similarly as in (35), we have

$$
\begin{aligned}
&2k|||\bar{\partial}\vartheta^n|||^2+a(U^n;\vartheta^n,\vartheta^n)-a(U^{n-1};\vartheta^{n-1},\vartheta^{n-1})+\alpha k^2\|\nabla\bar{\partial}\vartheta^n\|^2 \\
&\le 2k\delta(t^n;U^{n-1},\bar{\partial}\vartheta^n)+2k\varepsilon_a(U^{n-1};\vartheta^n,\bar{\partial}\vartheta^n) \\
&\quad+\{a(U^n;\vartheta^n,\vartheta^n)-a(U^{n-1};\vartheta^n,\vartheta^n)\}=I.
\end{aligned}
\tag{51}
$$

Thus, in a similar way that we obtained (36)–(38), we have

$$
\begin{aligned}
|I|\le 2k|||\bar{\partial}\vartheta^n|||^2+Ck\|\nabla(U^{n-1}-R_hu^n)\|^2+C(k+h^2)\|\nabla\vartheta^n\|^2 \\
+\alpha k^2\|\nabla\bar{\partial}\vartheta^n\|^2+CkE,
\end{aligned}
$$

with $E=O(k^2+k^{-1}h^4)$. Combining these in (51), using the fact that $U^{n-1}-R_hu^n=\vartheta^{n-1}-kR_h\bar{\partial}u^n$ and the stability of R_h, we obtain for k sufficiently small

$$
a(U^n;\vartheta^n,\vartheta^n)\le(1+Ck)a(U^{n-1};\vartheta^{n-1},\vartheta^{n-1})+CkE.
$$

Therefore, since $\vartheta^0=0$, we obtain

$$
\alpha\|\nabla\vartheta^n\|^2\le a(U^n;\vartheta^n,\vartheta^n)\le CkE\sum_{\ell=0}^{n}(1+Ck)^{n-\ell+1}\le C(k^2+k^{-1}h^4),
$$

which gives the desired bound. Finally, this estimate, the inverse inequality (19), and the fact that $k=O(h^{1+\varepsilon})$ give, for sufficiently small h, that $U^n\in\mathcal{B}_M$, which completes the proof. □

6 Numerical Examples

In this section we give numerical examples to illustrate the error estimates presented in the previous sections. Let $\{\phi_i\}_{i=1}^d$ be the standard piecewise linear basis functions of $\mathcal{X}_h$ and for $\chi\in\mathcal{X}_h$, let $\tilde{\chi}=(\tilde{\chi}_1,\dots,\tilde{\chi}_d)\in\mathbb{R}^d$ be the vector such that $\chi=\sum_{i=1}^d\tilde{\chi}_i\phi_i$. Then the backward Euler method (11) can be written as

$$
(D+kS(\tilde{U}^n))\tilde{U}^n=D\tilde{U}^{n-1}+kQ^n,
$$

where D is the mass matrix with elements $D_{ij}=\int_{V_i}\phi_j\,dx$, Q the vector with entries $Q_i=\int_{V_i}f\,dx$, and $S(\tilde{\chi})$ the resulting stiffness matrix for $\chi\in\mathcal{X}_h$, i.e.,

$$
S_{ij}(\tilde{\chi})=-\int_{\partial V_i}A(\chi)\nabla\phi_j\cdot n\,ds,\qquad \text{for } \chi\in\mathcal{X}_h.
$$

Table 1 Comparison of errors of backward Euler (BE) and LBE methods for various h with$k = h^{1.01}$

	BE				LBE			
h	$\|u-u_h\|$	Rate	$\|u-u_h\|_1$	Rate	$\|u-u_h\|$	Rate	$\|u-u_h\|_1$	Rate
0.125	3.6569e−03	–	8.8974e−02	–	4.9954e−03	–	8.8928e−02	–
0.0625	9.0420e−04	2.02	4.4710e−02	0.99	1.6205e−03	1.62	4.4763e−02	0.99
0.03125	2.0321e−04	2.15	2.2382e−02	1.00	6.4270e−04	1.33	2.2460e−02	1.00
0.015625	4.1362e−05	2.20	1.1194e−02	1.00	2.7213e−04	1.24	1.12480e−02	1.00
0.0078125	8.3814e−06	2.30	5.5974e−03	1.00	1.2512e−04	1.12	5.6268e−03	1.00

Since, this is a nonlinear problem, we employ the following iteration: Set $\xi^0 = \tilde{U}^{n-1}$ and for $m = 1,2,\dots,$ we solve

$$(D + kS(\tilde{\xi}^{m-1}))\tilde{\xi}^m = D\tilde{U}^{n-1} + kQ^n,$$

until some specified convergence. We note that if the iteration is stopped at $m = 1$, we recover the LBE method. For all examples below, we use as a stopping criteria

$$\|(D + kS(\tilde{\xi}^{m-1}))\tilde{\xi}^m - D\tilde{U}^{n-1} - kQ^n\|_{l_\infty} \leq \varepsilon,$$

for some preassigned small number ε, with $\|\tilde{\chi}\|_{l_\infty} = \max_i |\tilde{\chi}_i|$.

We consider $\Omega = [0,1] \times [0,1]$ and partition $[0,1]$ into N equidistant intervals; thus, N^2 squares are formed and divide each one into two triangles, which results in a mesh with size $h = \sqrt{2}/N$. Once the spatial mesh size is determined, the time step k is computed in such a way that $k = h^{1.01}$. Note that our numerical examples indicate that we could choose $k = h$; however, we do not know at this point how to proceed with the analysis under this assumption. We consider $u(x,y,t) = 8e^{-t}(x-x^2)(y-y^2)$ and use the nonlinear coefficient $A(u) = 1/(1-0.8\sin^2(4u))$, with forcing function f such that u satisfies the parabolic equation (1). We compute the error at final time $T = 1$ and the results are shown in Table 1. In both methods, the error convergence rate does follow the a priori estimates. We also see that in the LBE, that as we decrease h, the error contribution from k starts to dominate. This is indicated by the decrease of the convergence order in the L_2-norm.

Acknowledgements The research of P. Chatzipantelidis was partly supported by the FP7-REGPOT-2009-1 project "Archimedes Center for Modeling Analysis and Computation," funded by the European Commission. The research of V. Ginting was partially supported by the grants from DOE (DE-FE0004832 and DE-SC0004982), the Center for Fundamentals of Subsurface Flow of the School of Energy Resources of the University of Wyoming (WYDEQ49811GNTG, WYDEQ49811PER), and from NSF (DMS-1016283).

References

1. Chatzipantelidis, P., Ginting, V., Lazarov, R.D.: A finite volume element method for a nonlinear elliptic problem. Numer. Linear Algebra Appl. **12**, 515–546 (2005)
2. Chatzipantelidis, P., Lazarov, R.D., Thomée, V.: Error estimates for a finite volume element method for parabolic equations in convex polygonal domains. Numer. Methods Partial Differ. Equ. **20**, 650–674 (2004)
3. Chavent, G., Jaffré, J.: Mathematical Models and Finite Elements for Reservoir Simulation. Elsevier Science Publisher, B.V. Amsterdam, (1986)
4. Chou, S.-H., Li, Q.: Error estimates in L^2, H^1 and L^∞ in covolume methods for elliptic and parabolic problems: a unified approach. Math. Comp. **69**, 103–120 (2000)
5. Eymard, R., Gallouët, T., Herbin, R.: Finite volume methods. In: Ciarlet, P.G., Lions, J.L. (eds.) Handbook of Numerical Analysis, vol. VII, pp. 713–1020. North-Holland, Amsterdam (2000)
6. Keller, E., Segel, L.: Initiation of slime mold aggregation viewed as an instability. J. Theor. Biol. **26**, 399–415 (1970)
7. Ladyženskaja, O.A., Solonnikov, V.A., Uralćeva, N.N.: Linear and Quasilinear Equations of Parabolic Type. Translated from the Russian by S. Smith. American Mathematical Society, Providence (1968)
8. Li, R.: Generalized difference methods for a nonlinear Dirichlet problem. SIAM J. Numer. Anal. **24**, 77–88 (1987)
9. Li, R., Chen, Z., Wu, W.: Generalized Difference Methods for Differential Equations. Marcel Dekker, New York (2000)
10. Richards, L.A.: Capillary conduction of liquids through porous mediums. Physics **1**, 318–333 (1931)
11. Thomée, V.: Galerkin Finite Element Methods for Parabolic Problems. Springer, Berlin (2006)
12. Zhang, T., Zhong, H., Zhao, J.: A fully discrete two–grid finite–volume method for a nonlinear parabolic problem. Int. J. Comput. Math. **88**, 1644–1663 (2011)

Multidimensional Sensitivity Analysis of Large-Scale Mathematical Models

Ivan Dimov and Rayna Georgieva

Abstract Sensitivity analysis (SA) is a procedure for studying how sensitive are the output results of large-scale mathematical models to some uncertainties of the input data. The models are described as a system of partial differential equations. Often such systems contain a large number of input parameters. Obviously, it is important to know how sensitive is the solution to some uncontrolled variations or uncertainties in the input parameters of the model. Algorithms based on analysis of variances technique for calculating numerical indicators of sensitivity and computationally efficient Monte Carlo integration techniques have recently been developed by the authors. They have been successfully applied to sensitivity studies of air pollution levels calculated by the Unified Danish Eulerian Model with respect to several important input parameters. In this paper a comprehensive theoretical and experimental study of the Monte Carlo algorithm based on *symmetrised shaking* of Sobol sequences has been done. It has been proven that this algorithm has an optimal rate of convergence for functions with continuous and bounded second derivatives in terms of probability and mean square error. Extensive numerical experiments with Monte Carlo, quasi-Monte Carlo (QMC) and scrambled QMC algorithms based on Sobol sequences are performed to support the theoretical studies and to analyze applicability of the algorithms to various classes of problems. The numerical tests show that the Monte Carlo algorithm based on *symmetrised shaking* of Sobol sequences gives reliable results for multidimensional integration problems under consideration.

Keywords Global sensitivity analysis • Analysis for independent inputs • Monte Carlo and quasi-Monte Carlo algorithms • Optimal Monte Carlo methods

---

I. Dimov (✉) • R. Georgieva
Department of Parallel Algorithms, IICT, Bulgarian Academy of Sciences, Acad. G. Bonchev 25 A, 1113 Sofia, Bulgaria
e-mail: ivdimov@bas.bg; rayna@parallel.bas.bg

O.P. Iliev et al. (eds.), *Numerical Solution of Partial Differential Equations: Theory, Algorithms, and Their Applications*, Springer Proceedings in Mathematics & Statistics 45,
DOI 10.1007/978-1-4614-7172-1_8,

Mathematics Subject Classification (2010): 49Q12, 65C05, 65B99

1 Introduction

Most existing methods for providing SA rely on special assumptions connected to the behavior of the model (such as linearity, monotonicity and additivity of the relationship between model input and model output) [22]. Such assumptions are often applicable to a large range of mathematical models. At the same time there are models that include significant nonlinearities and/or stiffness. For such models assumptions about linearity and additivity are not applicable. This is especially true when one deals with nonlinear systems of partial differential equations. The numerical study and results reported in this paper have been done by using a large-scale mathematical model called Unified Danish Eulerian Model (UNI-DEM) [33, 34]. The model enables us to study the transport of air pollutants and other species over a large geographical region. The system of partial differential equations describes the main physical processes, such as advection, diffusion, deposition as well as chemical and photochemical processes between the studied species. The emissions and the quickly changing meteorological conditions are also described. The nonlinearity of the equations is mainly introduced when modeling chemical reactions [33]. If the model results are sensitive to a given process, one can describe it mathematically in a more adequate way or more precisely. Thus, the goal of our study is to increase the reliability of the results produced by the model and to identify processes that must be studied more carefully, as well as to find input parameters that need to be measured with a higher precision. A careful sensitivity analysis is needed in order to decide where and how simplifications of the model can be made. That is why it is important to develop and study more adequate and reliable methods for sensitivity analysis. A good candidate for reliable sensitivity analysis of models containing nonlinearity is the variance-based method [22]. The idea of this approach is to estimate how the variation of an input parameter or a group of inputs contributes into the variance of the model output. As a measure of this analysis we use the *total sensitivity indices (TSI)* (see, Sect. 2) described as multidimensional integrals:

$$I = \int_{\Omega} g(\mathrm{x})p(\mathrm{x})\,\mathrm{dx}, \;\; \Omega \subset \mathbf{R}^d, \tag{1}$$

where $g(\mathrm{x})$ is a square integrable function in Ω and $p(\mathrm{x}) \geq 0$ is a *p*robability *d*ensity *f*unction (p.d.f.), such that $\int_{\Omega} p(\mathrm{x})\,\mathrm{dx} = 1$.

Clearly, the progress in the area of sensitivity analysis is closely related to the progress in reliable algorithms for multidimensional integration.

2 Problem Setting

2.1 Modeling and Sensitivity

Assume that the mathematical model can be presented as a function

$$\mathrm{u} = f(\mathrm{x}), \quad \text{where} \quad \mathrm{x} = (x_1, x_2, \ldots, x_d) \in U^d \equiv [0;1]^d \tag{2}$$

is the vector of input parameters with a joint p.d.f. $p(\mathrm{x}) = p(x_1, \ldots, x_d)$. Assume also that the input variables are independent (noncorrelated) and the density function $p(\mathrm{x})$ is known, even if x_i are not actually random variables (r.v.). The TSI of an input parameter $x_i, i \in \{1, \ldots, d\}$ is defined in the following way [9, 26]:

$$S_i^{tot} = S_i + \sum_{l_1 \neq i} S_{il_1} + \sum_{l_1, l_2 \neq i, l_1 < l_2} S_{il_1 l_2} + \ldots + S_{il_1 \ldots l_{d-1}}, \tag{3}$$

where S_i is called *the main effect (first-order sensitivity index)* of x_i and $S_{il_1 \ldots l_{j-1}}$ is the j-th order sensitivity index. The higher-order terms describe the interaction effects between the unknown input parameters $x_{i_1}, \ldots, x_{i_\nu}, \nu \in \{2, \ldots, d\}$ on the output variance.

The method of global SA used in this work is based on a decomposition of an integrable model function f in the d-dimensional factor space into terms of increasing dimensionality [26]:

$$f(\mathrm{x}) = f_0 + \sum_{\nu=1}^{d} \sum_{l_1 < \ldots < l_\nu} f_{l_1 \ldots l_\nu}(x_{l_1}, x_{l_2}, \ldots, x_{l_\nu}), \tag{4}$$

where f_0 is a constant. The representation (4) is referred to as the ANOVA representation of the model function $f(\mathrm{x})$ if each term is chosen to satisfy the following condition [26]:

$$\int_0^1 f_{l_1 \ldots l_\nu}(x_{l_1}, x_{l_2}, \ldots, x_{l_\nu}) \mathrm{d}x_{l_k} = 0, \quad 1 \leq k \leq \nu, \quad \nu = 1, \ldots, d.$$

Let us mention the fact that if the whole presentation (4) of the right-hand side is used, this does not make the problem simpler. The hope is that a truncated sequence $f_0 + \sum_{\nu=1}^{d_{tr}} \sum_{l_1 < \ldots < l_\nu} f_{l_1 \ldots l_\nu}(x_{l_1}, x_{l_2}, \ldots, x_{l_\nu})$, where $d_{tr} < d$ (or even $d_{tr} \ll d$), can be considered as a good approximation to the model function f.

The quantities

$$\mathbf{D} = \int_{U^d} f^2(\mathrm{x}) \mathrm{d}\mathrm{x} - f_0^2, \quad \mathbf{D}_{l_1 \ldots l_\nu} = \int f_{l_1 \ldots l_\nu}^2 \mathrm{d}x_{l_1} \ldots \mathrm{d}x_{l_\nu} \tag{5}$$

are the so-called total and partial variances, respectively, and are obtained after squaring and integrating over U^d the equality (4) on the assumption that $f(\mathrm{x})$ is a square integrable function (thus, all terms in (4) are also square integrable functions). Therefore, the total variance of the model output is split into partial variances in the analogous way as the model function, that is, the unique ANOVA-decomposition: $\mathbf{D} = \sum_{\nu=1}^{d} \sum_{l_1<\ldots<l_\nu} \mathbf{D}_{l_1 \ldots l_\nu}$. The use of probability theory concepts is based on the assumption that the input parameters are random variables distributed in U^d that defines $f_{l_1 \ldots l_\nu}(x_{l_1}, x_{l_2}, \ldots, x_{l_\nu})$ also as random variables with variances (5). For example, f_{l_1} is presented by a conditional expectation: $f_{l_1}(x_{l_1}) = \mathbf{E}(\mathrm{u}|x_{l_1}) - f_0$ and, respectively, $\mathbf{D}_{l_1} = \mathbf{D}[f_{l_1}(x_{l_1})] = \mathbf{D}[\mathbf{E}(\mathrm{u}|x_{l_1})]$. Based on these assumptions about the model function and the output variance, the following quantities

$$S_{l_1 \ldots l_\nu} = \frac{\mathbf{D}_{l_1 \ldots l_\nu}}{\mathbf{D}}, \quad \nu \in \{1, \ldots, d\} \tag{6}$$

are referred to as the global sensitivity indices [26]. Based on the formulas (5)–(6), it is clear that the mathematical treatment of the problem of providing global sensitivity analysis consists in evaluating total sensitivity indices (3) of corresponding order that, in turn, leads to computing multidimensional integrals of the form (1). It means that to obtain S_i^{tot} in general, one needs to compute 2^d (or $2^{d_{tr}}$, with $d_{tr} \ll d$) integrals of type (5).

The procedure for computing global sensitivity indices (see [26]) is based on the following representation of the variance:

$$\mathbf{D}_\mathrm{y} : \mathbf{D}_\mathrm{y} = \int f(\mathrm{x})\, f(\mathrm{y}, \mathrm{z}')\mathrm{d}\mathrm{x}\mathrm{d}\mathrm{z}' - f_0^2, \tag{7}$$

where $\mathrm{y} = (x_{k_1}, \ldots, x_{k_m}),\ 1 \leq k_1 < \ldots < k_m \leq d,$ is an arbitrary set of m variables $(1 \leq m \leq d-1)$ and z is the set of $d-m$ complementary variables, i.e. $\mathrm{x} = (\mathrm{y}, \mathrm{z})$. The equality (7) enables the construction of a Monte Carlo algorithm for evaluating $f_0, \mathbf{D}$ and $\mathbf{D}_\mathrm{y}$:

$$\frac{1}{n}\sum_{j=1}^{n} f(\xi_j) \xrightarrow{P} f_0, \qquad \frac{1}{n}\sum_{j=1}^{n} f(\xi_j)\, f(\eta_j, \zeta_j') \xrightarrow{P} \mathbf{D}_\mathrm{y} + f_0^2,$$

$$\frac{1}{n}\sum_{j=1}^{n} f^2(\xi_j) \xrightarrow{P} \mathbf{D} + f_0^2, \qquad \frac{1}{n}\sum_{j=1}^{n} f(\xi_j)\, f(\eta_j', \zeta_j) \xrightarrow{P} \mathbf{D}_\mathrm{z} + f_0^2,$$

where $\xi = (\eta, \zeta)$ is a random sample and η corresponds to the input subset denoted by y.

Instead of randomized (Monte Carlo) algorithms for computing the above sensitivity parameters, one can use deterministic quasi-Monte Carlo (QMC) algorithms or randomized QMC [13, 14]. Randomized (Monte Carlo) algorithms have proven to be very efficient in solving multidimensional integrals in composite domains [3, 23]. At the same time the QMC based on well-distributed Sobol sequences can be considered as a good alternative to Monte Carlo algorithms, especially for smooth integrands and not very high *effective dimensions* (up to $d = 15$) [12]. Sobol $\Lambda\Pi_\tau$ are good candidates for efficient QMC algorithms. Algorithms based on $\Lambda\Pi_\tau$ sequences while being deterministic mimic the pseudorandom sequences used in Monte Carlo integration. One of the problems with $\Lambda\Pi_\tau$ sequences is that they may have *bad* two-dimensional projection. In this context *bad* means that the distribution of the points is far from being a uniform distribution. If such projections are used in a certain computational problem, then the lack of uniformity may provoke a substantial lost of accuracy. To overcome this problem randomized QMC can be used. There are several ways of randomization and *scrambling* is one of them. The original motivation of scrambling [10, 19] aims toward obtaining more uniformity for quasi-random sequences in high dimensions, which can be checked via two-dimensional projections. Another way of randomisation is to *shake* the quasi-random points according to some procedure. Actually, the scrambled algorithms obtained by *shaking* the quasi-random points can be considered as Monte Carlo algorithms with a special choice of the density function. It is a matter of definition. Thus, there is a reason to be able to compare two classes of algorithms: *deterministic* and *randomized.*

3 Complexity in Classes of Algorithms

One may pose the task to consider and compare two classes of algorithms: *deterministic algorithms* and *randomized (Monte Carlo) algorithms*. Let I be the desired value of the integral. Assume for a given r.v. θ one can prove that the mathematical expectation satisfies $\mathbf{E}\theta = I$. Suppose that the mean value of n values of θ: $\theta^{(i)}$, $i = 1,\dots,n$ is considered as a Monte Carlo approximation to the solution: $\bar{\theta}_n = 1/n\sum_{i=1}^{n}\theta^{(i)} \approx I$, where $\theta^{(i)}(i = 1,2,\dots,n)$ correspond to values (realizations) of a r.v. θ. In general, a certain randomized algorithm can produce the result with a given probability error. So, dealing with randomized algorithms one has to accept that the result of the computation can be true only with a certain (although high) probability. In most practical computations it is reasonable to accept an error estimate with a probability smaller than 1.

Consider the following integration problem:

$$S(f) := I = \int_{U^d} f(\mathrm{x})d\mathrm{x}, \tag{8}$$

where $\mathrm{x} \equiv (x_1,\dots,x_d) \in U^d \subset \mathbf{R}^d$ and $f \in C(U^d)$ is an integrable function on U^d. The computational problem can be considered as a mapping of function $f : \{[0,1]^d \to \mathbf{R}\}$ to $\mathbf{R}$: $S(f) : f \to \mathbf{R}$, where $S(f) = \int_{U^d} f(\mathrm{x})d\mathrm{x}$ and $f \in F_0 \subset C(U^d)$. We refer to S as the solution operator. The elements of F_0 are the data, for which the problem has to be solved, and for $f \in F_0$, $S(f)$ is the exact solution. For a given f, we want to compute exactly or approximately $S(f)$. One may be interested in cases when the integrand f has a higher regularity. It is because in many cases of practical computations f is smooth and has high-order bounded derivatives. If this is the case, then is it reasonable to try to exploit such a smoothness. To be able to do that we need to define the functional class $F_0 \equiv \mathbf{W}^k(\|f\|;U^d)$ in the following way:

Definition 3.1. Let d and k be integers, $d,k \geq 1$. We consider the class $\mathbf{W}^k(\|f\|;U^d)$ (sometimes abbreviated to $\mathbf{W}^k$) of real functions f defined over the unit cube $U^d = [0,1)^d$, possessing all the partial derivatives $\frac{\partial^r f(\mathrm{x})}{\partial x_1^{\alpha_1} \ldots \partial x_d^{\alpha_d}}$, $\alpha_1 + \cdots + \alpha_d = r \leq k$, which are continuous when $r < k$ and bounded in sup norm when $r = k$. The seminorm $\|\cdot\|$ on $\mathbf{W}^k$ is defined as

$$\|f\| = \sup\left\{\left|\frac{\partial^k f(\mathrm{x})}{\partial x_1^{\alpha_1} \ldots \partial x_d^{\alpha_d}}\right|, \ \alpha_1 + \cdots + \alpha_d = k, \ \mathrm{x} \equiv (x_1, \ldots, x_d) \in U^d\right\}.$$

We keep the seminorm $\|f\|$ into the notation for the functional class $\mathbf{W}^k(\|f\|;U^d)$ since it is important for our further consideration. We call a *quadrature formula* any expression of the form

$$A^D(f,n) = \sum_{i=1}^{n} c_i f(\mathrm{x}^{(i)}),$$

which approximates the value of the integral $S(f)$. The real numbers $c_i \in \mathbf{R}$ are called weights and the d-dimensional points $\mathrm{x}^{(i)} \in U^d$ are called nodes. It is clear that for fixed weights c_i and nodes $\mathrm{x}^{(i)} \equiv (x_{i,1}, \ldots, x_{i,d})$, the quadrature formula $A^D(f,n)$ may be used to define an algorithm with an integration error $err(f,A^D) \equiv \int_{U^d} f(\mathrm{x})d\mathrm{x} - A^D(f,n)$. We call a *randomized quadrature formula* any formula of the following kind: $A^R(f,n) = \sum_{i=1}^{n} \sigma_i f(\xi^{(i)})$, where σ_i and $\xi^{(i)}$ are random weights and nodes, respectively. The algorithm $A^R(f,n)$ belongs to the class of randomized (Monte Carlo) denoted by $\mathscr{A}^{\mathscr{R}}$.

Definition 3.2. Given a randomized (Monte Carlo) integration formula for the functions from the space $\mathbf{W}^k$, we define the integration error

$$err(f,A^R) \equiv \int_{U^d} f(\mathrm{x})d\mathrm{x} - A^R(f,n)$$

by the probability error $\varepsilon_P(f)$ in the sense that $\varepsilon_P(f)$ is the least possible real number, such that

$$Pr\left(\left|err(f,A^R)\right| < \varepsilon_P(f)\right) \geq P,$$

and the mean square error

$$r(f) = \left\{E\left[err^2(f,A^R)\right]\right\}^{1/2}.$$

We assume that it suffices to obtain an $\varepsilon_P(f)$-approximation to the solution with a probability $0 < P < 1$. If we allow equality, i.e. $0 < P \leq 1$ in Definition 3.2, then

$\varepsilon_P(f)$ can be used as an accuracy measure for both randomized and deterministic algorithms. In such a way it is consistent to consider a wider class $\mathscr{A}$ of algorithms that contains both classes: randomized and deterministic algorithms.

Definition 3.3. Consider the set $\mathscr{A}$ of algorithms A:

$$\mathscr{A} = \{A : Pr(|err(f,A)| \leq \varepsilon) \geq c\}, \quad A \in \{A^D, A^R\}, \quad 0 < c < 1$$

that solve a given problem with an integration error $err(f,A)$.

In such a setting it is correct to compare randomized algorithms with algorithms based on low-discrepancy sequences like Sobol $\Lambda\Pi_\tau$ sequences.

4 The Algorithms

The algorithms we study are based on Sobol $\Lambda\Pi_\tau$ sequences.

4.1 $\Lambda\Pi_\tau$ Sobol Sequences

$\Lambda\Pi_\tau$ sequences are *uniformly distributed sequences* (u.d.s.) The term *u.d.s.* was introduced by Hermann Weyl in 1916 [30]. For practical purposes a u.d.s. should satisfy the following three requirements [23, 25]: (i) the best asymptote as $n \to \infty$, (ii) well-distributed points for small n and (iii) a computationally inexpensive algorithm.

All $\Lambda\Pi_\tau$ sequences given in [25] satisfy the first requirement. Suitable distributions such as $\Lambda\Pi_\tau$ sequences are also called (t,m,s)-nets and (t,s)-sequences in base $b \geq 2$. To introduce them, define first an elementary s-interval in base b as a subset of U^s of the form $E = \prod_{j=1}^{s} \left[\frac{a_j}{b^{d_j}}, \frac{a_j+1}{b^{d_j}}\right]$, where $a_j, d_j \geq 0$ are integers and $a_j < b^{d_j}$ for all $j \in \{1,\ldots,s\}$. Given two integers $0 \leq t \leq m$, a (t,m,s)-net in base b is a sequence $\mathrm{x}^{(i)}$ of b^m points of U^s such that $Card\, E \cap \{\mathrm{x}^{(1)},\ldots,\mathrm{x}^{(b^m)}\} = b^t$ for any elementary interval E in base b of hypervolume $\lambda(E) = b^{t-m}$. Given a non-negative integer t, a (t,s)-sequence in base b is an infinite sequence of points $\mathrm{x}^{(i)}$ such that for all integers $k \geq 0, m \geq t$, the sequence $\{\mathrm{x}^{(kb^m)},\ldots,\mathrm{x}^{((k+1)b^m-1)}\}$ is a (t,m,s)-net in base b.

Sobol [23] defines his Π_τ-meshes and $\Lambda\Pi_\tau$ sequences, which are (t,m,s)-nets and (t,s)-sequences in base 2, respectively. The terms (t,m,s)-nets and (t,s)-sequences in base b (also called Niederreiter sequences) were introduced in 1988 by Niederreiter [18].

To generate the j-th component of the points in a Sobol sequence, we need to choose a primitive polynomial of some degree s_j over the Galois field of two elements GF(2) $P_j = x^{s_j} + a_{1,j}x^{s_j-1} + a_{2,j}x^{s_j-2} + \ldots + a_{s_j-1,j}x + 1$, where the coefficients $a_{1,j},\dots,a_{s_j-1,j}$ are either 0 or 1. A sequence of positive integers $\{m_{1,j},m_{2,j},\dots\}$ are defined by the recurrence relation

$$m_{k,j} = 2a_{1,j}m_{k-1,j} \oplus 2^2 a_{2,j}m_{k-2,j} \oplus \dots \oplus 2^{s_j} m_{k-s_j,j} \oplus m_{k-s_j,j},$$

where $\oplus$ is the bit-by-bit *exclusive-or* operator. The values $m_{1,j},\dots,m_{s_j,j}$ can be chosen freely provided that each $m_{k,j}, 1 \leq k \leq s_j$, is odd and less than 2^k. Therefore, it is possible to construct different Sobol sequences for the fixed dimension s. In practice, these numbers must be chosen very carefully to obtain really efficient Sobol sequence generators [27]. The so-called direction numbers $\{v_{1,j},v_{2,j},\dots\}$ are defined by $v_{k,j} = \frac{m_{k,j}}{2^k}$. Then the j-th component of the i-th point in a Sobol sequence is given by $x_{i,j} = i_1 v_{1,j} \oplus i_2 v_{2,j} \oplus \dots$, where i_k is the k-th binary digit of $i = (\dots i_3 i_2 i_1)_2$. Subroutines to compute these points can be found in [2, 24]. The work [15] contains more details.

4.2 The Monte Carlo Algorithms Based on Modified Sobol Sequences: MCA-MSS

One of the algorithms based on a procedure of *shaking* was proposed recently in [6]. The idea is that we take a Sobol $\Lambda\Pi_\tau$ point (vector) x of dimension d. Then x is considered as a centrum of a sphere with a radius $\rho < 1$. A random point $\xi \in U^d$ uniformly distributed on the sphere is taken. Consider a random variable θ defined as a value of the integrand at that random point, i.e. $\theta = f(\xi)$. Consider random points $\xi^{(i)}(\rho) \in U^d, i = 1,\dots,n$. Assume $\xi^{(i)}(\rho) = \mathrm{x}^{(i)} + \rho\omega^{(i)}$, where $\omega^{(i)}$ is a unique uniformly distributed vector in U^d. The radius ρ is relatively small $\rho \ll \frac{1}{2^{d_j}}$, such that $\xi^{(i)}(\rho)$ is still in the same elementary i-th interval $E_i^d = \prod_{j=1}^d \left[\frac{a_j^{(i)}}{2^{d_j}}, \frac{a_j^{(i)}+1}{2^{d_j}}\right]$, where the pattern $\Lambda\Pi_\tau$ point $\mathrm{x}^{(i)}$ is. We use a subscript i in E_i^d to indicate that the i-th $\Lambda\Pi_\tau$ point $\mathrm{x}^{(i)}$ is in it. So, we assume that if $\mathrm{x}^{(i)} \in E_i^d$, then $\xi^{(i)}(\rho) \in E_i^d$ too.

It was proven in [6] that the mathematical expectation of the random variable $\theta = f(\xi)$ is equal to the value of the integral (8), that is, $\mathbf{E}\theta = S(f) = \int_{U^d} f(\mathrm{x})d\mathrm{x}$. This result allows for defining a randomized algorithm. One can take the Sobol $\Lambda\Pi_\tau$ point $\mathrm{x}^{(i)}$ and *shake* it somewhat. *Shaking* means to define random points $\xi^{(i)}(\rho) = \mathrm{x}^{(i)} + \rho\omega^{(i)}$ according to the procedure described above. For simplicity the algorithm described above is abbreviated as MCA-MSS-1.

The probability error of the algorithm MCA-MSS-1 was analysed in [7]. It was proved that for integrands with continuous and bounded first derivatives, i.e. $f \in \mathbf{W}^1(L;U^d)$, where $L = \|f\|$, it holds

$$err(f,d) \leq c_d^{'} \|f\| \, n^{-\frac{1}{2}-\frac{1}{d}} \quad \text{and} \quad r(f,d) \leq c_d^{''} \|f\| \, n^{-\frac{1}{2}-\frac{1}{d}},$$

where the constants $c_d^{'}$ and $c_d^{''}$ do not depend on n.

In this work a modification of algorithm MCA-MSS-1 is proposed and analysed. The new algorithm will be called MCA-MSS-2.

It is assumed that $n = m^d$, $m \geq 1$. The unit cube U^d is divided into m^d disjoint subdomains, such that they coincide with the elementary d-dimensional subintervals defined in Sect. 4.1 $U^d = \bigcup_{j=1}^{m^d} K_j$, where $K_j = \prod_{i=1}^{d} [a_i^{(j)}, b_i^{(j)})$, with $b_i^{(j)} - a_i^{(j)} = \frac{1}{m}$ for all $i = 1, \dots, d$.

In such a way in each d-dimensional subdomain K_j, there is exactly one $\Lambda\Pi_\tau$ point $\mathrm{x}^{(j)}$. Assuming that after shaking, the random point stays inside K_j, i.e. $\xi^{(j)}(\rho) = \mathrm{x}^{(j)} + \rho\omega^{(j)} \in K_j$, one may try to exploit the smoothness of the integrand in case if the integrand f belongs to $\mathbf{W}^2(L; U^d)$.

Then, if $p(x)$ is a p.d.f., such that $\int_{U^d} p(\mathrm{x})d\mathrm{x} = 1$, then

$$\int_{K_j} p(\mathrm{x})d\mathrm{x} = p_j \leq \frac{c_1^{(j)}}{n},$$

where $c_1^{(j)}$ are constants. If d_j is the diameter of K_j, then

$$d_j = \sup_{x_1, x_2 \in K_j} |x_1 - x_2| \leq \frac{c_2^{(j)}}{n^{1/d}},$$

where $c_2^{(j)}$ are another constants.

In the particular case when the subintervals are with edge $1/m$ for all constants, we have $c_1^{(j)} = 1$ and $c_2^{(j)} = \sqrt{d}$. In each subdomain K_j the central point is denoted by $\mathrm{s}^{(j)}$, where $\mathrm{s}^{(j)} = (s_1^{(j)}, s_2^{(j)}, \dots, s_d^{(j)})$.

Suppose two random points $\xi^{(j)}$ and $\xi^{(j)'}$ are chosen, such that $\xi^{(j)}$ is selected during our procedure used in MCA-MSS-1. The second point $\xi^{(j)'}$ is chosen to be symmetric to $\xi^{(j)}$ according to the central point $s^{(j)}$ in each cube K_j. In such a way the number of random points is $2m^d$. One may calculate all function values $f(\xi^{(j)})$ and $f(\xi^{(j)'})$, for $j = 1, \dots, m^d$, and approximate the value of the integral in the following way:

$$I(f) \approx \frac{1}{2m^d} \sum_{j=1}^{2n} \left[f(\xi^{(j)}) + f(\xi^{(j)'}) \right]. \tag{9}$$

This estimate corresponds to MCA-MSS-2. We prove later on that this algorithm has an optimal rate of convergence for functions with bounded second derivatives, i.e. for functions $f \in \mathbf{W}^2(L; U^d)$, while the algorithm MCA-MSS-1 has an optimal rate of convergence for functions with bounded first derivatives: $f \in \mathbf{W}^1(L; U^d)$.

One can prove the following:

Theorem 1. *The quadrature formula (9) constructed above for integrands f from* $\mathbf{W}^2(L;U^d)$ *satisfies*

$$err(f,d) \le \tilde{c}_d' \, \|f\| \, n^{-\frac{1}{2}-\frac{2}{d}}$$

and

$$r(f,d) \le \tilde{c}_d'' \, \|f\| \, n^{-\frac{1}{2}-\frac{2}{d}},$$

where the constants $\tilde{c}_d'$ *and* $\tilde{c}_d''$ *do not depend on n.*

Proof. One can see that

$$\mathbf{E}\left\{\frac{1}{2m^d}\sum_{j=1}^{2n}\left[f(\xi^{(j)})+f(\xi^{(j)'})\right]\right\}=\int_{U^d} f(\mathrm{x})d\mathrm{x}.$$

For the fixed $\Lambda\Pi_\tau$ point $\mathrm{x}^{(j)} \in K_j$ one can use the d-dimensional Taylor formula to present the function $f(\mathrm{x}^{(j)})$ in K_j around the central point $\mathrm{s}^{(j)}$. Since $f \in \mathbf{W}^2(L;U^d)$, there exists a d-dimensional point $\eta^{(j)} \in K_j$ lying between $\mathrm{x}^{(j)}$ and $\mathrm{s}^{(j)}$ such that

$$f(\mathrm{x}^{(j)}) = f(\mathrm{s}^{(j)}) + \nabla f(\mathrm{s}^{(j)})\,(\mathrm{x}^{(j)}-\mathrm{s}^{(j)}) + \frac{1}{2}(\mathrm{x}^{(j)}-\mathrm{s}^{(j)})^T\,[D^2 f(\eta^{(j)})](\mathrm{x}^{(j)}-\mathrm{s}^{(j)}), \qquad (10)$$

where $\nabla f(\mathrm{x}) = \left[\frac{\partial f(\mathrm{x})}{\partial x_1},\ldots,\frac{\partial f(\mathrm{x})}{\partial x_d}\right]$ and $[D^2 f(\mathrm{x})] = \left[\frac{\partial^2 f(\mathrm{x})}{\partial x_i \partial x_k}\right]_{i,k=1}^{d}$. For simplicity the superscript of the argument (j) in the last two formulas is omitted assuming that the formulas are written for the j-th cube K_j. Now, we can write formula (10) at previously defined random points ξ and ξ' both belonging to K_j. In such a way we have

$$f(\xi) = f(\mathrm{s}) + \nabla f(\mathrm{s})\,(\xi-\mathrm{s}) + \frac{1}{2!}(\xi-\mathrm{s})^T[D^2 f(\eta)](\xi-\mathrm{s}), \qquad (11)$$

$$f(\xi') = f(\mathrm{s}) + \nabla f(\mathrm{s})\,(\xi'-\mathrm{s}) + \frac{1}{2!}(\xi'-\mathrm{s})^T[D^2 f(\eta')](\xi'-\mathrm{s}), \qquad (12)$$

where η' is another d-dimensional point lying between ξ' and s. Adding (11) and (12), we get

$$f(\xi)+f(\xi') = 2f(\mathrm{s}) + \frac{1}{2}\,\{(\xi-\mathrm{s})^T[D^2 f(\eta)](\xi-\mathrm{s}) + (\xi'-\mathrm{s})^T[D^2 f(\eta')]\,(\xi'-\mathrm{s})\}.$$

Because of the symmetry there is no term depending on the gradient $Df(\mathrm{s})$ in the previous formula. If we consider the variance $\mathbf{D}[f(\xi)+f(\xi')]$ taking into account that the variance of the constant $2f(\mathrm{s})$ is zero, then we get

$$\mathbf{D}[f(\xi) + f(\xi')] =$$

$$= \mathbf{D}\left\{\frac{1}{2}\left[(\xi-\mathrm{s})^T[D^2 f(\eta)](\xi-\mathrm{s})+(\xi'-\mathrm{s})^T[D^2 f(\eta')](\xi'-\mathrm{s})\right]\right\}$$

$$\leq \mathbf{E}\left\{\tfrac{1}{2}\left[(\xi-\mathrm{s})^T[D^2 f(\eta)](\xi-\mathrm{s})+(\xi'-\mathrm{s})^T[D^2 f(\eta')](\xi'-\mathrm{s})\right]\right\}^2.$$

Since $f \in \mathbf{W}^2(L;U^d)$, we can strengthen the last inequality if the terms $[D^2 f(\eta)]$ and $[D^2 f(\eta')]$ are substituted by the seminorm L (and removing front bracket) and the products $(\xi-\mathrm{s})^T$ $(\xi-\mathrm{s})$ and $(\xi'-\mathrm{s})^T$ $(\xi'-\mathrm{s})$ by the squared diameter of the subdomain K_j. Now we return back to the notation with superscript, taking into account that the above consideration is just for an arbitrary subdomain K_j. The variance can be estimated from above in the following way:

$$\mathbf{D}[f(\xi)+f(\xi')] \leq L^2 \sup_{x_1^{(j)},x_2^{(j)}} \left|x_1^{(j)}-x_2^{(j)}\right|^4 \leq L^2 (c_2^{(j)})^4 n^{-4/d}.$$

Now the variance of $\theta_n = \sum_{j=1}^{n} \theta^{(j)}$ can be estimated:

$$\mathbf{D}\theta_n = \sum_{j=1}^{n} p_j^2 \mathbf{D}[f(\xi)+f(\xi')] \leq \sum_{j=1}^{n} (c_1^{(j)})^2 n^{-2} L^2 (c_2^{(j)})^4 n^{-4/d}$$

$$\leq \left(L c_1^{(j)} c_2^{(j)2}\right)^2 n^{-1-4/d}. \tag{13}$$

Therefore, $r(f,d) \leq \tilde{c}_d'' \|f\| n^{-\frac{1}{2}-\frac{2}{d}}$. The application of Tchebycheff's inequality to the variance (13) yields

$$\varepsilon(f,d) \leq \tilde{c}_d' \|f\| n^{-\frac{1}{2}-\frac{2}{d}}$$

for the probable error ε, where $\tilde{c}_d' = \sqrt{2d}$, which concludes the proof.

One can see that the Monte Carlo algorithm MCA-MSS-2 has an optimal rate of convergence for functions with continuous and bounded second derivative [3]. This means that the rate of convergence ($n^{-\frac{1}{2}-\frac{2}{d}}$) cannot be improved for the functional class $\mathbf{W}^2$ in the class of the randomized algorithms $\mathcal{A}^{\mathcal{R}}$.

Note that both MCA-MSS-1 and MCA-MSS-2 have one control parameter, that is, the radius ρ of the *sphere of shaking*. At the same time, to be able to efficiently use this control parameter, one should increase the computational complexity. The problem is that after *shaking* the random point may leave the multidimensional subdomain. That is why after each such a procedure, one should be checking if the random point is still in the same subdomain. It is clear that the procedure of checking if a random point is inside the given domain is a computationally expensive procedure when one has a large number of points. A small modification of MCA-MSS-2 algorithm allows to overcome this difficulty. If we just generate a random point $\xi^{(j)} \in K_j$ uniformly distributed inside K_j and after that take the symmetric point $\xi^{(j)'}$ according to the central point $s^{(j)}$, then this procedure will simulate the algorithm MCA-MSS-2. Such a completely randomized approach simulates algorithm MCA-MSS-2, but the *shaking* is with different radiuses ρ in each subdomain. We call this algorithm MCA-MSS-2-S, because this approach looks like the stratified symmetrised Monte Carlo. Obviously, MCA-MSS-2-S is less expensive than MCA-MSS-2, but there is not such a control parameter like the radius ρ, which can be considered as a parameter randomly chosen in each subdomain K_j.

It is important to notice that all three algorithms MCA-MSS-1, MCA-MSS-2 and MCA-MSS-2-S have optimal (unimprovable) rate of convergence for the corresponding functional classes, that is, MCA-MSS-1 is optimal in $\mathbf{W}^1(L;U^d)$ and both MCA-MSS-2 and MCA-MSS-2-S are optimal in $\mathbf{W}^2(L;U^d)$.

We also consider the known Owen nested scrambling algorithm [19] for which it is proved that the rate of convergence is $n^{-3/2}(log\ n)^{(d-1)/2}$, which is very good but still not optimal even for integrands in $\mathbf{W}^1(L;U^d)$. One can see that if the *logarithmic function* from the estimate can be omitted, then the rate will become optimal. Let us mention that it is still not proven that the above estimate is exact, that is, we do not know if the *logarithm* can be omitted. It should be mentioned that the proved convergence rate for the Owen nested scrambling algorithm improves significantly the rate for the unscrambled nets, which is $n^{-1}(log\ n)^{d-1}$. That is why it is important to compare numerically our algorithms MCA-MSS with the Owen nested scrambling. The idea of Owen nested scrambling is based on randomization of a single digit at each iteration. Let $x^{(i)} = (x_{i,1}, x_{i,2}, \ldots, x_{i,s}),\ i = 1, \ldots, n$ be quasi-random numbers in $[0,1)^s$, and let $z^{(i)} = (z_{i,1}, z_{i,2}, \ldots, z_{i,s})$ be the scrambled version of the point $x^{(i)}$. Suppose that each $x_{i,j}$ can be represented in base b as $x_{i,j} = (0.x_{i1,j}\ x_{i2,j} \ldots x_{iK,j} \ldots)_b$ with K being the number of digits to be scrambled. Then nested scrambling proposed by Owen [19, 20] can be defined as follows: $z_{i1,j} = \pi_{\bullet}(x_{i1,j})$, and $z_{il,j} = \pi_{\bullet x_{i1,j} x_{i2,j} \ldots x_{il-1,j}}(x_{il,j})$, with independent permutations $\pi_{\bullet x_{i1,j} x_{i2,j} \ldots x_{il-1,j}}$ for $l \geq 2$. Of course, (t,m,s)-net remains (t,m,s)-net under nested scrambling. However, nested scrambling requires b^{l-1} permutations to scramble the l-th digit. Owen scrambling (nested scrambling), which can be applied to all (t,s)-sequences, is powerful; however, from the implementation point of view, nested scrambling or so-called path-dependent permutations require a considerable amount of bookkeeping and lead to more problematic implementation. There are various versions of scrambling methods based on digital permutation, and the differences among those methods are based on the definitions of the π_l's. These include Owen nested scrambling [19, 20], Tezuka's generalized Faure sequences [29] and Matousek's linear scrambling [17].

5 Case Study: Variance-Based Sensitivity Analysis of the Unified Danish Eulerian Model

The input data for the sensitivity analysis performed in this paper has been obtained during runs of a large-scale mathematical model for remote transport of air pollutants (UNI-DEM, [33]). The model enables us to study concentration variations in time of a high number of air pollutants and other species over a large geographical region (4,800 $\times$ 4,800 km), covering the whole of Europe, the Mediterranean and some parts of Asia and Africa. Such studies are important for environmental protection, agriculture and health care. The model presented as a system of partial differential equations describes the main processes in the atmosphere including photochemical processes between the studied species, the emissions and the quickly changing meteorological conditions. Both nonlinearity and stiffness of the equations are mainly introduced when modeling chemical reactions [33]. The chemical scheme used in the model is the well-known condensed CBM-IV (Carbon Bond Mechanism). Thus, the motivation to choose UNI-DEM is that it is one of the models of atmospheric chemistry, where the chemical processes are taken into account in a very accurate way.

This large and complex task is not suitable for direct numerical treatment. For the purpose of numerical solution, it is split into submodels, which represent the main physical and chemical processes. The sequential splitting [16] is used in the production version of the model, although other splitting methods have also been considered and implemented in some experimental versions [4, 5]. Spatial and time discretization makes each of the above submodels a huge computational task, challenging for the most powerful supercomputers available nowadays. That is why parallelization has always been a key point in the computer implementation of DEM since its very early stages.

Our main aim here is to study the sensitivity of the ozone concentration according to the rate variation of some chemical reactions. We consider the chemical rates to be the input parameters and the concentrations of pollutants to be the output parameters.

6 Numerical Results and Discussion

Some numerical experiments are performed to study experimentally various properties of the algorithms. We are interested in both smooth and non-smooth integrands. The reason to consider both cases is that we deal with many different output functions using the UNI-DEM model. Formally the output functions should have enough smoothness, because the solution has bounded second derivatives by definition. Nevertheless, some functions of concentrations that depend on photochemical reactions in the air have computational irregularities. It means that the derivative of

Table 1 Relative error and computational time for numerical integration of a smooth function ($S(f_2) \approx 0.10897$)

n	SFMT Rel. error	SFMT Time (s)	Sobol QMCA Rel. error	Sobol QMCA Time (s)	Owen scrambling Rel. error	Owen scrambling Time (s)	MCA-MSS-1 ρ $\times 10^3$	MCA-MSS-1 Rel. error	MCA-MSS-1 Time (s)
10^2	0.0562	0.002	0.0365	<0.001	0.0280	0.001	3.9	0.0363	0.001
							13	0.0036	0.001
10^3	0.0244	0.004	0.0023	0.001	0.0016	0.001	1.9	0.0038	0.010
							6.4	0.0019	0.010
10^4	0.0097	0.019	0.0009	0.002	0.0003	0.003	0.8	0.0007	0.070
							2.8	0.0006	0.065

the function is very high by modulo and it causes computational difficulties—the function behaves as a non-smooth function.

The expectations based on theoretical results are that for non-smooth functions MCA-MSS algorithms based on the *shaking* procedures outperform the QMC even for relatively low dimensions. It is also interesting to observe how behave the randomized QMC based on scrambled Sobol sequences.

For our numerical tests we use the following non-smooth integrand:

$$f_1(x_1,x_2,x_3,x_4) = \sum_{i=1}^{4} |(x_i - 0.8)^{-1/3}|, \qquad (14)$$

for which even the first derivative does not exist. Such kinds of applications appear also in some important problems in financial mathematics. The referent value of the integral $S(f_1)$ is approximately equal to 7.22261. To make a comparison we also consider an integral with a smooth integrand:

$$f_2(x_1,x_2,x_3,x_4) = x_1\, x_2^2\, \mathbf{e}^{x_1 x_2} \sin x_3 \cos x_4. \qquad (15)$$

The second integrand (15) is a function $f_2 \in C^\infty(U^d)$ with a referent value of the integral $S(f_2)$ approximately equal to 0.10897. The integration domain in both cases is $U^4 = [0,1]^4$.

Some results from the numerical integration tests with a smooth (15) and a non-smooth (14) integrand are presented in Tables 1 and 2, respectively. As a measure of the efficiency of the algorithms, both the relative error (defined as the absolute error divided by the referent value) and computational time are shown. For generating Sobol quasi-random sequences, the algorithm with Gray code implementation [1] and sets of direction numbers proposed by Joe and Kuo [11] are used. The MCA-MSS-1 algorithm [6] involves generating random points uniformly distributed on a sphere with radius ρ. One of the best available random number generators, SIMD-oriented Fast Mersenne Twister (SFMT) [21, 32] 128-bit pseudorandom number generator of period $2^{19937}-1$ has been used to generate the required random points. SFMT algorithm is a very efficient implementation of the plain Monte

Table 2 Relative error and computational time for numerical integration of a non-smooth function ($S(f_1) \approx 7.22261$)

n	SFMT Rel. error	SFMT Time (s)	Sobol QMCA Rel. error	Sobol QMCA Time (s)	Owen scrambling Rel. error	Owen scrambling Time (s)	MCA-MSS-1 ρ $\times 10^3$	MCA-MSS-1 Rel. error	MCA-MSS-1 Time (s)
10^3	0.0010	0.011	0.0027	0.001	0.0021	0.002	1.9	0.0024	0.020
							6.4	0.0004	0.025
7.10^3	0.0009	0.072	0.0013	0.009	0.0003	0.011	1.0	0.0004	0.110
							3.4	0.0005	0.114
3.10^4	0.0005	0.304	0.0003	0.032	0.0003	0.041	0.6	0.0001	0.440
							1.9	0.0002	0.480
5.10^4	0.0007	0.513	0.0002	0.053	2e-05	0.066	0.4	7e-05	0.775
							1.4	0.0001	0.788

Carlo method [23]. The radius ρ depends on the integration domain, number of samples and minimal distance between Sobol deterministic points δ. We observed experimentally that the behavior of the relative error of numerical integration is significantly influenced by the fixed radius of spheres. That is why the values of the radius ρ are presented according to the number of samples n used in our experiments, as well as to a fixed coefficient, *radius coefficient* $\kappa = \rho/\delta$. The latter parameter gives the ratio of the radius to the minimal distance between Sobol points. The code of scrambled quasi-random sequences used in our studies is taken from the collection of NAG C Library [31]. This implementation of scrambled quasi-random sequences is based on TOMS Algorithm 823 [10]. In the implementation of the scrambling, there is a possibility to make a choice of three methods of scrambling: the first is a restricted form of Owen scrambling [19], the second is based on the method of Faure and Tezuka [8] and the last method combines the first two (it is referred to as a *combined* approach).

Random points for the MCA-MSS-1 algorithm have been generated using the original Sobol sequences and modeling a random direction in d-dimensional space. The computational time of the calculations with pseudorandom numbers generated by SFMT (see columns labeled as *SFMT* and *MCA-MSS* in Tables 1 and 2) has been estimated for all 10 algorithm runs.

Comparing the results in Tables 1 and 2 one observes that:

- All algorithms under consideration are efficient and converge with the expected rate of convergence.
- In the case of *smooth functions*, the Sobol algorithm is better than SFMT (the relative error is up to 10 times smaller than for SFMT).
- The scrambled QMC and MCA-MSS-1 are much better than the classical Sobol algorithm; in many cases even the simplest *shaking* algorithm MCA-MSS-1 gives a higher accuracy than the scrambled algorithm.
- In the case of *non-smooth functions*, SFMT algorithm implementing the plain Monte Carlo method is better than the Sobol algorithm for relatively small samples (n).

Table 3 Relative error and computational time for numerical integration of a smooth function ($S(f) \approx 0.10897$)

No. of points n	Sobol QMCA		MCA-MSS-1			MCA-MSS-2		MCA-MSS-2-S	
(No. of double points $2n$)	Rel. error	Time (s)	ρ $\times 10^3$	Rel. error	Time (s)	Rel. error	Time (s)	Rel. error	Time (s)
2^9	0.0059	< 0.001	2.1	0.0064	0.009	0.0033	0.010	0.0016	0.005
(2×2^9)			6.4	0.0061	0.010	0.0032	0.010		
2^{10}	0.0035	0.002	1.9	0.0037	0.010	9e-05	0.020	0.0002	0.007
(2×2^{10})			6.4	0.0048	0.010	0.0002	0.020		
2^{16}	2e-05	0.027	0.4	3e-05	1.580	7 e-06	1.340	9e-06	0.494
(2×2^{16})			1.2	0.0001	1.630	5e-06	1.380		

- In the case of *non-smooth functions*, our Monte Carlo *shaking* algorithm MCA-MSS-1 gives similar results as the scrambled QMC; for several values of n, we observe advantages for MCA-MSS-1 in terms of accuracy.
- Both MCA-MSS-1 and scrambled QMC are better than SFMT and Sobol quasi MC algorithm in the case of non-smooth functions.

Another observation is that for the chosen integrands the scrambling algorithm does not outperform the algorithm with the original Sobol points, but the scrambled algorithm and Monte Carlo algorithm MCA-MSS-1 are more stable with respect to relative errors for relatively small values of n.

In Table 3 we compare Sobol QMCA with MCA-MSS-2 and MCA-MSS-2-S, as well as with simplest *shaking* algorithm MCA-MSS-1. The results show that the simplest *shaking* algorithm MCA-MSS-1 gives relative errors similar to errors of the Sobol QMCA, which is expected since the $\Lambda\Pi_\tau$ Sobol sequences are already quite well distributed. That is why one should not expect improvement for a very smooth integrand. But the *symmetrised shaking* algorithm MCA-MSS-2 improves the relative error. The effect of this improvement is based on the fact that the second derivatives of the integrand exists, they are bounded and the construction of the MCA-MSS-2 algorithm gives a better convergence rate of order $O(n^{-1/2-2/d})$. The same convergence rate has the algorithm MCA-MSS-2-S, but the latter one does not allow to control the value of the radius of *shaking*. As expected MCA-MSS-2-S gives better results than MCA-MSS-1. The relative error obtained by MCA-MSS-2 and MCA-MSS-2-S are of the same magnitude (see Table 3). The advantage of MCA-MSS-2-S is that its computational complexity is much smaller. A comparison of the relative error and computational complexity for different values of n is presented in Table 4. To have a fair comparison we have to consider again a smooth function (15). The observation is that MCA-MSS-2-S algorithm outperforms the simplest *shaking* algorithm MCA-MSS-1 in terms of relative error and complexity.

After testing the algorithms under consideration on the *smooth* and *non-smooth functions*, we studied the efficiency of the algorithms on *real-life* functions obtained after running UNI-DEM. Polynomials of 4th degree with 35 unknown coefficients are used to approximate the *mesh functions* containing the model outputs.

Table 4 Relative error and computational time for numerical integration of a smooth function ($S(f) \approx 0.10897$) (comparison between MCA-MSS-1 and MCA-MSS-2-S algorithms)

	Sobol QMCA		MCA-MSS-1			MCA-MSS-2-S	
	Rel.	Time	ρ	Rel.	Time	Rel.	Time
n	$\times 10^3$	error	(s)	error	(s)	error	(s)
2×4^4	0.0076	< 0.001	2.1	0.0079	< 0.001	0.0016	0.005
(512)			6.4	0.0048	< 0.001		
2×6^4	0.0028	0.001	1.2	0.0046	0.030	0.0004	0.009
(2,592)			4.1	0.0046	0.030		
2×8^4	0.0004	0.004	0.9	0.0008	0.090	0.0002	0.025
(8,192)			2.9	0.0024	0.090		
2×10^4	0.0002	0.008	0.6	0.0001	0.220	5e-05	0.070
(20,000)			2.0	0.0013	0.230		
2×13^4	0.0001	0.022	0.4	0.0001	0.630	4e-06	0.178
(57,122)			1.2	0.0007	0.640		
2×14^4	5e-06	0.029	0.4	1e-05	0.860	1e-05	0.237
(76,832)			1.2	0.0005	0.880		
2×15^4	8e-06	0.036	0.4	0.0001	1.220	9e-07	0.313
(101,250)			1.2	0.0005	1.250		

We use various values of the number of points that corresponds to situations when one needs to compute the sensitivity measures with different accuracy. We have computed results for g_0 (g_0 is the integral over the integrand $g(x) = f(x) - c$, $f(x)$ is the approximate model function of UNI-DEM and c is a constant obtained as a Monte Carlo estimate of f_0, [28]), the total variance **D** as well as total sensitivity indices $S_i^{tot}, i = 1,2,3$. The above-mentioned parameters are presented in Table 5. Table 5 presents the results obtained for a relatively low sample size $n = 6,600$.

One can notice that for most of the sensitivity parameters, the simplest *shaking* algorithm MCA-MSS-1 outperforms the scrambled Sobol sequences, as well as the algorithm based on the $\Lambda\Pi_\tau$ Sobol sequences in terms of accuracy. For higher values of sample sizes this effect is even stronger.

One can clearly observe that the simplest *shaking* algorithm MCA-MSS-1 based on modified Sobol sequences improves the error estimates for non-smooth integrands. For smooth functions modified algorithms MCA-MSS-2 and MCA-MSS-2-S give better results than MCA-MSS-1. Even for relatively large radiuses ρ the results are good in terms of accuracy. The reason is that centers of spheres are very well uniformly distributed by definition. So that even for large values of radiuses of *shaking* the generated random points continue to be well distributed. We should stress on the fact that for relatively low number of points ($< 1,000$) the algorithm based on modified Sobol sequences gives results with a high accuracy.

Table 5 Relative error (in absolute value) and computational time for estimation of sensitivity indices of input parameters using various Monte Carlo and quasi-Monte Carlo approaches ($n = 6,600, c \approx 0.51365, \delta \approx 0.08$)

Estimated quantity	Sobol QMCA	Owen scrambling	MCA-MSS-1 ρ	MCA-MSS-1 Rel. error
g_0	1e-05	0.0001	0.0007	0.0001
			0.007	6e-05
D	0.0007	0.0013	0.0007	0.0003
			0.007	0.0140
S_1^{tot}	0.0036	0.0006	0.0007	0.0009
			0.007	0.0013
S_2^{tot}	0.0049	6e-05	0.0007	2e-05
			0.007	0.0034
S_3^{tot}	0.0259	0.0102	0.0007	0.0099
			0.007	0.0211

7 Conclusions

A comprehensive theoretical and experimental study of the Monte Carlo algorithm MCA-MSS-2 based on *symmetrised shaking* of Sobol sequences has been done. The algorithm combines properties of two of the best available approaches—Sobol QMC integration and a high-quality SFMT pseudorandom number generator. It has been proven that this algorithm has an optimal rate of convergence for functions with continuous and bounded second derivatives in terms of probability and mean square error.

A comparison with the scrambling approach, as well as with the Sobol QMC algorithm and the algorithm using SFMT generator, has been provided for numerical integration of smooth and non-smooth integrands. The algorithms mentioned above are tested numerically also for computing sensitivity measures for UNI-DEM model to study sensitivity of ozone concentration according to variation of chemical rates. All algorithms under consideration are efficient and converge with the expected rate of convergence. It is important to notice that the Monte Carlo algorithm MCA-MSS-2 based on modified Sobol sequences when *symmetrised shaking* is used has a unimprovable rate of convergence and gives reliable numerical results.

Acknowledgements The research reported in this paper is partly supported by the Bulgarian NSF Grants DTK 02/44/2009 and DMU 03/61/2011.

References

1. Antonov, I., Saleev, V.: An economic method of computing LP_τ-sequences. USSR Comput. Math. Math. Phys. **19**, 252–256 (1979)
2. Bradley, P., Fox, B.: Algorithm 659: implementing Sobol's quasi random sequence generator. ACM Trans. Math. Software **14**(1), 88–100 (1988)
3. Dimov, I.T.: Monte Carlo Methods for Applied Scientists. World Scientific, London (2008)
4. Dimov, I.T., Farago, I., Havasi, A., Zlatev, Z.: Operator splitting and commutativity analysis in the Danish Eulerian Model. Math. Comput. Simulat. **67**, 217–233 (2004)
5. Dimov, I.T., Ostromsky, Tz., Zlatev, Z.: Challenges in using splitting techniques for large-scale environmental modeling. In: Farago, I., Georgiev, K., Havasi, A. (eds.) Advances in Air Pollution Modeling for Environmental Security. NATO Science Series 54, pp. 115–132. Springer, Dordrecht (2005)
6. Dimov, I.T., Georgieva, R.: Monte Carlo method for numerical integration based on Sobol's sequences. In: Dimov, I., Dimova, S., Kolkovska, N. (eds.) Numerical Methods and Applications. Lecture Notes in Computer Science 6046, pp. 50–59. Springer, Berlin (2011)
7. Dimov, I.T., Georgieva, R., Ostromsky, Tz., Zlatev, Z.: Advanced algorithms for multidimensional sensitivity studies of large-scale air pollution models based on Sobol sequences. Comput. Math. Appl. **65**(3), 338–351 (2013). doi: 10.1016/j.camwa.2012.07.005
8. Faure, H., Tezuka, S.: Another random scrambling of digital (t,s)-sequences. In: (Fang, K., Hickernell, F., Niederreiter, H. (eds.) Monte Carlo and Quasi-Monte Carlo Methods. Springer, Berlin, pp. 242–256 (2000)
9. Homma, T., Saltelli, A.: Importance measures in global sensitivity analysis of nonlinear models. Reliab. Eng. Syst. Saf. **52**, 1–17 (1996)
10. Hong, H., Hickernell, F.: Algorithm 823: implementing scrambled digital sequences. ACM Trans. Math. Software **29**(2), 95–109 (2003)
11. Joe, S., Kuo, F.: Constructing Sobol' sequences with better two-dimensional projections. SIAM J. Sci. Comput. **30**, 2635–2654 (2008)
12. Kucherenko, S., Feil, B., Shah, N., Mauntz, W.: The identification of model effective dimensions using global sensitivity analysis. Reliab. Eng. Syst. Saf. **96**, 440–449 (2011)
13. L'Ecuyer, P., Lecot, C., Tuffin, B.: A randomized quasi-Monte Carlo simulation method for Markov chains. Oper. Res. **56**(4), 958–975 (2008)
14. L'Ecuyer, P., Lemieux, C.: Recent advances in randomized quasi-Monte Carlo methods. In: Dror, M., L'Ecuyer, P., Szidarovszki, F. (eds.) Modeling Uncertainty: An Examination of Stochastic Theory, Methods, and Applications, pp. 419–474. Kluwer, Boston (2002)
15. Levitan, Y., Markovich, N., Rozin, S., Sobol, I.: On quasi-random sequences for numerical computations. USSR Comput. Math. Math. Phys. **28**(5), 755–759 (1988)
16. Marchuk, G.I.: Mathematical Modeling for the Problem of the Environment, Studies in Mathematics and Applications, No. 16. North-Holland, Amsterdam (1985)
17. Matousek, J.: On the L_2-discrepancy for anchored boxes. J. Complex. **14**, 527–556 (1998)
18. Niederreiter, H.: Low-discrepancy and low-dispersion sequences. J. Number Theor. **30**, 51–70 (1988)
19. Owen, A.: Randomly permuted (t,m,s)-nets and (t,s)-sequences. In: Niederreiter, H., Shiue, P.J.-S. (eds.) Monte Carlo and Quasi-Monte Carlo Methods in Scientific Computing. Lecture Notes in Statistics, vol. 106, pp. 299–317. Springer, New York (1995)
20. Owen, A.: Variance and discrepancy with alternative scramblings. ACM Trans. Comput. Logic. **V**, 1–16 (2002)
21. Saito, M., Matsumoto, M.: SIMD-oriented Fast Mersenne Twister: a 128-bit pseudorandom number generator. In: Keller, A., Heinrich, S., Niederreiter, H. (eds.) Monte Carlo and Quasi-Monte Carlo Methods 2006, pp. 607–622. Springer, Heidelberg (2008)
22. Saltelli, A., Tarantola, S., Campolongo, F., Ratto, M.: Sensitivity Analysis in Practice: A Guide to Assessing Scientific Models. Halsted Press, New York (2004)
23. Sobol, I.M.: Monte Carlo Numerical Methods. Nauka, Moscow (1973) (in Russian)

24. Sobol, I.M.: On the systematic search in a hypercube. SIAM J. Numer. Anal. **16**, 790–793 (1979)
25. Sobol, I.M.: On quadratic formulas for functions of several variables satisfying a general Lipschitz condition. USSR Comput. Math. Math. Phys. **29**(6), 936–941 (1989)
26. Sobol, I.M.: Global sensitivity indices for nonlinear mathematical models and their Monte Carlo estimates. Math. Comput. Simulat. **55**(1–3), 271–280 (2001)
27. Sobol, I., Asotsky, D., Kreinin, A., Kucherenko, S.: Construction and comparison of high-dimensional Sobol' generators. Wilmott J. **2011**, 67–79 (2011)
28. Sobol, I., Myshetskaya, E.: Monte Carlo estimators for small sensitivity indices. Monte Carlo Meth. Appl. **13**(5–6), 455–465 (2007)
29. Tezuka, S.: Uniform Random Numbers, Theory and Practice. Kluwer Academic Publishers, Boston; IBM Japan, Tokyo (1995)
30. Weyl, H.: Ueber die Gleichverteilung von Zahlen mod Eins. Math. Ann. **77**(3), 313–352 (1916)
31. The NAG C Library (www.nag.co.uk/numeric/CL/CLdescription.asp.)
32. SIMD-oriented Fast Mersenne Twister (SFMT) (http://www.math.sci.hiroshima-u.ac.jp/~m-mat/MT/SFMT/).
33. Zlatev, Z., Dimov, I.T.: Computational and Numerical Challenges in Environmental Modelling. Elsevier, Amsterdam (2006)
34. Zlatev, Z., Dimov, I.T., Georgiev, K.: Three-dimensional version of the Danish Eulerian model. Z. Angew. Math. Mech. **76**(S4), 473–476 (1996)

Structures and Waves in a Nonlinear Heat-Conducting Medium

Stefka Dimova, Milena Dimova, and Daniela Vasileva

Abstract This paper is an overview of the main contributions of a Bulgarian team of researchers to the problem of finding the possible structures and waves in the open nonlinear heat-conducting medium, described by a reaction–diffusion equation. Being posed and actively worked out by the Russian school of A.A. Samarskii and S.P. Kurdyumov since the seventies of the last century, this problem still contains open and challenging questions.

Keywords Nonlinear heat-conducting medium • Self-organization • Reaction-diffusion equation • Self-similar solutions • Blow-up • Finite element method.

Mathematics Subject Classification (2010): 35K57, 35B40, 35B44, 65N30, 65M60, 65P40

1 Introduction

A very general form of the model of heat structures reads as follows:

$$u_t = \sum_{i=1}^{N} (k_i(u) u_{x_i})_{x_i} + Q(u),\ t > 0,\ x \in \mathbb{R}^N, \tag{1}$$

S. Dimova (✉)
Faculty of Mathematics and Informatics, University of Sofia,
5 James Bourchier Blvd., 1164 Sofia, Bulgaria
e-mail: dimova@fmi.uni-sofia.bg

M. Dimova • D. Vasileva
Institute of Mathematics and Informatics, Bulgarian Acad. Sci.,
Acad. G.Bonchev Str., bl. 8, 1113 Sofia, Bulgaria
e-mail: mkoleva@math.bas.bg; vasileva@math.bas.bg

O.P. Iliev et al. (eds.), *Numerical Solution of Partial Differential Equations: Theory, Algorithms, and Their Applications*, Springer Proceedings in Mathematics & Statistics 45, DOI 10.1007/978-1-4614-7172-1_9,

where the heat conductivity coefficients $k_i(u) \geq 0$ and the heat source $Q(u) \geq 0$ are nonlinear functions of the temperature $u(t,x) \geq 0$.

Models such as (1) are studied by many researchers in various contexts. A part of this research is devoted to semilinear equations: $k_i(u) \equiv 1$, $Q(u) = \lambda e^u$ (Frank–Kamenetskii equation) and $Q(u) = u^\beta$, $\beta > 1$. After the pioneer work of Fujita [28], these equations and some generalizations of theirs are studied intensively by many authors, including J. Bebernes, A. Bressan, H. Brezis, D. Eberly, A. Friedman, V.A. Galaktionov, I.M. Gelfand, M.A. Herrero, R. Kohn, L.A. Lepin, S.A. Posashkov, A.A. Samarskii, J.L. Vázquez, and J.J. L. Velázquez. The book [6] contains a part of these investigations and a large bibliography.

The quasilinear equation is studied by D.G. Aronson, A. Friedman, H.A. Levine, S. Kaplan, L.A. Peletier, J.L. Vázquez, and others. The contributions of the Russian school are significant. The unusual localization effect of the blow-up boundary regimes is discovered by numerical experiment in the work of A.A. Samarskii and M.I. Sobol in 1963 [49]. The problem of localization for quasilinear equations with a source is posed by S.P. Kurdyumov [42] in 1974. The works of I.M. Gelfand, A.S. Kalashnikov, and the scientists of the school of A.A. Samarskii and S.P. Kurdyumov are devoted to the challenging physical and mathematical problems, related with this model and its generalizations. Among them are localization in space of the process of burning, different types of blow-up, and arising of structures—traveling and standing waves, complex structures with varying degrees of symmetry. The combination of the computational experiment with the progress in the qualitative and analytical methods of the theory of ordinary and partial differential equations, the Lie and the Lie–Bäcklund group theory, has been crucial for the success of these investigations. The book [52] contains many of these results, achieved till 1986; in the review [32], there are citations of later works. A special part of these investigations is devoted to finding and studying different kinds of self-similar and invariant solutions of (1) with power nonlinearities:

$$k_i(u) = u^{\sigma_i}, \ \ Q(u) = u^\beta. \tag{2}$$

This choice is suggested by the following reasoning:

First, such temperature dependencies are usual for many real processes [5,54,57]. For example, when $\sigma_i = \sigma = 2.5$, $\beta \leq 5.2$, (1) describes thermonuclear combustion in plasma in the case of electron heat conductivity, the parameters $\sigma = 0$, $2 \leq \beta \leq 3$ correspond to the models of autocatalytic processes with diffusion in the chemical reactors, $\sigma \approx 6.5$ corresponds to the radiation heat conductivity of the high-temperature plasma in the stars, and so on.

Second, it is shown in [25] that in the class of power functions, the symmetry of (1) is maximal in some sense—the equation admits a rich variety of invariant solutions. In general, almost all of the dissipative structures known so far are invariant or partially invariant solutions of nonlinear equations. The investigations of the dissipative structures provide reasons to believe that the invariant solutions describe the attractors of the dissipative structures' evolution and thus they characterize important internal properties of the nonlinear dissipative medium.

Third, this rich set of invariant solutions of (1) with power nonlinearities is necessary for the successful application of the methods for investigating the same equation in the case of more general dependencies $k_i(u)$, $Q(u)$. By using the methods of operator comparison [29] and stationary states [34], it is possible to analyze the properties of the solutions (such as localization, blow-up, asymptotic behavior) of whole classes general nonlinear equations. The method of approximate self-similar solutions [33], developed in the works of A.A. Samarskii and V.A. Galaktionov, makes it possible to put in accordance with such general equations some other, basic equations. The latter could have invariant solutions even if the original equations do not have such. Moreover, the original equations may significantly differ from the basic equations, and nevertheless their solutions tend to the invariant solutions of the basic equations at the asymptotic stage.

Finally, in the case of power coefficients, the dissipation and the source are coordinated so that complex structures arise; moreover, a spectrum of structures, burning consistently, occurs.

Below we report about the main contributions of the Bulgarian research team, S.N. Dimova, M.S. Kaschiev, M.G. Koleva, D.P. Vasileva, and T.P. Chernogorova, to the problem of finding the possible evolution patterns in the heat-conducting medium, described by the reaction–diffusion Eq. (1), (2). The outline of this paper is as follows. The main notions, needed further, are introduced in Sect. 2 on the simplest and the most-studied radially symmetric case. The specific peculiarities of the numerical methods, developed and applied to solve the described problems, are systematized in Sect. 3. A brief report on the main achievements of the team is made in Sect. 4. Section 5 contains some open problems.

2 The Radially Symmetric Case, the Main Notions

Let us introduce the main notions to be used further on the Cauchy problem for (1) with initial data:

$$u(0,x) = u_0(x) \geq 0,\ x \in \mathbb{R}^N,\ \sup u_0(x) < \infty.$$

This problem could have global or blow-up solutions. The global in time solution is defined and bounded in $\mathbb{R}^N$ for every t. *The unbounded (blow-up) solution* is defined in $\mathbb{R}^N$ on a finite interval $[0,T_0)$, moreover

$$\overline{\lim}_{t\to T_0^-} \sup_{x\in\mathbb{R}^N} u(t,x) = +\infty.$$

The time T_0 is called *blow-up time*.

The unbounded solution of the Cauchy problem with finite support initial data $u_0(x)$ is called *localized (in a strong sense)*, if the set

$$\Omega_L = \{x \in \mathbb{R}^N : u(T_0^-, x) := \overline{\lim}_{t \to T_0^-} u(t,x) > 0\}$$

is bounded in $\mathbb{R}^N$. The set Ω_L is called *localization region*. The solution localized in a strong sense grows infinitely for $t \to T_0^-$ in a finite region

$$\omega_L = \{x \in \mathbb{R}^N : u(T_0^-, x) = \infty\}$$

in general different from Ω_L.

If for $k_i(u) \equiv 0$ the condition

$$\int_1^\infty \frac{du}{Q(u)} < +\infty \qquad (3)$$

holds, then the solution of the Cauchy problem is unbounded [52]. The heating of the medium happens in a blow-up regime; moreover, the blow-up time of every point of the medium is different, depending on its initial temperature.

If for $Q(u) \equiv 0$ the condition

$$\int_0^1 \frac{k_i(u)}{u} du < +\infty, \quad i = 1, 2, \ldots, N, \qquad (4)$$

holds, then a **finite speed** of heat propagation takes place for a finite support initial perturbation in an absolutely cold medium [52].

In the case (2) of power nonlinearities, it is sufficient to have $\sigma_i > 0, \beta > 1$ for the conditions (3) and (4) to be satisfied. Then $k_i(0) = 0$ and (1) degenerates. In general it has a generalized solution, which could have discontinuous derivatives on the surface of degeneration $\{u = 0\}$.

2.1 The Basic Blow-Up Regimes

The basic blow-up regimes will be explained on the radially symmetric version of the Cauchy problem for (1):

$$u_t = \frac{1}{x^{N-1}} (x^{N-1} u^\sigma u_x)_x + u^\beta, \quad x \in \mathbb{R}_+^1, \quad t > 0, \quad \sigma > 0,\ \beta > 1, \qquad (5)$$

$$u_x(t,0) = 0, \quad u(0,x) = u_0(x) \geq 0,\ 0 \leq x < l, \quad u_0(x) \equiv 0, \quad x \geq l. \qquad (6)$$

If $u_0(x)$ satisfies the additional conditions $u_0(x) \in C(\mathbb{R}_+^1)$, $(u_0^\sigma u_0')(0) = 0$, there exists unique local (in time) generalized solution $u = u(t,x)$ of problem (5)–(6), which is a nonnegative continuous function in $\mathbb{R}_+^1 \times (0,T)$, where $T \in (0,\infty]$ is the finite or infinite time of existence of the solution (see the bibliography in the review [39]). Moreover $u(t,x)$ is a classical solution in a vicinity of every point (t,x), where $u(t,x)$ is strictly positive. It could not have the necessary smoothness at the points of degeneracy, but the heat flux $-x^{N-1}u^{\sigma}u_x$ must be continuous. It means that $u^{\sigma}u_x = 0$ everywhere $u = 0$. Equation (5) admits a *self-similar solution (s.-s.s.)* [52]:

$$u_s(t,x) = \varphi(t)\theta_s(\xi) = \left(1 - \frac{t}{T_0}\right)^{\frac{-1}{\beta-1}} \theta_s(\xi), \tag{7}$$

$$\xi = x/\psi(t) = x/\left(1 - \frac{t}{T_0}\right)^{\frac{m}{\beta-1}}, \quad m = \frac{\beta - \sigma - 1}{2}. \tag{8}$$

The s.-s.s. corresponds to initial data $u_s(0,x) = \theta_s(x)$. The function $\varphi(t)$ determines the amplitude of the solution. The *self-similar function (s.-s.f.)* $\theta_s(\xi) \geq 0$ determines the space-time structure of the s.-s.s. (7). This function satisfies the degenerate ordinary differential equation in $\mathbb{R}^1_+$:

$$L(\theta_s) \equiv -\frac{1}{\xi^{N-1}}(\xi^{N-1}\theta_s^{\sigma}\theta_s')' + \frac{\beta-\sigma-1}{2(\beta-1)T_0}\xi\theta_s' + \frac{1}{(\beta-1)T_0}\theta_s - \theta_s^{\beta} = 0 \tag{9}$$

and the boundary conditions:

$$\theta_s'(0) = 0, \;\; \theta_s(\infty) = 0, \;\; \theta_s^{\sigma}\theta_s'(\xi_0) = 0, \text{ if } \; \theta_s(\xi_0) = 0. \tag{10}$$

Equation (9) has two constant solutions: $\theta_s(\xi) \equiv \theta_H = (T_0(\beta-1))^{\frac{-1}{\beta-1}}$ and $\theta_s(\xi) \equiv 0$. These two solutions play an important role in the analysis of the different solutions of (9). For blow-up regimes we assume $T_0 > 0$. Without loss of generality we set

$$T_0 = 1/(\beta-1), \;\; \text{then} \;\; \theta_H \equiv 1. \tag{11}$$

The analysis of the solutions of problem (9), (10), carried out in the works [1,26, 50,51], (see also [52], Chap. IV), gives the following results:

- For arbitrary $1 < \beta \leq \sigma + 1$ there exist a finite support solution $\theta_s(\xi) \geq 0$.
- For $\beta < \sigma + 1$, $N \geq 1$ and $\beta = \sigma + 1$, $N > 1$ the problem has no nonmonotone solutions. The uniqueness is proved only for $\beta < \sigma + 1$, $N = 1$.
  The graphs of the s.-s.f. $\theta_s(\xi)$ for $\beta = \sigma + 1 = 3$, $N = 1,2,3$ are shown in Fig. 1, the graphs of the s.-s.f. $\theta_s(\xi)$ for $\beta = 2.4 < \sigma + 1 = 3$, $N = 1,2,3$ in Fig. 2.
- For $\beta > \sigma + 1$, $N \geq 1$ the problem has no finite support solutions.
- If $\sigma + 1 < \beta < \beta_s = (\sigma+1)(N+2)/(N-2)_+$, ($\beta_s$—*the critical Sobolev exponent)*, the problem has at least one solution $\theta_s(\xi) > 0$ in $\mathbb{R}^1_+$, strictly monotone decreasing in ξ and having the asymptotics

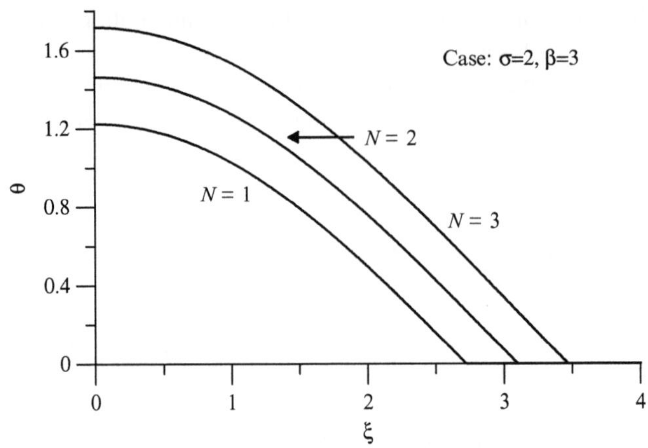

Fig. 1 $\beta = \sigma + 1 = 3$, *S*-regime

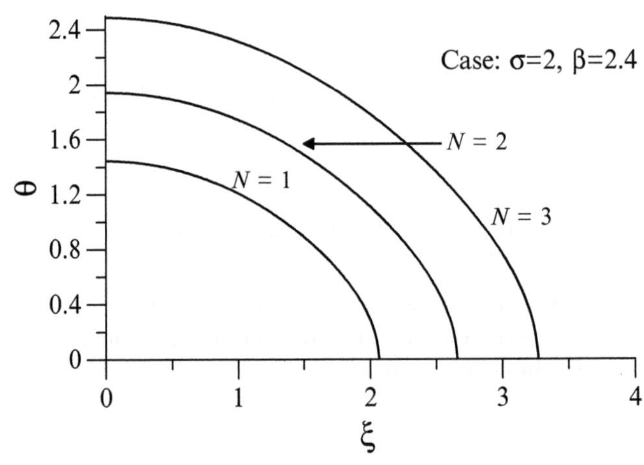

Fig. 2 $\beta = 2.4 < \sigma + 1 = 3$, *HS*-regime

$$\theta_s(\xi) = C_s\xi^{-2/(\beta-\sigma-1)}[1+\omega(\xi)], \qquad \omega(\xi) \to 0, \qquad \xi \to \infty, \tag{12}$$

$C_s = C_s(\sigma,\beta,N)$ is a constant. Later on in [31] the interval in β has been extended.

- For $N = 1,\ \beta > \sigma + 1$ the problem has at least

$$K = -[-a] - 1, \ a = \frac{\beta - 1}{\beta - \sigma - 1} > 1 \tag{13}$$

different solutions [1,26,52]. Let us introduce the notations $\theta_{s,i}(\xi)$, $i = 1,2,\dots,K$ for them. On the basis of linear analysis and some numerical results in the works

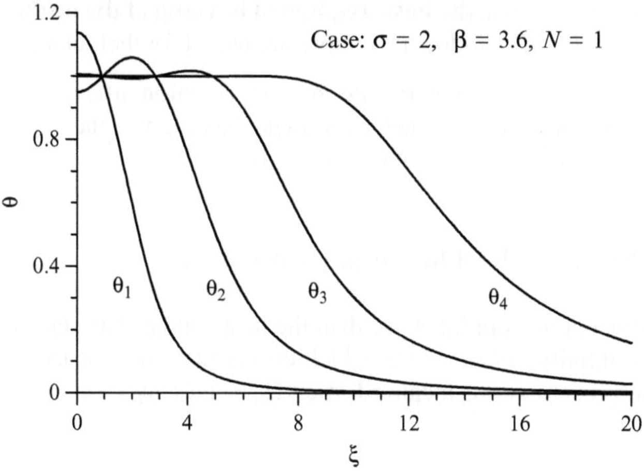

Fig. 3 $\beta = 3.6 > \sigma + 1 = 3$, *LS*-regime, $N = 1$

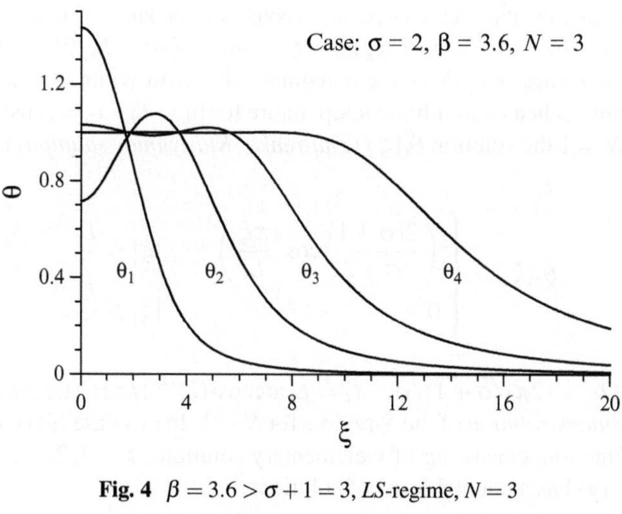

Fig. 4 $\beta = 3.6 > \sigma + 1 = 3$, *LS*-regime, $N = 3$

[43, 44], it has been supposed that the number of different solutions $\theta_{s,i}(\xi)$ for $\beta > \sigma + 1$ and $N \geq 1$ is $K + 1$. For $N = 1$ this result was refined [45] by using bifurcation analysis: the number of solutions is $K = [a]$, if a is not an integer, and $K = a - 1$, if a is an integer. For $N = 2, 3$ the bifurcation analysis gives the same estimate for the number of different solutions, but for $\beta \approx \sigma + 1$, $\beta > \sigma + 1$ it is violated (see Sect. 4.1).

The graphs of the four self-similar functions, existing for $\sigma = 2, \beta = 3.6 > \sigma + 1$ ($K = 4$), are shown in Fig. 3 ($N = 1$) and Fig. 4 ($N = 3$).

These results determine **the basic regimes of burning of the medium**, described by the s.-s.s. (7), (8). The following notions are useful for their characterization:

- *Semi-width* $x_s = x_s(t)$, determined by the equation $u(t,x_s) = u(t,0)/2$ for solutions, monotone in x and having a single maximum at the point $x = 0$.
- *Front-point* x_f: $u(t,x_f) = 0, \ u^\sigma u_x(t,x_f) = 0$.

2.1.1 *HS*-Evolution, Total Blow-Up, $1 < \beta < \sigma + 1$

The heat diffusion is more intensive than the heat source. The semi-width and the front tend to infinity; a *heat wave*, which covers the whole space for time T_0, is formed. The process is not localized: mes $\Omega_L =$ mes $\omega_L = \infty, \ \ x_s \to \infty, \ \ x_f \to \infty, \ t \to T_0^-$.

2.1.2 *S*-Evolution, Regional Blow-Up, $\beta = \sigma + 1$

The heat diffusion and the source are correlated in such a way that leads to localization of the process in a region $\Omega_L = \omega_L = \{|x| < L_s/2\}$ of diameter L_s, called a *fundamental length* of the *S*-regime. The semi-width is constant; inside Ω_L the medium is heated to infinite temperature for time $T_0: \ x_s = \text{const}, \ x_f = L_s/2$. In the case $N = 1$ the solution $\theta_s(\xi)$ (*Zmitrenko–Kurdyumov solution*) is found [50] explicitly:

$$\theta_s(\xi) = \begin{cases} \left(\dfrac{2(\sigma+1)}{\sigma+2} \cos^2 \dfrac{\pi\xi}{L_s} \right)^{1/\sigma}, & |\xi| \leq \dfrac{L_s}{2} \\ 0 & |\xi| > \dfrac{L_s}{2}, \end{cases} \tag{14}$$

$L_s = \text{diam } \Omega_L = (2\pi\sqrt{\sigma+1})/\sigma, \ \ x_s = L_s \arccos((2^{-\frac{\sigma}{2}})/\pi)$. The solution (14) is called *elementary solution* of the *S*-regime for $N = 1$. In this case (9) is autonomous, and every function, consisting of k elementary solutions, $k = 1,2,\ldots$, is a solution as well, i.e., (9) has a countable set of solutions.

2.1.3 *LS*-Evolution, Single Point Blow-Up, $\sigma + 1 < \beta < \beta_f = \sigma + 1 + \frac{2}{N}$

Here β_f *is the critical Fujita exponent* [46]. The intensity of the source is bigger than the diffusion. The front of the s.-s.s. is at infinity (12), the semi-width decreases, and the medium is heated to infinite temperature in a single point:

$$\text{mes } \omega_L = 0, \ \ x_s \to 0, \ \ t \to T_0^-.$$

According to the different s.-s.f. $\theta_{s,i}(\xi)$, $i = 1,2,\ldots$, the medium burns as a *simple* structure $(i = 1)$ and as *complex* structures $(i > 1)$ with the same blow-up time.

2.2 Stability of the Self-similar Solutions

To show the important property of the s.-s.s. as attractors of wide classes of other solutions of the same equation, we will need of additional notions.

In the case of arbitrary finite support initial data $u_0(x)$ (6), the so-called *self-similar representation* [26] of the solution $u(t,x)$ of problem (5), (6) is defined. It is determined at every time t according to the structure of the s.-s.s. (7), (8):

$$\Theta(t,\xi) = (1 - t/T_0)^{\frac{1}{\beta-1}}\, u\left(t, \xi\,(1 - t/T_0)^{\frac{m}{\beta-1}}\right) = \varphi^{-1}(t) u(t, \xi\,\psi(t)). \qquad (15)$$

The s.-s.s. $u_s(t,x)$ is called *asymptotically stable* [52] if there exists a sufficiently large class of solutions $u(t,x)$ of problem (5), (6) for initial data $u_0(x) \not\equiv \theta_s(x)$, whose self-similar representations $\Theta(t,\xi)$ tend in some norm to $\theta_s(\xi)$ when $t \to T_0^-$:

$$\|\Theta(t,\xi) - \theta_s(\xi)\| \to 0,\ t \to T_0^-. \qquad (16)$$

The definition of the self-similar representation (15) contains the blow-up time T_0. For theoretical investigations this is natural, but for numerical investigations definition (15) is unusable since for arbitrary initial data $u_0(x) \neq \theta_s(x)$ T_0 is not known. Therefore, another approach has been proposed and numerically implemented (for $N = 1$) in [26, 51]. This approach gives a possibility to investigate **the structural stability** of the unbounded solutions in a special **"self-similar" norm**, consistent for every t with the geometric form of the solution and not using explicitly the blow-up time T_0. A new self-similar representation, consistent with the structure of the s.-s.s. (7), (8), is introduced:

$$\Theta(t,\xi) = u(t, \xi(\gamma(t))^{-m})/\gamma(t), \qquad \gamma(t) = \frac{\max_x u(t,x)}{\max_\xi \theta_s(\xi)}. \qquad (17)$$

If the limit (16) takes place for $\Theta(t,\xi)$, given in (17), then the self-similar solution $u_s(t,x)$ is called *structurally stable*.

The notion of structural stability, i.e., the preservation in time of some characteristics of the structures, such as geometric form, rate of growth, and localization in space, is tightly connected with the notion invariance of the solutions with respect to the transformations, involving the time [30]. This determines its advisability for investigating the asymptotic behavior of the blow-up solutions.

In the case of complex structures, another notion of stability is needed, namely, **metastability**. The self-similar solution $u_s(t,x)$ is called *metastable* if for every $\varepsilon > 0$ there exists a class of initial data $u(0,x) \approx \theta_s(x)$ and a time T, $T_0 - T \ll T_0$ such that

$$\| \Theta(t,\xi) - \theta_s(\xi) \| \leq \varepsilon, \quad \text{for } 0 \leq t \leq T$$

holds for the self-similar representations (17) of the corresponding solutions. This means, that the metastable s.-s.s. preserves its complex space-time structure during the evolution up to time T, very close to the blow-up time T_0. After that time the complex structure could degenerate into one or several simple structures.

3 Numerical Methods

To solve the reaction–diffusion problem (5), (6) and the corresponding self-similar problem (9), (10), as well as their generalizations both for systems of such equations and for the 2D case, appropriate numerical methods and algorithms were developed.

The difficulties, common for the nonstationary and for the self-similar problems, were the nonlinearity, the dependence on a number of parameters (not less than 3), and insufficient smoothness of the solutions on the degeneration surface, where the solutions vanish. In the case of radial symmetry for $N > 1$ and polar coordinates in the 2D case, additional singularity at $x = 0$ ($\xi = 0$) occurs.

The main challenge in solving the self-similar problems is the nonuniqueness of their solutions for some ranges of the parameters. The following problems arise: to find a "good" approximation to each of the solutions; to construct an iteration process, converging fast to the desired solution (corresponding to the initial approximation) and ensuring sufficient accuracy; to construct a computational process, which enables finding all different solutions for given parameters (σ, β, N) in one and the same way; and to determine in advance where to translate the boundary conditions from infinity, for example (10), for the asymptotics (12) to be fulfilled.

The main difficulty in solving the nonstationary problems is the blow-up of their solutions: blow-up in a single point, in a finite region, and in the whole space. Two other difficulties are connected with—the moving front of the solution, where it is often not sufficiently smooth, and the instability of the blow-up solutions.

3.1 *Initial Approximations to the Different s.-s.f. for a Given Set of Parameters*

To overcome the difficulty with the initial approximations to the different s.-s.f., we have used the approach proposed and used in the works [43, 44]. Based on the hypothesis that in the region of their nonmonotonicity the s.-s.f. have small oscillations around the homogeneous solution θ_H, this approach consists of "linearization" of the self-similar equation around θ_H and followed by "sewing" the solutions of the resulting linear equation with the known asymptotics at infinity, e.g., (12).

Our experiments showed [19] that when $\beta \to \sigma + 1 + 0$, the hypothesis about small oscillations of the s.-s.f. around θ_H is not fulfilled. The detailed analytical and numerical investigations [12, 19, 38] of the "linear approximations" in the radially symmetric case for $N = 1, 2, 3$ showed that even in the case that these approximations take negative values in vicinity of the origin, they still give the true number of crossings with θ_H and, thus, the character of nonmonotonicity of the s.-s. functions. Recommendations of how to use the linear approximations in these cases are made in [19]. Let us note, the "linear approximations" are expressed by different special functions: the confluent hypergeometric function ${}_1F_1(a,b;z)$ and the Bessel function $J_k(z)$ for different parameter ranges within the complex plane for a and b and different ranges of the variable z. To compute these special functions, various methods were used: Taylor series expansions, expansions in ascending series of Chebyshev polynomials, rational approximations, and asymptotic series [15, 17].

3.2 *Numerical Method for the Self-similar Problems*

To solve the self-similar problem (9), (10) and its generalization for systems of ODE and for the 2D case, the continuous analog of the Newton's method (CANM) was used [12–16, 19–23, 38, 40, 41]. Proposed by Gavurin [35], this method was further developed in [48, 56] and used for solving many nonlinear problems. The idea behind it is to reduce the stationary problem $L(\theta) = 0$ to the evolution one:

$$L'(\theta)\frac{\partial \theta}{\partial t} = -L(\theta), \quad \theta(\xi,0) = \theta_0(\xi), \tag{18}$$

by introducing a continuous parameter t, $0 < t < \infty$, on which the unknown solution depends: $\theta = \theta(\xi,t)$. By setting $v = \partial\theta/\partial t$ and applying the Euler's method to the Cauchy problem (18), one comes to the iteration scheme:

$$L'(\theta_n)v_n = -L(\theta_n), \tag{19}$$

$$\begin{gathered} \theta_{n+1} = \theta_n + \tau_n v_n, \quad 0 < \tau_n \leq 1, \quad n = 0, 1, \ldots, \\ \theta_n = \theta_n(\xi) = \theta(\xi, t_n), \quad v_n = v_n(\xi) = v(\xi, t_n), \\ \theta_0(\xi) \text{ being the initial approximation.} \end{gathered} \tag{20}$$

The linear equations (19) (or the system of such equations in the case of a two-component medium) are solved by the Galerkin finite element method (GFEM) at every iteration step. The combination of the CANM and the GFEM turned out to be very successful. The linear system of the FEM with nonsymmetric matrix is solved by using the LU decomposition. The iteration process (20) converges very fast—usually less than 15–16 iterations are sufficient for stop-criterium $\|L(\theta_n)\| < 10^{-7}$.

The numerical investigation of the accuracy of the method being implemented shows errors (a) of order $O(h^4)$ when using quadratic elements in the radially symmetric case and (b) of optimal-order $O(h^2)$ when using linear elements in the same case or bilinear ones in the 2D case. To achieve the same accuracy in vicinity of the origin in the radially symmetric case for $N \geq 3$, the nonsymmetric Galerkin method [27] was developed [14, 38].

The computing of the solutions of the linearized self-similar equation and their sewing with the known asymptotics is implemented in a software, so the process is fully automatized. The software enables the computing of the self-similar functions for all of the blow-up regimes; moreover, in the case of *LS*-regime, only the number k of the self-similar function $\theta_{s,k}(\xi)$ must be given.

3.3 Numerical Method for the Reaction–Diffusion Problems

The GFEM, based on the Kirchhoff transformation of the nonlinear heat-conductivity coefficient:

$$G(u) = \int_0^u s^\sigma \, ds = u^{\sigma+1}/(\sigma+1), \tag{21}$$

was used [3, 4, 12–15, 17, 18, 20–23, 53] for solving the reaction–diffusion problems. This transformation is crucial for the further interpolation of the nonlinear coefficients on the basis of the finite element space and for optimizing of the computational process:

Here below we point out the main steps of the method on the problem (5), (6) in a finite interval $[0, X(t)]$ under the boundary condition $u(X(t)) = 0$. Because of the finite speed of heat propagation we choose $X = X(t)$ so as to avoid the influence of the boundary condition on the solution. The discretization is made on the Galerkin form of the problem:

Find a function $u(t,x) \in D$,

$$D = \{u : x^{(N-1)/2} u,\ x^{(N-1)/2} \partial u^{(\sigma+1)/2}/\partial x \in L_2,\ u(X(t)) = 0\},$$

which for every fixed t satisfies the integral identity

$$(u_t, v) = A(t; u, v), \quad \forall v \in H^1(0, X(t)), \quad 0 < t < T_0, \tag{22}$$

and the initial condition (6).

Here

$$(u,v) = \int_0^{X(t)} x^{N-1} u(x) v(x) dx, \quad A(t; u, v) = \int_0^{X(t)} \left[x^{N-1} \frac{\partial G(u)}{\partial x} \frac{\partial v}{\partial x} + x u^\beta v \right] dx,$$

$$H^1(0, X(t)) = \{v : x^{(N-1)/2} v,\ x^{(N-1)/2} v' \in L_2(0, X(t)),\ v(X(t)) = 0.\}$$

The lumped mass finite element method [55] with interpolation of the nonlinear coefficients $G(u)$ (21) and $q(u) = u^\beta$:

$$G(u) \sim G_I = \sum_{i=1}^{n} G(u_i)\varphi_i(x), \qquad q(u) \sim q_I = \sum_{i=1}^{n} q(u_i)\varphi_i(x)$$

on the basis $\{\varphi_i\}, i = 1,\ldots,n$, of the finite element space is used for discretization of (22). The resulting system of ordinary differential equations with respect to the vector $U(t) = (u_1(t), u_2(t), \ldots, u_n(t))^T$ of the nodal values of the solution $u(t,x)$ at time t is:

$$\dot{U} = \tilde{M}^{-1}(-KG(U)) + q(U), \quad U(0) = U_0. \tag{23}$$

Here the following denotations are used: $G(U) = (G(u_1),\ldots,G(u_n))^T$, $q(U) = (q(u_1),\ldots,q(u_n))^T$, $\tilde{M}$ is the lumped mass matrix, and K is the stiffness matrix. Let us mention, thanks to the Kirchhoff transformation and the interpolation of the nonlinear coefficients, only the two vectors $G(U)$ and $q(U)$ contain the nonlinearity of the problem, while the matrix K does not depend on the unknown solution.

To solve the system (23) an explicit Runge–Kutta method [47] of second order of accuracy and an extended region of stability was used. A special algorithm for choosing the time step τ ensures the validity of the weak maximum principle and, in the case of smooth solutions, the achievement of a given accuracy ε up to the end of the time interval. The stop criterion is $\tau < 10^{-16}$, and then $\tilde{T}_0$ is the approximate blow-up time, found in the computations.

It is worth mentioning that the nonlinearity has changed the prevailing opinion about the explicit methods. Indeed, there are at least two reasons for an explicit method to be preferred over the implicit one for solving the system (23):

- The condition for solvability of the nonlinear discrete system on the upper time level imposes the same restriction on the relation "time step—step in space", as does the condition for validity of the weak maximum principle for the explicit scheme (see [52], Chap. VII, Sect. 5).
- The explicit method for solving large discrete systems has a significant advantage over the implicit one with respect to the computational complexity.

Let us also mention that in the case of blow-up solutions, the discrete system on the upper time level would connect solution values differing by 6–12 orders of magnitude, which causes additional difficulties to overcome. Finally the explicit methods allow easy parallelization.

The special achievements of the proposed methods are the **adaptive meshes** in the *LS*-regime (refinement of the mesh) and in the *HS*-regime (stretching meshes with constant number of mesh points), consistent with the self-similar low. Let us briefly describe this adaptation idea on the differential problem

$$u_t = Lu, \;\; u = u(t,x), \;\; x \in \mathbb{R}^{\mathrm{N}}, \;\; t > 0, \tag{24}$$

which admits a self-similar solution of the kind

$$u_s(t,x) = \varphi(t)\theta_s(\xi), \qquad \xi = x/\psi(t), \;\; u_s(0,x) = \theta_s(x). \tag{25}$$

Since the invariant solution $u_s(x,t)$ is an attractor of the solutions of (24) for large classes of initial data different from $\theta_s(x)$, it is important to incorporate the structure (25) in the numerical method for solving (24). The relation (25) between ξ and x gives the idea how to adapt the mesh in space. Let $\Delta x^{(0)}$ be the initial step in space, $\Delta x^{(k)}$—the step in space at $t = t^k$. Then $\Delta x^{(k)}$ must be chosen so that $\Delta\xi^{(k)}$ is bounded from below and from above

$$\Delta x^{(0)}/\lambda \leq \Delta\xi^{(k)} \leq \lambda\Delta x^{(0)}$$

for an appropriate λ (usually $\lambda = 2$).

Further, by using the relation between $\psi(t)$ and $\varphi(t)$, it is possible to incorporate the structure (25) of the s.-s.s. in the adaptation procedure. In the case of (5), we have

$$\xi = x\Gamma(t)^m, \quad \Delta\xi = \Delta x\Gamma(t)^m, \quad m = (\beta - \sigma - 1)/2, \quad \Gamma(t) = \frac{\max_x u(t,x)}{\max_x u_0(x)}. \tag{26}$$

On the basis of the relations (26), the following strategy is accepted.

In the case of a single point blow-up, $m > 0$, we choose the step $\Delta x^{(k)}$ so that the step $\Delta\xi^{(k)}$ be bounded from above:

$$\Delta\xi^{(k)} = \Delta x^{(k)}\Gamma(t)^m \leq \lambda\Delta x^{(0)}. \tag{27}$$

When condition (27) is violated, the following procedure is carried out: every element in the region, in which the solution is not established with a given accuracy δ_u (usually $\delta_u = 10^{-7}$), is divided into two equal elements, and the values of the solution in the new mesh points are found by interpolating from the old values; the elements, in which the solution is established with a given accuracy δ_u, are neglected.

In the case of a total blow-up, $m < 0$, we choose the step $\Delta x^{(k)}$ so that the step $\Delta\xi^{(k)}$ be bounded from below:

$$\Delta\xi^{(k)} = \Delta x^{(k)}\Gamma(t)^m \geq \lambda\Delta x^{(0)}. \tag{28}$$

When condition (28) is violated, the lengths of the elements are doubled, and so is the interval in x: $X(t_{k+1}) = 2X(t_k)$; thus, the number of mesh points remains constant.

This adaptation procedure makes it possible to compute efficiently the single point blow-up as well as the total blow-up solutions up to amplitudes 10^6–10^{12}, depending on the medium parameters. It ensures the authenticity of the results of investigation of the structural stability and the metastability of the self-similar solutions. Let us note, this approach does not require an auxiliary differential problem for the mesh to be solved, unlike the moving mesh methods [10]. The idea to use the invariant properties of the differential equations and their solutions [7] and to incorporate the structural properties (e.g., geometry, different kind of symmetries, the conservation laws) of the continuous problems in the numerical method lies at the basis of an important direction of the computational mathematics—geometric integration, to which many works and monographs are devoted—see [9, 24, 37] and the references therein.

The numerous computational experiments carried out with the exact self-similar initial data (14) for $N = 1, \beta = \sigma + 1$, as well as with the computed self-similar initial data, show a good blow-up time restoration (set into the self-similar problem) in the process of solving the reaction–diffusion problem. The preservation of the self-similarity and the restoration of the blow-up time demonstrate the high quality of the numerical methods for solving both the self-similar and the nonstationary problems.

4 Results and Achievements

The developed numerical technique was used to analyze and solve a number of open problems. Below we present briefly some of them.

4.1 *The Transition LS- to S-Regime in the Radially Symmetric Case*

The investigation of the limit case $\beta \to \sigma + 1 + 0$ resolved the following paradox for $N > 1$: there exists one simple-structure s.-s. function in S-regime ($\beta = \sigma + 1$), whereas in LS-regime for $\beta \to \sigma + 1 + 0$, their number tends to infinity according to formula (13). The detailed numerical experiment in [19, 38] yielded the following results. First, it was shown that the structure of the s.-s.f. for $N > 1$ and $\beta \sim \sigma + 1$, $\beta > \sigma + 1$ is substantially different from the one for $N = 1$. Second, for $N = 1$ the transition $\beta \to \sigma + 1 + 0$ is "continuous"—the self-similar function $\theta_{s,k}(\xi)$ for the LS-regime tends to a s.-s.f. of the S-regime, consisting of k elementary solutions. For $N > 1$ the transition behaves very differently.

For σ fixed and $\beta \to \sigma + 1 + 0$ the central minimum of the "even" s.-s.f. $\theta_{2j}^{(N)}$, $j = 1, 2, \ldots$, decreases and surprisingly becomes zero for some $\beta = \beta_j^*(\sigma, N)$ (Fig. 5, left). For $\beta < \beta_j^*(\sigma, N)$ all of the s.-s.f. have zero region around the center of symmetry, and the radius of this region tends to infinity for $\beta \to \sigma + 1 + 0$. All of the maxima of $\theta_{s,2j}^{(N)}(\xi)$ tend to the maximum of the s.-s.f. of the S-regime for the corresponding σ and for $N = 1$. Thus, the s.-s.f. $\theta_{s,2j}^{(N)}(\xi)$, "going to infinity" when $\beta \to \sigma + 1 + 0$, tends to a s.-s.f. of the S-regime for $N = 1$ and the same σ, consisting of j elementary solutions.

For fixed σ there exists such a value $\beta_j^{**}(\sigma, N)$ that for $\sigma + 1 < \beta < \beta_j^{**}$ the "odd" s.-s.f. $\theta_{s,2j+1}^{(N)}(\xi), j = 1, 2, \ldots$ split into two parts: a central one, tending to the s.-s.f. of the S-regime for the same N, and second one, coinciding with the s.-s.f. $\theta_{s,2j}^{(N)}(\xi)$, "going to infinity" when $\beta \to \sigma + 1 + 0$ (Fig. 5, right).

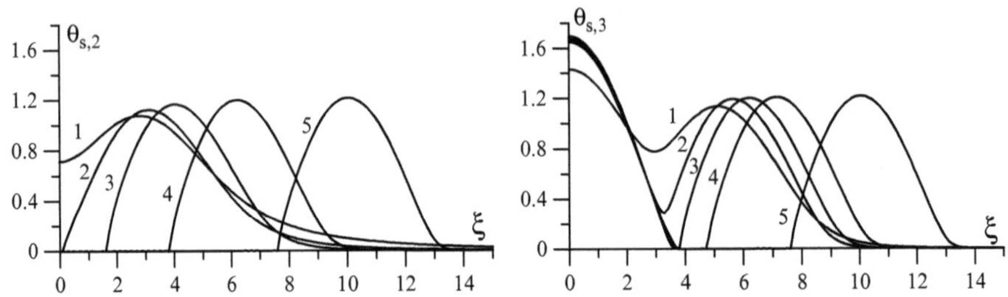

Fig. 5 Graphs of the s.-s.f. $\theta_{s,2}$ for $N=3$, $\sigma=2$, $\beta=\{3.6(1);\ 3.38(2);\ 3.2(3);\ 3.08(4);\ 3.03(5)\}$ (*left*) and $\theta_{s,3}$ for $N=3$, $\sigma=2$, $\beta=\{3.2(1);\ 3.1(2);\ 3.08(3);\ 3.06(4);\ 3.03(5)\}$ (*right*)

According to the described "scenario", when $\beta \to \sigma+1+0$ only the first s.-s.f. of the *LS*-regime remains, and it tends to the unique s.-s.f. of the *S*-regime.

As a result of this investigation, new-structure s.-s.f. were found—s.-s.f. with **a left front**. The existence of such s.-s. functions was confirmed by an asymptotic analysis, i.e., the asymptotics in the neighborhood of the left front-point were found analytically [12]. This new type of solutions initiated investigations of other authors [36, 45] by other methods (the method of dynamical analogy, bifurcation analysis). Their investigations confirmed our results.

4.2 *The Asymptotic Behavior of the Blow-Up Solutions of Problem (5), (6) Beyond the Critical Fujita Exponent*

For $\beta > \beta_f = \sigma+1+2/N$ the problem (5), (6) could have blow-up or global solutions depending on the initial data. For the s.-s. blow-up solution (7), (8) it holds $u_s(t,r) \notin L_1(\mathbb{R}^N)$. The qualitative theory of nonstationary averaging "amplitude-semi-width" predicts a self-similar behavior of the amplitude of the blow-up solutions and a possible non-self-similar behavior of the semi-width [52]. A question was posed there: what kind of invariant or approximate s.-s.s. describes the asymptotic stage $(t \to T_0^-)$ of the blow-up process?

The detailed numerical experiment carried out in [18, 53] showed that the s.-s.s. (7), (8), corresponding to $\theta_{s,1}(\xi)$, is structurally stable: all of the numerical experiments with finite support initial data (6), ensuring blow-up, yield solutions tending to the self-similar one on the asymptotic stage.

4.3 *Asymptotically Self-similar Blow-Up Beyond Some Other Critical Exponents*

The numerical investigation of the blow-up processes in the radially symmetric case for high space dimensions N was carried out in [13, 14]. The aim was to check some hypotheses [31] about the solutions of the s.-s. problem (9), (10) and about the asymptotic stability of the corresponding s.-s. solutions for parameters beyond the following critical exponents:

$\beta_s = (\sigma+1)(N+2)/(N-2)$, $N \geq 3$ (Sobolev's exponent).

$\beta_u = (\sigma+1)(1+4/(N-4-2\sqrt{N-1}))$, $N \geq 11$.

$\beta_p = 1+3(\sigma+1)+(\sigma^2(N-10)^2+2\sigma(5\sigma+1)(N-10)+9(\sigma+1)^2)^{1/2}/(N-10)$, $N \geq 11$.

Self-similar functions, monotone in space, were constructed numerically for all of these cases, thus confirming the hypotheses of their existence (not proved for $\beta > \beta_u$). It was also shown that the corresponding s.-s.s. are structurally stable, thus confirming another hypothesis of [31]. Due to the strong singularity at the origin, the nonsymmetric Galerkin method and the special refinement of the finite element mesh [14] were crucial for the success of these investigations.

4.4 Two-Component Nonlinear Medium

The methods, developed for the radially symmetric problems (5), (6) and (9), (10), were generalized in [22, 40, 53] for the case of two-component nonlinear medium, described by the system:

$$\left| \begin{aligned} u_{1t} &= \frac{1}{x^{N-1}}(x^{N-1}u_1^{\sigma_1}u_{1x})_x + u_1^{\beta_1}u_2^{\gamma_2}, \quad x \in \mathbb{R}^1_+,\ N=1,2,3, \\ u_{2t} &= \frac{1}{x^{N-1}}(x^{N-1}u_2^{\sigma_2}u_{2x})_x + u_1^{\gamma_1}u_2^{\beta_2}, \quad \sigma_i > 0,\ \beta_i > 1,\ \gamma_i \geq 0, i=1,2. \end{aligned} \right. \tag{29}$$

This system admits blow-up s.-s.s. of the form

$$\begin{aligned} u_{1s} &= (1-t/T_0)^{m_1}\theta_{1s}(\xi),\ \xi = x/(1-t/T_0)^n, \\ u_{2s} &= (1-t/T_0)^{m_2}\theta_{2s}(\xi),\ m_i < 0,\ i=1,2, \end{aligned} \tag{30}$$

where

$$m_i = \frac{\alpha_i}{p},\quad \alpha_i = \gamma_i + 1 - \beta_i,\ i=1,2,\quad p = (\beta_1-1)(\beta_2-1) - \gamma_1\gamma_2,$$

$$n = \frac{m_1\sigma_1+1}{2} = \frac{m_2\sigma_2+1}{2},\quad \sigma_1(\gamma_2+1-\beta_2) = \sigma_2(\gamma_1+1-\beta_1).$$

The s.-s.f. satisfy the system of nonlinear ODE:

$$\left| \begin{array}{l} L_1(\theta_{1s},\theta_{2s}) \equiv -\frac{1}{\xi^{N-1}}(\xi^{N-1}\theta_{1s}^{\sigma_1}\theta_{1s}')' + n\xi\theta_{1s}' - m_1\theta_{1s} - \theta_{1s}^{\beta_1}\theta_{2s}^{\gamma_2} = 0, \\ L_2(\theta_{1s},\theta_{2s}) \equiv -\frac{1}{\xi^{N-1}}(\xi^{N-1}\theta_{2s}^{\sigma_2}\theta_{2s}')' + n\xi\theta_{2s}' - m_2\theta_{2s} - \theta_{1s}^{\gamma_1}\theta_{2s}^{\beta_2} = 0 \end{array} \right. \tag{31}$$

and the boundary conditions

$$\lim_{\xi\to 0}\xi^{N-1}\theta_{is}^{\sigma_1}\theta_{is}' = 0, \ \lim_{\xi\to\infty}\theta_{is} = 0, \ i = 1,2. \tag{32}$$

Superconvergence of the FEM (of order $O(h^4)$) for solving the s.-s. problem (31)–(32) by means of quadratic elements and optimal-order convergence ($O(h^2)$) by means of linear elements was achieved. The structural stability of the s.-s.s. (30) for parameters $\sigma_i,\ \beta_i,\ \gamma_i,\ i = 1,2$, corresponding to the *LS*-regime ($n > 0$), was analyzed in [22,40,53]. It was shown that only the s.-s.s. of systems (29) with strong feedback ($p < 0$), corresponding to the s.-s.f. with two simple-structure components, were structurally stable. All the other s.-s.s. were metastable—self-similarity was preserved to times not less than $99.3\,\%\tilde{T}_0$.

The proposed computational technique can be applied to investigating the self-organization processes in wide classes of nonlinear dissipative media described by nonlinear reaction–diffusion systems.

4.5 Directed Heat Diffusion in a Nonlinear Anisotropic Medium

Historically the first Bulgarian contribution to the topic under consideration was the numerical realization of the self-similar solutions, describing directed heat diffusion and burning of a two-dimensional nonlinear anisotropic medium. It was shown in [25] that the model of heat structures in the anisotropic case

$$u_t = (u^{\sigma_1}u_{x_1})_{x_1} + (u^{\sigma_2}u_{x_2})_{x_2} + u^\beta,\ x = (x_1,x_2) \in \mathbb{R}^2,\ \sigma_1 > 0,\ \sigma_2 > 0,\ \beta > 1 \tag{33}$$

admits invariant solutions of the kind

$$u_s(t,x_1,x_2) = \left(1 - \frac{t}{T_0}\right)^{-\frac{1}{\beta-1}} \theta_s(\xi),\ \xi = (\xi_1,\xi_2) \in \mathbb{R}^2,$$

$$\xi_i = x_i/\left(1 - \frac{t}{T_0}\right)^{\frac{m_i}{\beta-1}},\ m_i = \frac{\beta-\sigma_i-1}{2},\ i = 1,2.$$

The self-similar function $\theta_s(\xi_1,\xi_2)$ satisfies the nonlinear elliptic problem

$$L(\theta_s) \equiv \sum_{i=1}^{2}\left(-\frac{\partial}{\partial\xi_i}\left(\theta_s^{\sigma_i}\frac{\partial\theta_s}{\partial\xi_i}\right) + \frac{\beta-\sigma_i-1}{2}\xi_i\frac{\partial\theta_s}{\partial\xi_i}\right) + \theta_s - \theta_s^\beta = 0, \tag{34}$$

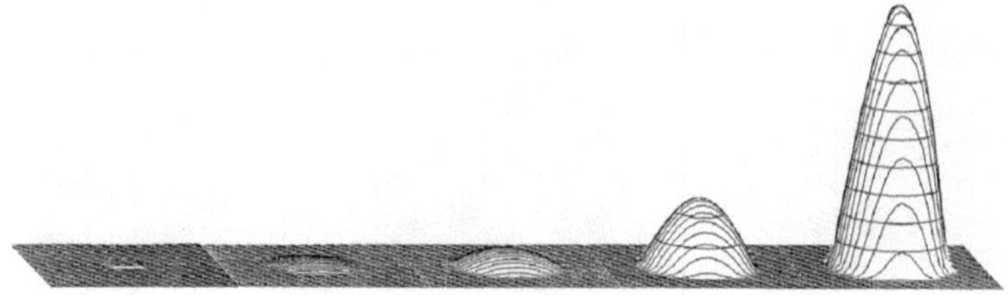

Fig. 6 S-regime: $\sigma_1 = 2,\ \sigma_2 = 2,\ \beta = 3$

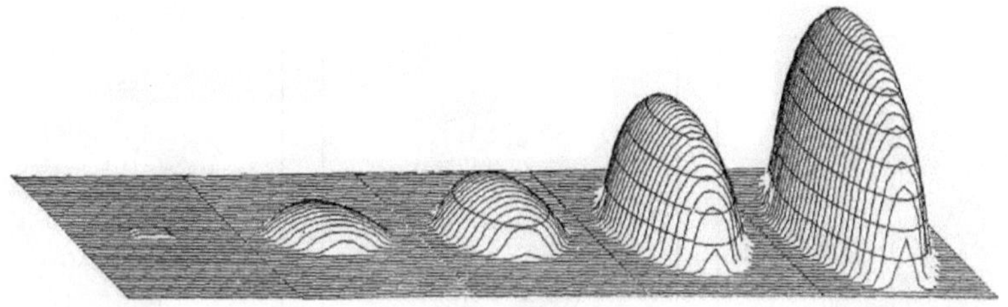

Fig. 7 HS–S-regime: $\sigma_1 = 3,\ \sigma_2 = 2,\ \beta = 3$

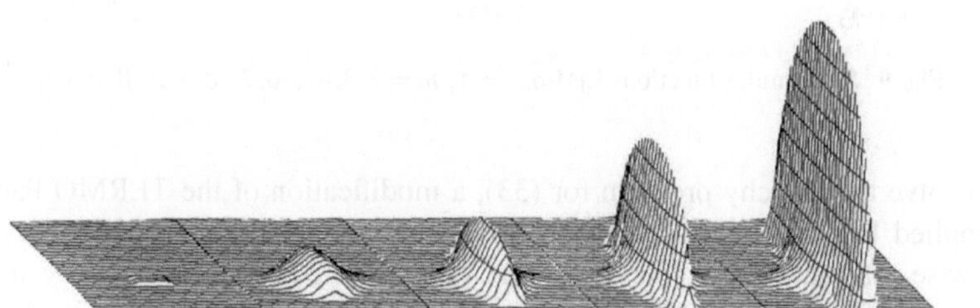

Fig. 8 HS–LS-regime: $\sigma_1 = 3,\ \sigma_2 = 1,\ \beta = 3$

$$\left.\frac{\partial \theta_s}{\partial \xi_i}\right|_{\xi_i=0} = 0,\ i = 1,2;\ \ \theta_s(\xi) \to 0,\ \ |\xi| \to \infty. \tag{35}$$

The Cauchy problem for equation (33) was investigated in the works [3, 4] for different parameters σ_1, σ_2 and β. Depending on the parameters, different mixed regimes S–HS, HS–LS, and S–LS of heat transfer and burning were implemented numerically. The evolution in time of one and the same initial perturbation is shown for the cases of the 2D radially symmetric S-regime (Fig. 6), the mixed HS–S-regime (Fig. 7), and the mixed HS–LS-regime (Fig. 8).

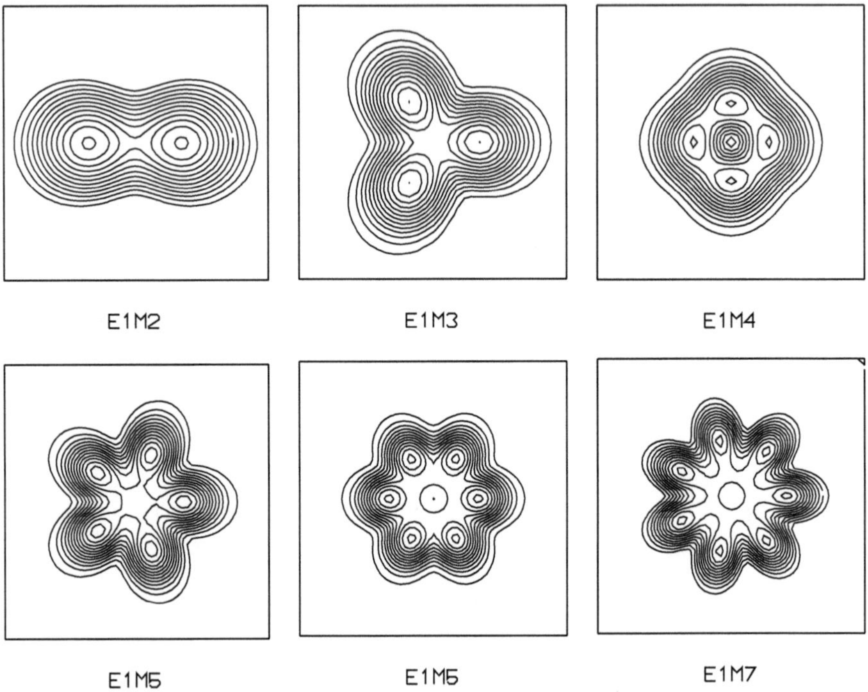

Fig. 9 Self-similar functions EjMm, $j = 1$, $m = 2,3,4,5,6,7$, $\sigma = 2$, $\beta = 3.25$

To solve the Cauchy problem for (33), a modification of the TERMO Package of Applied Programs [8], designed initially for solving isotropic problems with piecewise constant coefficients, was done. TERMO had been worked out by an IMI-BAS team, after the idea of Raytcho Lazarov and under his guidance, a merit worth mentioning here.

The s.-s. functions for the corresponding mixed regimes were found in [3] by self-similar processing of the solution of the Cauchy problem for (33). Later, in [16], they were found as solutions of the self-similar problem (34), (35). The self-similar functions of complex symmetry for the isotropic case in Cartesian coordinates (denoted in [44] as Ei/j) were found as a special case.

Later on, in [40, 41], the numerical methods were modified for the isotropic 2D self-similar problem in polar coordinates to construct numerically another class of self-similar functions of complex symmetry (denoted in [44] as $EjMm$) in *LS*-regime and to investigate their structural stability.

Graphical representations of the evolution of the anisotropic invariant solutions, as well as the s.-s. functions $EjMm$ for some different values of j, m, σ, β, are included in the Handbook [11]. We show some of the s.-s.f. $EjMm$ in Fig. 9.

4.6 Spiral Waves in HS-Regime

The numerical realization of the invariant solutions, describing "spiral" propagation of the nonhomogeneities in two-dimensional isotropic medium, appears to be one of the most interesting contributions of ours. The mathematical model in polar coordinates reads:

$$u_t = \frac{1}{r}(ru^\sigma u_r)_r + \frac{1}{r^2}(u^\sigma u_\varphi)_\varphi + u^\beta, \ \sigma > 0, \ \beta > 1. \tag{36}$$

It admits s.-s.s. of the kind [4, 30]:

$$u_s(t,r,\varphi) = \left(1 - \frac{t}{T_0}\right)^{-\frac{1}{\beta-1}} \theta_s(\xi,\phi), \tag{37}$$

$$\xi = r/\left(1 - \frac{t}{T_0}\right)^{\frac{m}{\beta-1}}, \quad \phi = \varphi + \frac{c_0}{\beta-1}\ln\left(1 - \frac{t}{T_0}\right), \quad m = \frac{\beta-\sigma-1}{2}. \tag{38}$$

The self-similar function $\theta_s(\xi,\phi)$ satisfies the nonlinear elliptic equation

$$\begin{aligned} L(\theta_s) \equiv & -\frac{1}{\xi}\frac{\partial}{\partial\xi}\left(\xi\theta_s^\sigma \frac{\partial\theta_s}{\partial\xi}\right) - \frac{1}{\xi^2}\frac{\partial}{\partial\phi}\left(\theta_s^\sigma \frac{\partial\theta_s}{\partial\phi}\right) + \frac{\beta-\sigma-1}{2}\xi\frac{\partial\theta_s}{\partial\xi} \\ & -c_0\frac{\partial\theta_s}{\partial\phi} + \theta_s - \theta_s^\beta = 0, \quad T_0 = \frac{1}{\beta-1}. \end{aligned} \tag{39}$$

Here $c_0 \neq 0$ is the parameter of the family of solutions. From (38) it follows

$$\xi e^{s\phi} = re^{s\varphi} = \text{const}, \quad s = \frac{\beta-\sigma-1}{2c_0}.$$

This means that the trajectories of the nonhomogeneities in the medium (say local maxima) are logarithmic spirals for $\beta \neq \sigma+1$ or circles for $\beta = \sigma+1$. The direction of movement for fixed c_0, for example, $c_0 > 0$, depends on the relation between σ and β : for $\beta > \sigma+1$ towards the center (twisting spirals) and for $\beta < \sigma+1$ from the center (untwisting spirals).

The problem for the numerical realization of the spiral s.-s.s. (37), (38) was posed in 1984, when the possibility for their existence has been established by the method of invariant group analysis in the Ph.D. Thesis of S.R. Svirshchevskii. As it was stated in [2], there were significant difficulties for finding such solutions. First, the linearization of the self-similar Eq. (39) was not expected to give the desired result, because it is not possible to separate the variables in the linearized equation. Second, the asymptotics at infinity of the solutions of the self-similar equation were not known.

The first successful step was the appropriate (complex) separation of variables in the linearized equation. Using the assumption for small oscillations of the s.-s.f. $\theta_s(\xi,\phi)$ around $\theta_H^1 \equiv 1$, i. e., $\theta_s(\xi,\phi) = 1+\alpha y(\xi,\phi)$, $\alpha = \text{const}$, $|\alpha y| \ll 1$ and the idea of linearization around it, the following linear equation for $y(\xi,\phi)$ was found [17]:

$$-\frac{1}{\xi}\frac{\partial}{\partial \xi}\left(\xi\frac{\partial y}{\partial \xi}\right)-\frac{1}{\xi^2}\frac{\partial^2 y}{\partial \phi^2}+\frac{\beta-\sigma-1}{2}\xi\frac{\partial y}{\partial \xi}-c_0\frac{\partial y}{\partial \phi}+(1-\beta)y=0.$$

Seeking for particular solutions $Y_k(\xi,\phi) = R_k(\xi)e^{\mathrm{i},k,\phi}$, $k \in \mathbb{N}$, it was found: for $\beta = \sigma+1$

$$R_k(\xi) = J_k(z), \quad z = (\sigma + c_0 k\mathrm{i})^{1/2}\xi,$$

where $J_k(z)$ is the first kind Bessel function of order k, and for $\beta \neq \sigma+1$

$$R_k(\xi) = \xi^k {}_1F_1(a,b;z), \quad a = -\frac{\beta-1+c_0 k\mathrm{i}}{\beta-\sigma-1}+\frac{k}{2}, \quad b = 1+k, \quad z = \frac{\beta-\sigma-1}{4}\xi^2,$$

where ${}_1F_1(a,b,z)$ is the confluent hypergeometric function. The detailed analytical and numerical investigation [17] of the functions $y_k(\xi,\phi) = \Re(Y_k(\xi,\phi))$ showed that their asymptotics at infinity are self-similar as well as that the functions

$$\tilde{\theta}_{s,k}(\xi,\phi) = 1+\alpha y_k(\xi,\phi), \; |\alpha y_k| \ll 1 \tag{40}$$

are very close to the sought-after solutions $\theta_s(\xi,\phi)$. Moreover, the amplitude of the linear approximations $y_k(\xi,\phi)$ tends to zero for $\xi \to \infty$ in the case of HS-regime and to infinity in the case of LS-regime. This gave the idea to seek for s.-s.s. of the HS-regime, tending to the nontrivial constant solution $\theta_s \equiv \theta_H$, i.e., to generalize the notion of s.-s. functions, and, consequently, the notion of the structures and waves, which arise and preserve themselves in the absolutely cold medium. Although not exploited earlier, this change is reasonable and fully adequate to the real systems.

All stated above enabled us to predict the asymptotics of the solutions of (39), to derive a boundary condition by using this asymptotics and to close the s.-s. problem by the following boundary and periodic conditions [15, 21, 23]:

$$\begin{gathered}
\lim_{\xi\to 0}\xi\theta_{s,k}^{\sigma}\frac{\partial\theta_{s,k}}{\partial\xi} = 0, \; \phi \in \left[0,\frac{2\pi}{k}\right], \\
\frac{\partial\theta_{s,k}}{\partial\xi} = \frac{\theta_{s,k}-1}{\bar{m}\xi}-\frac{\gamma k}{s\xi^{1-\frac{1}{\bar{m}}}}\sin(k\phi+\frac{k}{s}\ln\xi+\mu), \; \xi = l \gg 1, \; \phi \in \left[0,\frac{2\pi}{k}\right], \\
\theta_{s,k}(\xi,0) = \theta_{s,k}\left(\xi,\frac{2\pi}{k}\right), \quad \frac{\partial\theta_{s,k}}{\partial\phi}(\xi,0) = \frac{\partial\theta_{s,k}}{\partial\phi}\left(\xi,\frac{2\pi}{k}\right), \quad 0 \le \xi \le l,
\end{gathered} \tag{41}$$

Fig. 10 One-armed spiral solution ($k=1$), two-armed spiral solution ($k=2$), three-armed spiral solution ($k=3$), $c_0=1$, $\sigma=3$, $\beta=3.6$

where $\bar{m}=m/(\beta-1)$, γ and μ are constants, depending on σ,β,c_0. The numerical solving of the problem (39), (41) for $c_0\neq 0$ with initial approximations (40) gives "the spiral" s.-s. functions of the *HS*-regime; some of them and their evolution in time by solving (36) were investigated in [15, 21, 23]. The graphs of the s.-s.f. for $k=1$ (one-armed spiral), $k=2$ (two-armed spiral), and $k=3$ (three-armed spiral) are shown in Fig. 10. The rest of the parameters are $\sigma=3$, $\beta=3.6$, $c_0=1$.

The evolution of three-armed s.-s.f. for parameters $\sigma=2$, $\beta=2.4$, $k=3$, $c_0=1$ is shown in Fig. 11. The exact blow-up time is (11) $T_0=1/(\beta-1)=0.(714285)$. Similarly to all complex-symmetry s.s.-s., the three-armed spiral one is metastable—at $T\to T_0$ it degenerates into the simplest radially symmetric s.-s.s. for the same parameters σ,β. The mesh adaptation in r-direction when solving (36) is realized again in consistency with the self-similar law, keeping the same number of mesh points during the whole process of evolution. Having to solve two nonlinear 2D problems (elliptic and reaction–diffusion), going through a number of approximations, it is astonishing that the restoration (with accuracy 10^{-6}) of the exact blow-up time is practically perfect (Fig. 11, right most bottom).

The existence of spiral structures in *LS*-regime is an open question by now. The "ridges" of the linear approximations of the s.-s.f. in the *LS*-regime tend also to the self-similar ones for $\xi\to\infty$ [17], but their amplitudes tend to infinity, and it is not clear by what asymptotics to sew the linear approximations.

4.7 Complex-Symmetry Waves in HS-Regime and S-Regime

In the process of solving the problem for the spiral s.-s.s., an idea arose to seek for complex nonmonotone waves, tending to the nonzero homogeneous solution for $\xi\to\infty$ in *HS*-regime and $c_0=0$.

Figure 12 shows a complex-symmetry s.-s.f. and its evolution in time for the same parameters σ,β, as in Fig. 11, and $k=2$, $c_0=0$. Note the same perfect restoration of the blow-up time.

The results about the spiral and the complex-symmetry s.-s.f. are included in the book [54].

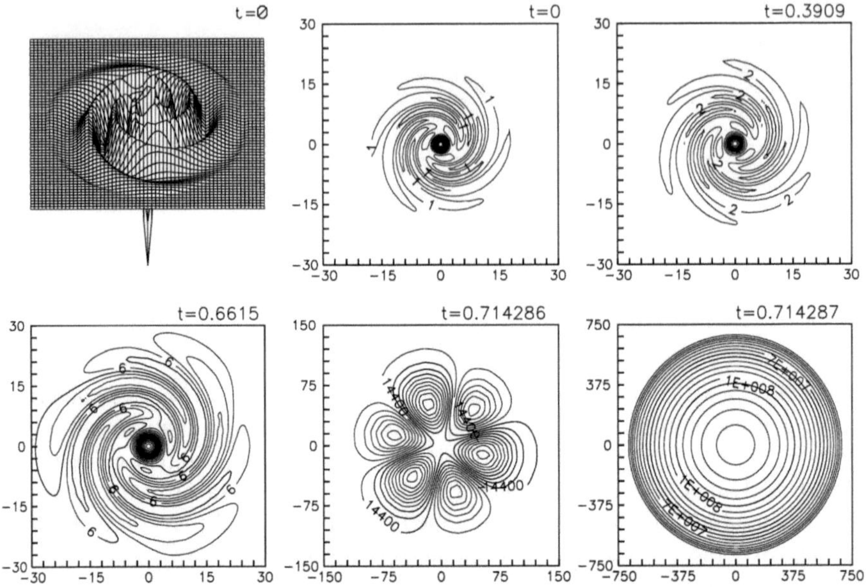

Fig. 11 Evolution of three-armed spiral wave: $\sigma = 2,\ \beta = 2.4,\ c_0 = 1,\ k = 3,\ T_0 = 0.(714285)$

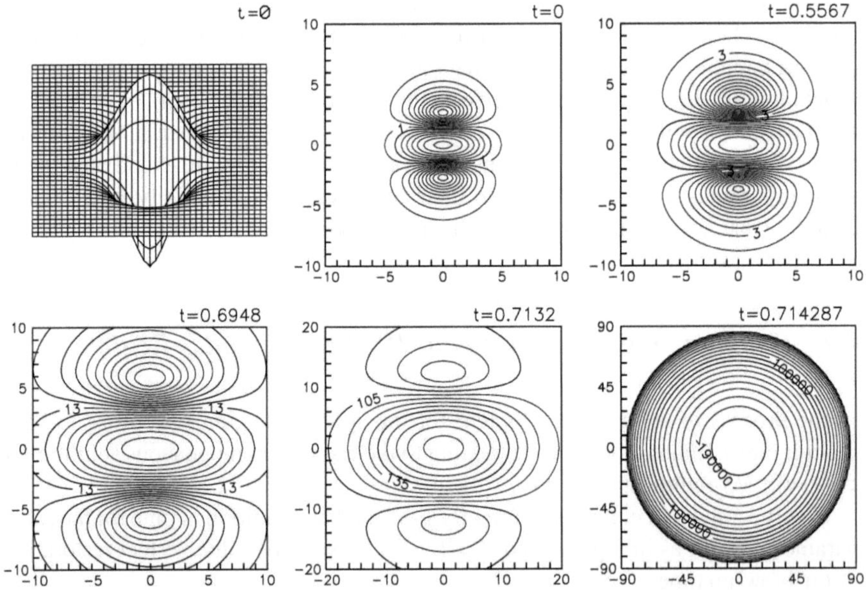

Fig. 12 Evolution of a complex wave in *HS*-regime: $\sigma = 2,\ \beta = 2.4,\ c_0 = 0,\ k = 2,\ T_0 = 0.(714285)$

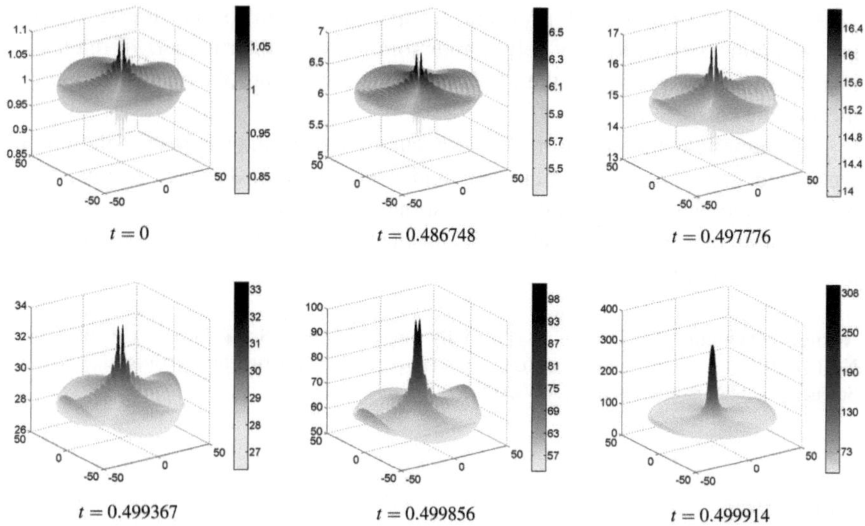

Fig. 13 Evolution of a complex wave in S-regime: $\sigma = 2,\ \beta = 3,\ c_0 = 0,\ k = 2,\ T_0 = 0.5$

Later the problem of finding nonmonotone s.-s.f. in the S-regime, tending to the nontrivial homogeneous solution θ_H, was posed and successfully solved [20]. The self-similar equation was solved with boundary conditions

$$\frac{\partial \theta_{s,k}}{\partial \xi} = \frac{1-\theta_{s,k}}{2\xi} - \gamma\sqrt{\frac{2}{\pi\sqrt{\sigma}\xi}}\sin\left(\sqrt{\sigma}\xi - \frac{k\pi}{2} - \frac{\pi}{4}\right)\cos(k\phi),\ \xi = l \gg 1, \phi \in \left[0, \frac{2\pi}{k}\right].$$

The s.-s.f. for $\beta = \sigma + 1 = 3$ and its evolution in time are shown in Fig. 13.

Let us mention that the existence of continuum of solutions to the radially symmetric s.-s. problem in HS- and S-regimes, which tend to the nontrivial homogeneous solution θ_H for $\xi \to \infty$, was mentioned in [52], but this result has remained without attention. As it turned out, it is these solutions that determine the spiral structures and the complex structures in HS- and S-regimes.

5 Open Problems

In spite of the numerous achievements related to the problems considered here, many interesting questions are still open: how many complex-symmetry s.-s.f. of the kind Ei/j and $EiMj$, tending to the trivial constant solution $\theta_s \equiv 0$, exist; how can the self-similar problem for the spiral s.-s.f. in the LS-regime be closed, and therefore how to construct these spiral s.-s.f. numerically; how wide are the different classes of spiral s.-s.f. for $\beta < \sigma + 1$ and the different classes of complex-symmetry s.-s.f. in S- and HS-regimes.

Although the method of invariant group analysis shows that the differential problem admits some kind of self-similar solutions and their numerical realization is a constructive "proof" of their existence, theoretical proofs are still missing in most of the cases.

The numerical investigation of the accuracy of the approximate solutions on embedded grids shows optimal-order and even superconvergence results, but it would be interesting to find theoretical estimates as well.

All these questions pose challenging problems both from theoretical and computational points of view.

Acknowledgements The first author is partially supported by the Sofia University research grant No 181/2012; the second and the third authors are partially supported by the Bulgarian National Science Foundation under Grant DDVU02/71.

References

1. Adyutov, M.M., Klokov, Yu.A., Mikhailov, A.P.: Self-similar heat structures with reduced half width. Differ. Equ. **19**(7), 805–815 (1983)
2. Akhromeeva, T.S., Kurdyumov, S.P., Malinetzkii, G.G., Samarskii, A.A.: Nonstationary Structures and Diffusion-Induced Chaos. Nauka, Moscow (1992) (in Russian)
3. Bakirova, M.I., Borshukova, S.N., Dorodnicyn, V.A., Svirshchevskii, S.: Directed heat diffusion in a nonlinear anisotropic medium. Preprint IPM AN SSSR, vol. 182 (1985) (in Russian)
4. Bakirova, M.I., Dimova, S.N., Dorodnicyn, V.A., Kurdyumov, S.P., Samarskii, A.A., Svirshchevskii, S.: Invariant solutions of heat-transfer equation, describing directed heat diffusion and spiral waves in nonlinear medium. Sov. Phys. Dokl. **33**(3), 187–189 (1988)
5. Barenblatt, I.G.: Scaling, Self-similarity and Intermediate Asymptotics. Cambridge University Press, Cambridge (1996)
6. Bebernes, J., Eberly, D.: Mathematical problems from combustion theory. In: Applied Mathematical Sciences, vol. 83. Springer, New York (1989)
7. Berger, M., Kohn, R.: A rescaling algorithm for the numerical calculation of blowing up solutions. Commun. Pure Appl. Math. **41**, 841–863 (1988)
8. Borshukova, S.N., Yotova, A.I., Lazarov, R.D.: PAP TERMO for solving parabolic problems by the FEM. In: Numerical Methods for Solving Problems of Mathematical Physics. Computer Center of the Siberian Branch of USSR Academy of Sciences, pp. 31–41 (1985) (in Russian)
9. Budd, C.J., Piggott, M.D.: Geometric integration and its applications, In: Handbook of Numerical Analysis, Elsevier, vol. 18, pp. 111–241 (2003)
10. Budd, C.J., Huang, W., Russell, R.D.: Addaptivity with moving grids. Acta Numerica, Elsevier, vol. 18, 111–241 (2009)
11. Ibragimov, N.H. (Ed.): CRC Handbook of Lie Group Analysis of Differential Equations, vol. 1, Symmetries, Exact Solutions and Conservation Laws. CRC Press Inc., Boca Raton (1993)
12. Dimova S.: Numerical Investigation of Nonstationary Heat Structures. Dissertation for Doctor of Sciences, Dubna, Russia (2004) (in Russian)
13. Dimova, S.N., Chernogorova, T.P.: Asymptotically self-similar blow-up for a quasilinear parabolic equation byond some critical exponents. C.R. Acad. Bulg. Sci. **53**(12), 21–24 (2000)

14. Dimova, S.N., Chernogorova, T.P.: Nonsymmetric Galerkin finite element method with dynamic mesh refinement for singular nonlinear problems. J. Comp. Appl. Math. (Kiev University) **92**, 3–16 (2005)
15. Dimova, M., Dimova, S.: Numerical investigation of spiral structure solutions of a nonlinaer elliptic problem. In: Dimov, I., Dimova, S., Kolkovska, N. (eds.) NMA 2010. Lecture Notes in Computer Science, Springer, vol. 6046, pp. 395–403 (2011)
16. Dimova, S.N., Kaschiev, M.S.: Numerical analysis of two-dimensional eigen functions of burning of a nonlinear anysotropic medium. Commun. JINR, Dubna **11-88-876**(9), (1988) (in Russian)
17. Dimova, S.N., Vasileva, D.P.: Numerical realization of blow-up spiral wave solutions of a nonlinear heat-transfer equation. Int. J. Numer. Methods Heat Fluid Flow **4**(6), 497–511 (1994)
18. Dimova, S.N., Vasileva, D.P.: Lumped-mass finite element method with interpolation of the nonlinear coefficients for a quasilinear heat transfer equation. Numer. Heat Transf. B **28**, 199–215 (1995)
19. Dimova, S.N., Kaschiev, M.S., Kurdyumov, S.P.: Numerical analysis of the eigenfunctions of combustion of a nonlinear medium in the radial-symmetric case. USSR Comp. Math. Math. Phys. **29**(6), 61–73 (1989)
20. Dimova, M., Dimova, S., Vasileva, D.: Numerical investigation of a new class of waves in an open nonlinear heat-conducting medium. Cent. Eur. J. Math.
21. Dimova, S.N., Kaschiev, M.S., Koleva, M.G., Vasileva, D.P.: Numerical analysis of radially nonsymmetric structures in a nonlinear heat-conducting medium. Sov. Phys. Dokl. **39**(10), 673–676 (1994)
22. Dimova, S.N., Kaschiev, M.S., Koleva, M.G., Vasileva, D.P.: Numerical analysis of the blow-up regimes of combustion of two-component nonlinear heat-conducting medium. Comput. Math. Math. Phys. **35**(3), 303–319 (1995)
23. Dimova, S.N., Kaschiev, M.S., Koleva, M.G., Vasileva, D.P.: Numerical analysis of radially nonsymmetric blow-up solutions of a nonlinear parabolic problem. J. Comp. Appl. Math. **97**, 81–97 (1998)
24. Dorodnitsyn, V.A.: Application of Lie groups to difference equations. CRC Press, Chapman & Hall (2010)
25. Dorodnitsyn, V.A., Knyazeva, I.V., Svirshchevskii, S.R.: Group properties of the nonlinear heat equation with a source in 2D and 3D case. Differ. Equ. **19**(7), 901–908 (1983)
26. Elenin, G.G., Kurdyumov, S.P., Samarskii, A.A.: Nonstationary dissipative structures in a nonlinear heat conducting medium, Zh. Vychisl. Mat. Mat. Fiz. **23**, 380–390 (1983) (in Russian)
27. Eriksson, K., Thomee, V.: Galerkin methods for singular boundary value problems in one space dimension. Math. Comp. **42**(166), 345–367 (1984)
28. Fujita, H.: On the blowup of solutions of the Cauchy problem for $u_t = \triangle u + u^{1+\alpha}$. J. Fac. Sci. Univ. Tokyo Sect. A Math. **16**, 105–113 (1966)
29. Galaktionov, V.A.: Two methods for comparing solutions of parabolic equations. Sov. Phys. Dokl. **25**, 250–251 (1980)
30. Galaktionov, V.A., Dorodnicyn, V.A., Elenin, G.G., Kurdyumov, S.P., Samarskii, A.A.: A quasilinear equation of heat conduction with a source: peaking, localization, symmetry, exact solutions, asymptotic behavior, structures. J. Math. Sci. **4**(5), 1222–1292 (1988)
31. Galaktionov, V.A., Vazquez, J.L.: Continuation of blow-up solutions of nonlinear heat equation in several space dimentions. Commun. Pure Appl. Math. **50**(1), 1–67 (1997)
32. Galaktionov, V.A., Vazquez, J.L.: The problem of blow-up in nonlinear parabolic equations. Discrete Contin. Dyn. Syst. **8**(2), 399–433 (2002)
33. Galaktionov, V.A., Kurdyumov S.P., Samarskii, A.A.: On appriximate self-similar solutions of a class of quasilinear heat equations with a source. Math. USSR Sb. **52**, 155–180 (1985)
34. Galaktionov, V.A., Kurdyumov S.P., Samarskii, A.A.: On the method of stationary states for quasilinear parabolic equations. Math. USSR Sb. **67**, 449–471 (1990)
35. Gavurin, M.K.: Nonlinear equations and continuous analogues of iteration methods. Izvestia Vuzov Math. **5**, 18–31 (1958) (in Russian)

36. Gurevich, M.I., Tel'kovskaya, O.V.: Existence of self-similar solutions of a nonlinear heat-transfer equation with zero region in the vicinity of the origin. Preprint IAE, Moscow, 5565/1 (1992) (in Russian)
37. Hairer, E., Lubich, C., Wanner, G.: Geometric numerical integration. In: Structure-Preserving Algorithms for ODE. Springer, Berlin (2002)
38. Ivanova, D.I., Dimova, S.N., Kaschiev, M.S.: Numerical analysis of the onedimensional eigen functions of combustion of a nonlinear medium. Commun. JINR, Dubna **11-90-11** (1990) (in Russian)
39. Kalashnikov, A.S.: Some problems of the qualitative theory of non-linear degenerate second order parabolic equations. Russ. Math. Surv. **42**, 135–176 (1987) (in Russian)
40. Koleva M.G.: Numerical Analysis of the Eigen Functions of Combustion of a Nonlinear Heat-conducting Medium. Ph.D. Thesis, Sofia (2000) (in Bulgarian)
41. Koleva, M.G., Dimova, S.N., Kaschiev M.S.: Analisys of the eigen functions of combustion of a nonlinear medium in polar coordinates. Math. Model. **3**, 76–83 (1992)
42. Kurdyumov, S.P.: Nonlinear processes in dense plasma. In: Proceedings of the II International Conference on Plasma Theory, Kiev 1974. Naukova Dumka, Kiev, pp. 278–287 (1976); Preprint IPM, Moscow, vol. 18 (1975) (in Russian)
43. Kurdyumov S.P., Kurkina E.S., Malinezkii G.G., Samarskii A.A.: Dissipative structures in a nonlinear nonhomogenious burning medium. Sov. Phys. Dokl. **251**(3), 587–591 (1980) (in Russian)
44. Kurdyumov, S.P., Kurkina E.S., Potapov A.B., Samarskii A.A.: Complex multidimensional structures of burning of a nonlinear medium. USSR Comp. Math. Math. Phys. **26**(8), 1189–1205 (1986) (in Russian)
45. Kurkina, E.S., Kurdyumov, S.P.: The spectrum of dissipative structures developing in a peaking regime. Dokl. Acad. Nauk **395**(6), 743–748 (2004) (in Russian)
46. Levin, H.A.: The role of critical exponents in blowup theorems. SIAM Rev. **32**, 262–288 (1990)
47. Novikov, V.A., Novikov, E.A.: Stability control for explicit one-step integration methods for ODE. Dokl. Akad. Nauk SSSR **272**, 1058-1062 (1984) (in Russian)
48. Puzynin, I.V., Boyadzhiev, T.L., Vinitskii, S.I., Zemlyanaya, E.V., Puzynina, T.P., Chuluun-baatar, O.: Methods of computational physics for investigation of models of complex physical systems. Phys. Particals Nucley **38**, 70–116 (2007)
49. Samarskii, A.A., Sobol', I.M.: Examples of numerical computation of temperature waves. USSR Comp. Math. Math. Phys. **3**, 945–970 (1963)
50. Samarskii, A.A., Zmitrenko, N.V., Kurdyumov, S.P., Mikhailov, A.P.: Thermal structures and fundamental length in a medium with nonlinear heat-conduction and volumetric heat source. Sov. Phys. Dokl. **21**, 141–143 (1976)
51. Samarskii, A.A., Elenin, G.G., Kurdyumov, S.P., Mikhailov, A.P.: The burning of nonlinear medium in the form of a complex structure. Sov. Phys. Dokl. **22**, 737–739 (1977)
52. Samarskii, A.A., Galaktionov, V.A., Kurdyumov, S.P., Mikhailov, A.P.: Blowup in Problems for Quasilinear Parabolic Equations. Nauka, Moscow (1987) (in Russian); Valter de Gruyter & Co., Berlin (1995)
53. Vasileva D.P.: Numerical Analysis of Highly Nonstationary Processes in Nonlinear Heat-conducting Medium. Ph.D. Thesis, Sofia (1997) (in Bulgarian)
54. Wilhelmsson, H., Lazaro, E.: Reaction-Diffusion Problems in the Physics of Hot Plasmas. IOP Publishing, Bristol (2001)
55. Yi-Yong-Nie, Tomee, V.: A lumped mass FEM with quadrature for a nonlinear parabolic problem. IMA J. Numer. Anal. **5**, 371–396 (1985)
56. Zhidkov, E.P., Makarenko, G.I., Puzynin, I.V.: Continuous analog of Newton method for nonlinear problems of physics. Particals Nucley **4**(1), 127–166 (1973)(in Russian)
57. Zmitrenko, N., Kurdyumov, S.P., Mickhailov, A.P., Samarskii, A.A.: Localization of termo-nuclear combustion in plasma with electron heat-conductivity. Lett. JETF **26**(9), 620–624 (1977) (in Russian)

Efficient Parallel Algorithms for Unsteady Incompressible Flows

Jean-Luc Guermond and Peter D. Minev

Abstract The objective of this paper is to give an overview of recent developments on splitting schemes for solving the time-dependent incompressible Navier–Stokes equations and to discuss possible extensions to the variable density/viscosity case. A particular attention is given to algorithms that can be implemented efficiently on large parallel clusters.

Keywords Navier-Stokes • Fractional Time-Stepping • Projection Methods • Direction Splitting • Variable Density • Variable Viscosity

Mathematics Subject Classification (2010): 65N12, 65N15, 35Q30

1 Introduction

Most of the algorithms that are currently used for solving the time-dependent incompressible fluid flows are based on three basic ideas:

(i) The advection operator is discretized explicitly.
(ii) The diffusion is treated implicitly to avoid the stability constraint on the time step induced by the second-order diffusion operator. (This rule does not apply to direct simulation of turbulence since in that case the stability restriction is set by the advection term.)

J.-L. Guermond
Department of Mathematics, Texas A&M University 3368 TAMU, College Station, TX 77843-3368, USA. On leave from CNRS, France
e-mail: guermond@math.tamu.edu

P.D. Minev (✉)
Department of Mathematical and Statistical Sciences, University of Alberta, Edmonton, Alberta, Canada T6G 2G1
e-mail: minev@ualberta.ca

O.P. Iliev et al. (eds.), *Numerical Solution of Partial Differential Equations: Theory, Algorithms, and Their Applications*, Springer Proceedings in Mathematics & Statistics 45, DOI 10.1007/978-1-4614-7172-1_10,

(iii) The pressure and velocity are decoupled to avoid solving the coupled saddle-point problem. In most cases this decoupling is done by using the so-called projection methods; see Chorin [5] and Temam [27] for the earliest examples and Guermond et al. [15] for a review on this class of techniques.

This paper reviews recent developments related to item (iii). In particular we discuss a method recently introduced in Guermond and Minev [8]. It is a fractional time stepping technique that departs from the projection paradigm in the sense that it uses a pressure equation derived from a perturbation of the incompressibility constraint induced by direction splitting. This approach is particularly suitable for implementation on parallel platforms. We also discuss some new implicit schemes for the approximation of the parabolic subproblem listed in item (ii) in case of flows with variable density and viscosity.

This paper is organized as follows. The pressure-velocity decoupling schemes that are the most widely used are summarized in Sect. 2, and the corresponding convergence results are recalled. We propose in Sect. 3 some new schemes for flows with variable density and viscosity, and we analyze their stability for first-order time discretization.

2 Pressure-Velocity Decoupling Schemes

2.1 *Notation and Preliminaries*

We consider the time-dependent Navier–Stokes equations on a finite time interval $[0,T]$ and in a domain $\Omega = (0,1)^3$.

As suggested in item (i) above, it is reasonable to discretize the nonlinear term in the Navier–Stokes equations explicitly. Even if it is treated semi-implicitly, this term has no significant influence on the pressure-velocity coupling, and we henceforth mainly consider the time-dependent Stokes equations written in terms of velocity $\mathfrak{u}$ and pressure $\mathfrak{p}$, restricting ourselves for the time being to the case of constant density and viscosity:

$$\begin{cases} \partial_t \mathfrak{u} - \nu \nabla^2 \mathfrak{u} + \nabla \mathfrak{p} = \mathbf{f} \quad \text{in } \Omega \times [0,T], \\ \nabla \cdot \mathfrak{u} = 0 \quad \text{in } \Omega \times [0,T], \\ \mathfrak{u}|_{\partial\Omega} = 0 \quad \text{in [0,T]}, \qquad \text{and } \mathfrak{u}|_{t=0} = \mathfrak{u}_0 \quad \text{in } \Omega, \end{cases} \tag{1}$$

where $\mathbf{f}$ is a smooth source term and $\mathfrak{u}_0$ is a solenoidal initial velocity field with zero normal trace at the boundary of Ω. We consider homogeneous Dirichlet boundary conditions on the velocity for the sake of simplicity.

Let $\Delta t > 0$ be a time step and set $t^k = k\Delta t$ for $0 \le k \le K = [T/\Delta t]$. Let $\phi^0, \phi^1, \dots \phi^K$ be some sequence of functions in a Hilbert space E. We denote by $\phi_{\Delta t}$ this sequence, and we define the following discrete norms:

$$\|\phi_{\Delta t}\|_{\ell^2(E)} := \left(\Delta t \sum_{k=0}^{K} \|\phi^k\|_E^2\right)^{1/2}, \; \|\phi_{\Delta t}\|_{\ell^\infty(E)} := \max_{0\le k\le K}\left(\|\phi^k\|_E\right). \tag{2}$$

In addition, we denote the first divided differences of the elements of the sequence by $\delta_t\phi^k = \Delta t^{-1}(\phi^k - \phi^{k-1})$ and the sequence of first divided differences by $\delta_t\phi_{\Delta t}$, i.e., $\delta_t\phi_{\Delta t} = \delta_t\phi^k, k = 1,\dots,K$.

Given the functional space $X(\Omega)$, we denote by $X_{\int=0}(\Omega)$ its subspace of functions with a zero mean, i.e., $X_{\int=0}(\Omega) = \{v \in X : \int_\Omega v d\Omega = 0\}$.

We denote by c a generic constant that is independent of Δt but possibly depends on the data, the domain, and the solution. We shall use the expression $A \lesssim B$ to say that there exists a generic constant c such that $A \le cB$.

2.2 Projection Schemes

Historically the oldest and probably the most widely used decoupling algorithms nowadays contain two basic steps. The first one consists of an implicit discretization of the momentum equation in which the pressure is approximated explicitly or just ignored. The second step updates the pressure by solving a Poisson equation with a Neumann boundary condition. This step is equivalent to a L^2-projection of the predicted velocity onto the divergence-free subspace of $H_0(\mathrm{div},\Omega)$. Depending on the way these two steps are organized, these schemes can be subdivided into two broad categories: pressure correction and velocity correction. The most accurate variants of these schemes are second-order accurate in time on the velocity in the L^2-norm.

2.2.1 Pressure-Correction Projection Algorithm

In this class of schemes the momentum equation is approximated first by solving

$$\tfrac{1}{2\Delta t}(3\tilde{u}^{k+1} - 4u^k + u^{k-1}) - \nu\nabla^2\tilde{u}^{k+1} + \nabla p^k = f(t^{k+1}), \quad \tilde{u}^{k+1}|_{\partial\Omega} = 0. \tag{3}$$

Then the predicted velocity, $\tilde{u}^{k+1}$, is projected onto the divergence-free subspace of $H_0(\mathrm{div},\Omega)$ by solving the following elliptic problem in mixed form:

$$\begin{cases} \tfrac{1}{2\Delta t}(3u^{k+1} - 3\tilde{u}^{k+1}) + \nabla\phi^{k+1} = 0, \\ \nabla\cdot u^{k+1} = 0, \quad u^{k+1}\cdot n|_{\partial\Omega} = 0. \end{cases} \tag{4}$$

Finally, the pressure is updated explicitly by setting

$$p^{k+1} = \phi^{k+1} + p^k - \chi\nu\nabla\cdot\tilde{u}^{k+1}. \tag{5}$$

The scheme is said to be in standard form if $\chi = 0$ and is said to be in rotational form if $0 < \chi \leq 1$. The rotational form of the algorithm was introduced in this form in Timmermans et al. [28] and in a somewhat different form in Kim and Moin [18]. It is also sometimes referred to in the literature as the Gauge method, E and Liu [29]. The classification in incremental and non-incremental form was introduced in Guermond and Quartapelle [14]. The classification in standard and rotational form was introduced in Guermond and Shen [13].

It is sometimes debated in the literature whether the projection step should be solved in mixed form or in primal form, i.e.,

$$\begin{cases} \frac{1}{2\Delta t}(3u^{k+1} - 3\tilde{u}^{k+1}) + \nabla\phi^{k+1} = 0, \\ \nabla\cdot u^{k+1} = 0, \quad u^{k+1}\cdot n|_{\partial\Omega} = 0. \end{cases} \tag{6}$$

or

$$\begin{cases} \Delta\phi^{k+1} = \frac{3}{2\Delta t}\nabla\cdot\tilde{u}^{k+1}, \quad \partial_n\phi^{k+1}|_{\partial\Omega} = 0, \\ u^{k+1} = \tilde{u}^{k+1} - \frac{2\Delta t}{3}\phi^{k+1}. \end{cases} \tag{7}$$

This issue has been thoroughly investigated in Guermond [7], and it is shown therein that all the discrete implementations of these two variants of the projection step are equivalent in terms of stability and approximation.

The stability and convergence properties of the scheme are stated in the following theorem:

Theorem 2.1. *Under suitable initialization hypothesis and provided that the solution to* (1) *is smooth enough in time and space, the solution* $(u_{\Delta t}, \tilde{u}_{\Delta t}, p_{\Delta t})$ *of* (3)–(5) *satisfies the estimate:*

$$\|\mathfrak{u}_{\Delta t} - u_{\Delta t}\|_{\ell^2(\mathbf{L}^2(\Omega))} + \|\mathfrak{u}_{\Delta t} - \tilde{u}_{\Delta t}\|_{\ell^2(\mathbf{L}^2(\Omega))} \lesssim \Delta t^2,$$

$$\|\mathfrak{u}_{\Delta t} - \tilde{u}_{\Delta t}\|_{\ell^\infty(\mathbf{H}^1(\Omega))} + \|\mathfrak{p}_{\Delta t} - p_{\Delta t}\|_{\ell^\infty(L^2(\Omega))} \lesssim \Delta t, \text{ if } \chi = 0$$

$$\|\mathfrak{u}_{\Delta t} - u_{\Delta t}\|_{\ell^2(\mathbf{H}^1(\Omega))} + \|\mathfrak{u}_{\Delta t} - \tilde{u}_{\Delta t}\|_{\ell^2(\mathbf{H}^1(\Omega))} + \|\mathfrak{p}_{\Delta t} - p_{\Delta t}\|_{\ell^2(L^2(\Omega))} \lesssim \Delta t^{\frac{3}{2}},$$

$$\text{if } 0 < \chi \leq 1$$

Proof. See Guermond and Shen [13].

2.2.2 Velocity-Correction Projection Algorithm

In this class of methods one first computes an implicit approximation for the pressure at each time step by using an explicit approximation for the velocity, and then one uses this pressure approximation to update the velocity implicitly. The second-order incremental velocity-correction algorithm in rotational form is given by

$$\begin{cases} \frac{1}{2\Delta t}(3u^{k+1} - 4\tilde{u}^k + \tilde{u}^{k-1}) + \nu\nabla\times\nabla\times\tilde{u}^{\star,k+1} + \nabla p^{k+1} = f(t^{k+1}), \\ \nabla\cdot u^{k+1} = 0, \quad u^{k+1}\cdot n|_{\partial\Omega} = 0, \end{cases} \tag{8}$$

and

$$\frac{1}{2\Delta t}(3\tilde{u}^{k+1} - 3u^{k+1}) - \nu\nabla^2\tilde{u}^{k+1} - \nu\nabla\times\nabla\times\tilde{u}^{\star,k+1} = 0, \quad \tilde{u}^{k+1}|_{\partial\Omega} = 0. \tag{9}$$

This scheme was introduced in this form in Guermond and Shen [11,12]. It has been introduced in a somewhat different (although equivalent) form by Orszag et al. [21] and Karniadakis et al. [17]. The difference between the standard and rotational forms of the velocity-correction algorithm is that in the standard form, the $\nabla\times\nabla\times$ operator in (8) and (9) is substituted by the ∇^2 operator. Similarly to the pressure-correction algorithm, the rotational form of the velocity-correction algorithm yields a better pressure approximation than the standard form as stated in the following theorem.

Theorem 2.2. *If the solution to* (1) *is smooth enough in time and space, and under suitable initialization hypothesis, the solution* $(u_{\Delta t}, \tilde{u}_{\Delta t}, p_{\Delta t})$ *to* (8)–(9) *satisfies the estimates:*

$$\|\mathfrak{u}_{\Delta t} - u_{\Delta t}\|_{\ell^2(\mathbf{L}^2(\Omega))} + \|\mathfrak{u}_{\Delta t} - \tilde{u}_{\Delta t}\|_{\ell^2(\mathbf{L}^2(\Omega))} \lesssim \Delta t^2,$$

$$\|\mathfrak{u}_{\Delta t} - \tilde{u}_{\Delta t}\|_{\ell^2(\mathbf{H}^1(\Omega))} + \|\mathfrak{p}_{\Delta t} - p_{\Delta t}\|_{\ell^2(L^2(\Omega))} \lesssim \Delta t^{\frac{3}{2}}.$$

Proof. We refer to Guermond and Shen [12].

Remark 2.1. Note that the Poisson problems in mixed form (4) and (8), arising in the two schemes above, are in fact a consistent perturbation of the incompressibility constraint. This fact has been used by Rannacher [22] and Shen [25,26] to analyze the first- and second-order pressure-correction schemes.

2.3 Direction-Splitting Schemes

The computational bottleneck that is common to both schemes discussed in the previous subsection is the solution of the Poisson problem (7) for the pressure which, if solved iteratively, requires many more iterations to converge than the problem for the velocity. This problem can be tackled using geometric or algebraic multigrid algorithms that scale well on large distributed clusters. Alternatively, taking into account that the pressure Poisson equation of the projection schemes is just a regularization of the incompressibility constraint, we can explore other regularization options. This idea was explored in Guermond and Salgado [10] and Guermond and Minev [8] where we proposed to abandon the L^2-projection paradigm, which yields the Poisson equation (4) or (8), and to use instead a perturbation of the incompressibility that allows for a faster computation of the pressure. It is shown in Guermond and Minev [8] that a general perturbation of the form

$$\frac{\nabla\cdot\tilde{u}^{k+1}}{\Delta t} = A\phi^{k+1}, \tag{10}$$

can be used if the operator $A : D(A) \subset L^2_{\int=0}(\Omega) \to L^2_{\int=0}(\Omega)$, with domain $D(A) \subset H^1_{\int=0}(\Omega)$, satisfies the following properties:

$$\begin{cases} \|\nabla q\|^2_{\mathbf{L}^2} \le \langle Aq, q\rangle, & \forall q \in D(A), \\ \langle Ap, q\rangle = \langle p, Aq\rangle, & \forall p, q \in D(A). \end{cases} \tag{11}$$

This is a natural generalization of the usual Poisson equation used in the classical projection schemes. If the domain Ω is a rectangle in 2D or a parallelepiped in 3D, a good choice for A is

$$\begin{cases} A := (1-\partial_{xx})(1-\partial_{yy}), \\ D(A) := \left\{ p \in H^1_{\int=0}(\Omega) : \partial_{yy}p, Ap \in L^2(\Omega) : \right. \\ \quad \left. \partial_y p|_{y=0,1} = 0,\ \partial_x(1-\partial_{yy})p|_{x=0,1} = 0 \right\}, \end{cases} \tag{12}$$

in two space dimensions and

$$\begin{cases} A := (1-\partial_{xx})(1-\partial_{yy})(1-\partial_{zz}), \\ D(A) := \left\{ p \in H^1_{\int=0}(\Omega) : \partial_{zz}p, (1-\partial_{yy})(1-\partial_{zz})p, Ap \in L^2(\Omega) : \right. \\ \quad \partial_z p|_{z=0,1} = 0,\ \partial_y(1-\partial_{zz})p|_{y=0,1} = 0, \\ \quad \left. \partial_x(1-\partial_{yy})(1-\partial_{zz})p|_{x=0,1} = 0 \right\}, \end{cases} \tag{13}$$

in three space dimensions. The most important advantage of this perturbation is that the computation of the pressure requires the solution of one-dimensional tridiagonal systems only, and this can be done efficiently by using the Thomas algorithm. Another possible choice is the BPX preconditioner (see Bramble et al. [4] and Bramble and Zhang [3, Chap. II, Sect. 4] for the proof of uniform spectral equivalence) or the multigrid $\mathscr{V}$-cycle with a variable number of smoothing steps per level (see Bramble and Zhang [3, Chap. II, Sect. 7.4]).

We concentrate in the rest of this section on the direction-splitting perturbation (12) and (13). When combined with a direction-splitting technique for the momentum equations, the resulting scheme is very simple and efficient, particularly on large distributed clusters. For instance, upon using the Douglas scheme for approximating the momentum equation (see Douglas [6]), the overall procedure in three space dimensions is given by:

- *Pressure predictor:* Denoting by $\mathfrak{p}_0$ the pressure field at $t = 0$ and $\phi^{\star,-\frac{1}{2}}$ an approximation of $\frac{1}{2}\Delta t \partial_t \mathfrak{p}(0)$, the algorithm is initialized by setting $p^{-\frac{1}{2}} = \mathfrak{p}_0$ and $\phi^{-\frac{1}{2}} = \phi^{\star,-\frac{1}{2}}$. Then for all $k \geq 0$ a pressure predictor is computed as follows:

$$p^{\star,k+\frac{1}{2}} = p^{k-\frac{1}{2}} + \phi^{k-\frac{1}{2}}. \tag{14}$$

- *Velocity update:* The velocity field is initialized by setting $u^0 = \mathfrak{u}_0$, and for all $k \geq 0$ the velocity update is computed by solving the following series of one-dimensional problems: Find ξ^{k+1}, η^{k+1}, ζ^{k+1}, and u^{k+1} such that

$$\frac{\xi^{k+1} - u^k}{\Delta t} - \nabla^2 u^k + \nabla p^{\star,k+\frac{1}{2}} = f^{k+\frac{1}{2}}, \quad \xi^{k+1}|_{\partial\Omega} = 0, \tag{15}$$

$$\frac{\eta^{k+1} - \xi^{k+1}}{\Delta t} - \frac{1}{2}\partial_{xx}(\eta^{k+1} - u^k) = 0, \quad \eta^{k+1}|_{x=0,1} = 0, \tag{16}$$

$$\frac{\zeta^{k+1} - \eta^{k+1}}{\Delta t} - \frac{1}{2}\partial_{yy}(\zeta^{k+1} - u^k) = 0, \quad \zeta^{k+1}|_{y=0,1} = 0, \tag{17}$$

$$\frac{u^{k+1} - \zeta^{k+1}}{\Delta t} - \frac{1}{2}\partial_{zz}(u^{k+1} - u^k) = 0, \quad u^{k+1}|_{z=0,1} = 0. \tag{18}$$

- *Penalty step:* The pressure-correction $\phi^{k+\frac{1}{2}}$ is computed by solving

$$A\phi^{k+\frac{1}{2}} = -\frac{1}{\Delta t}\nabla{\cdot}u^{k+1}. \tag{19}$$

- *Pressure update:* The last sub-step of the algorithm consists of updating the pressure as follows:

$$p^{k+\frac{1}{2}} = p^{k-\frac{1}{2}} + \phi^{k+\frac{1}{2}} - \frac{\chi}{2}\nabla{\cdot}\left(u^{k+1} + u^k\right). \tag{20}$$

The two-dimensional version of the algorithm is obtained by skipping the last step in (18) and setting $u^{k+1} = \zeta^{k+1}$. The best error estimate proven to date for this algorithm is stated in the following theorem:

Theorem 2.3 ($\ell^2(\mathbf{L}^2)$ **Velocity Estimate).** *Assume that the space dimension is two. If* $0 < \chi \leq 1$, $\mathfrak{u}, \mathfrak{p}$ *is smooth enough, and under suitable initialization assumptions, the solution* $(u_{\Delta t}, p_{\Delta t})$ *of the scheme* (14)–(20) *in two space dimensions satisfies*

$$\|\mathfrak{u}_{\Delta t} - u_{\Delta t}\|_{\ell^2(\mathbf{L}^2)} \leq c\Delta t^{\frac{3}{2}}.$$

Proof. See Theorem 4.2 in Guermond et al. [16].

Note that this is a suboptimal convergence estimate and it cannot be easily extended to 3D since the usual argument used in the estimation of the error for the rotational form of the projection schemes does not apply in this case. Nevertheless, various numerical tests suggest that the scheme is as accurate on both, pressure and velocity, as the classical incremental projection schemes (see, e.g., Guermond et al. [16], Fig. 2).

In this paper we do not pay much attention to the spatial approximation for the velocity and pressure, but we should mention that it needs to satisfy the usual $inf-sup$ condition. An obvious candidate that suits the needs of the direction-splitting algorithms is the staggered finite volume grid based on the so-called MAC stencil. All our numerical experience with direction-splitting schemes so far is based on this discretization although other options can certainly be exploited.

The scheme discussed in this section has several advantages. It is quite clear that on a staggered MAC grid it requires the storage of only $d(d+1)$ one-dimensional tridiagonal matrices. This is much better than storing the entire d-dimensional matrix and allows to solve significantly larger local problems (per processor) on a parallel cluster. In addition, it requires the solution of tridiagonal systems only, which can be performed very efficiently with the Thomas algorithm. Its implementation on a parallel cluster is also very efficient since the Schur complements for the interface unknowns are also tridiagonal and therefore can also be solved with the same algorithm. Probably the most important advantage of this scheme is that it has very low communication costs. Indeed, if we presume that the grid is partitioned into blocks of equal size, then the solution of the Navier–Stokes equations per time step would require only one communication, per internal interface, of the unknowns on this interface, for each of the velocity components and the pressure. Therefore, the parallel performance of the scheme is very efficient. More details on the parallel implementation of the algorithm can be found in Guermond and Minev [9].

2.4 Non-commutative One-Dimensional Operators

The algorithms and the results mentioned in the previous section are applicable only if the domain Ω is simple, i.e., a rectangle in 2D or a parallelepiped in 3D. Otherwise, the integration by parts of the mixed derivative that appear in (Aq,q) and in the momentum equation cannot be done. This difficulty is probably one of the main reasons direction-splitting schemes were abandoned after massive computer resources have become readily available. However, this problem can be tackled, at least partially, by the use of penalty or fictitious domain methods (see Korobytsina [19], Angot [1], Kuttykozhaeva et al. [20]). Unfortunately, the accuracy of the spatial approximation achievable by these approaches is usually suboptimal for the momentum equation. This problem can be solved by modifying the approximation of second-order spatial derivatives in the vicinity of the boundary of the domain to achieve optimal approximation there (see Angot et al. [2]). When all the fixes mentioned above are applied, the resulting one-dimensional discrete operators no longer commute. Nevertheless, as shown by Samarskii and

Vabishchevich [24], Sect. 2.2.3, the 2D version of the Douglas scheme (15)–(18) is still unconditionally stable even if the one-dimensional operators involved in the direction splitting do not commute. This proof does not generalize to the 3D case. In fact, it was experimentally verified in Angot et al. [2] that the 3D direction-splitting scheme (15)–(18) is unconditionally unstable in some cases of non-commutative operators. Therefore, the authors of Angot et al. [2] proposed the following modification of the Douglas scheme, which seems to be unconditionally stable for parabolic problems with non-commutative one-dimensional second-order operators (this property was verified only numerically):

$$\frac{\boldsymbol{\xi}^{n+1}-\mathbf{u}^n}{\Delta t}-(A_1\boldsymbol{\eta}^n+A_2\boldsymbol{\zeta}^n+A_3\mathbf{u}^n)=\mathbf{f}^{n+1/2}, \tag{21}$$

$$\frac{\boldsymbol{\eta}^{n+1}-\boldsymbol{\xi}^{n+1}}{\Delta t}-\frac{1}{2}A_1\left(\boldsymbol{\eta}^{n+1}-\boldsymbol{\eta}^n\right)=0, \tag{22}$$

$$\frac{\boldsymbol{\zeta}^{n+1}-\boldsymbol{\eta}^{n+1}}{\Delta t}-\frac{1}{2}A_2\left(\boldsymbol{\zeta}^{n+1}-\boldsymbol{\zeta}^n\right)=0, \tag{23}$$

$$\frac{\mathbf{u}^{n+1}-\boldsymbol{\zeta}^{n+1}}{\Delta t}-\frac{1}{2}A_3\left(\mathbf{u}^{n+1}-\mathbf{u}^n\right)=0. \tag{24}$$

Here A_1, A_2, A_3 are positive, possibly non-commutative operators, resulting from a penalty approximation of the original problem in a complex-shaped domain (see Angot et al. [2] for details). The above stability claim is based only on numerical evidence; a rigorous analysis of this scheme is yet to be done.

Another second-order scheme is provided by Samarskii and Vabishchevich [24], Sect. 4.3.2:

$$\frac{\boldsymbol{\eta}^{n+1}-\boldsymbol{\eta}^{n-1}}{\Delta t}+\mu A_1\left(\boldsymbol{\eta}^{n+1}-2\boldsymbol{\eta}^n+\boldsymbol{\eta}^{n-1}\right)+A_1\boldsymbol{\eta}^n+A_2\boldsymbol{\zeta}^n+A_3\mathbf{u}^n=\mathbf{f}^{n+1/2},$$

$$\frac{\boldsymbol{\zeta}^{n+1}-\boldsymbol{\zeta}^{n-1}}{\Delta t}+\mu A_2\left(\boldsymbol{\zeta}^{n+1}-2\boldsymbol{\zeta}^n+\boldsymbol{\zeta}^{n-1}\right)+A_1\boldsymbol{\eta}^n+A_2\boldsymbol{\zeta}^n+A_3\mathbf{u}^n=\mathbf{f}^{n+1/2},$$

$$\frac{\mathbf{u}^{n+1}-\mathbf{u}^{n-1}}{\Delta t}+\mu A_3\left(\mathbf{u}^{n+1}-2\mathbf{u}^n+\mathbf{u}^{n-1}\right)+A_1\boldsymbol{\eta}^n+A_2\boldsymbol{\zeta}^n+A_3\mathbf{u}^n=\mathbf{f}^{n+1/2}. \tag{25}$$

The scheme has been proved therein to be unconditional stable if the parameter μ is large enough.

For more details on the accuracy and stability of direction splittings with non-commutative operators, and on the spatial discretization of the Navier–Stokes equations in case of complex-shaped domains, the reader is referred to Angot et al. [2] and Samarskii and Vabishchevich [24].

3 Variable Density or Viscosity Flows

The scheme (14)–(20) is applicable only when the density and the viscosity are both constant. The computational practice often requires the solution of incompressible problems with variable density and viscosity, multicomponent flows being the most obvious example. Therefore, we propose in this note an extension of the direction-splitting scheme which is unconditionally stable in the variable density/viscosity case.

3.1 The Perturbation Algorithms

Since the major difficulty in this case arises when splitting the momentum equation, we simplify the analysis by focusing our attention on the heat equation with variable coefficients:

$$\begin{cases} \rho \partial_t \mathrm{u} - \nabla\cdot \nu \nabla \mathrm{u} = \mathbf{f} \quad \text{in } \Omega \times [0,T], \\ \mathrm{u}|_{\partial\Omega} = 0 \quad \text{in [0,T]}, \quad \text{and } \mathrm{u}|_{t=0} = \mathrm{u}_0 \quad \text{in } \Omega, \end{cases} \tag{26}$$

where ρ, ν are functions of the spatial variables and time. We further assume that there exist strictly positive numbers $\check{\rho}$, $\check{\nu}$, $\hat{\rho}$, and $\hat{\nu}$ so that $\check{\rho} \leq \rho(x,t) \leq \hat{\rho}$, $\check{\nu} \leq \nu(x,t) \leq \hat{\nu}$ for all $(x,t) \in \Omega \times [0,T]$. We also assume that $\nu \in W^{1,\infty}((0,T);L^\infty(\Omega))$ and we set $\hat{\nu}_t = \|\partial_t \nu\|_{L^\infty(\Omega\times(0,T))}$.

We start with the following implicit scheme with nonconstant coefficients

$$\rho^k \delta_t u^k = \nabla\cdot \left(\nu^{k+1} \nabla u^k\right), \qquad u^k|_{\partial\Omega} = 0$$

Next we perturb it so that the resulting fully discrete linear system has time-independent matrices:

$$\gamma \delta_t u^{k+1} - \sigma \nabla^2 u^{k+1} = \left(\gamma - \rho^k\right) \delta_t u^k + \nabla\cdot \left((\nu^{k+1} - \sigma)\nabla u^k\right), \tag{27}$$

where γ, σ are positive constants yet to be fully defined (see Theorem 3.1 below). Note that Samarskii [23] proposed to use similar perturbations in order to regularize unconditionally or conditionally stable schemes. In the present case we start with a stable scheme and employ Samarskii's trick solely to make the matrix of the discrete problem time independent and suitable for further direction splitting. Provided γ and σ are chosen appropriately, the scheme remains unconditionally stable as stated in the following theorem:

Theorem 3.1. *Assume that* $\gamma \geq \hat{\rho}$, $\sigma \geq 0.5\hat{\nu}$, $\|\partial_t \nu\|_{L^\infty(\Omega\times[0,T])} \leq \hat{\nu}_t < \infty$, *and choose* $\delta_t u^0$ *to satisfy* $\rho^0 \delta_t u^0 = \nabla\cdot \left(\nu^0 \nabla u^0\right)$. *Then if the solution of* (26) *is smooth enough* *and under suitable initialization assumptions, the following stability estimate for the solution of* (27) *holds:*

$$\begin{aligned}&\Delta t\gamma\|\delta_t u_{\Delta t}\|^2_{\ell^\infty(\mathbf{L}^2(\Omega))}+\check{\rho}\|\delta_t u_{\Delta t}\|^2_{\ell^2(\mathbf{L}^2(\Omega))}+\\&(2\sigma-\hat{\nu})\,\Delta t\|\nabla\delta_t u_{\Delta t}\|^2_{\ell^2(\mathbf{L}^2(\Omega))}+\|\sqrt{\nu}\nabla u_{\Delta t}\|^2_{\ell^\infty(\mathbf{L}^2(\Omega))}<\\&\Delta t\,(\gamma-\check{\rho})\,\|\delta_t u^0\|^2_{\mathbf{L}^2(\Omega)}+\|\sqrt{\nu^0}\nabla u^0\|^2_{\mathbf{L}^2(\Omega)}+\Delta t\hat{\nu}_t\sum_{k=0}^{K-1}\|\nabla u^k\|^2_{\mathbf{L}^2(\Omega)}.\end{aligned}$$

Proof. We first rewrite (27) as follows: $\gamma\delta_t u^{k+1}-\sigma\Delta t\nabla^2\delta_t u^{k+1}=\left(\gamma-\rho^k\right)\delta_t u^k+\nabla\cdot\left(\nu^{k+1}\nabla u^k\right)$, and then following an idea from Samarskii and Vabishchevich [24], Sect. 1.2.2, we multiply the equation by $\delta_t u^{k+1}$ and use the identity $u^k=0.5(u^{k+1}+u^k)-0.5\Delta t\delta_t u^{k+1}$ to obtain

$$\begin{aligned}&\gamma\|\delta_t u^{k+1}\|^2_{\mathbf{L}^2(\Omega)}+\sigma\Delta t\|\delta_t\nabla u^{k+1}\|^2_{\mathbf{L}^2(\Omega)}=\left((\gamma-\rho)\,\delta_t u^k,\delta_t u^{k+1}\right)+\\&\frac{\Delta t}{2}\left(\nu^{k+1}\nabla\delta_t u^{k+1},\nabla\delta_t u^{k+1}\right)-\frac{1}{2\Delta t}\left(\nu^{k+1}\nabla\left(u^{k+1}+u^k\right),\nabla\left(u^{k+1}-u^k\right)\right).\end{aligned}$$

The inequality $|\gamma-\rho^k|\leq\gamma-\check{\rho}$ and the conditions $\gamma\geq\hat{\rho}$, $\sigma\geq 0.5\hat{\nu}$ immediately yield

$$\begin{aligned}&\frac{1}{2}(\gamma+\check{\rho})\,\|\delta_t u^{k+1}\|^2_{\mathbf{L}^2(\Omega)}+\left(\sigma-\frac{\hat{\nu}}{2}\right)\Delta t\|\delta_t\nabla u^{k+1}\|^2_{\mathbf{L}^2(\Omega)}+\\&\frac{1}{2\Delta t}\|\sqrt{\nu^{k+1}}\nabla u^{k+1}\|^2_{\mathbf{L}^2(\Omega)}\leq\frac{1}{2}(\gamma-\check{\rho})\,\|\delta_t u^k\|^2_{\mathbf{L}^2(\Omega)}+\\&\frac{1}{2\Delta t}\|\sqrt{\nu^k}\nabla u^k\|^2_{\mathbf{L}^2(\Omega)}+\frac{1}{2\Delta t}\left(\left(\nu^{k+1}-\nu^k\right)\nabla u^k,\nabla u^k\right),\end{aligned}$$

which after summing for $k=0,\dots K-1$, taking into account that $\gamma-\check{\rho}<\gamma$, and setting $\hat{\nu}_t:=\|\partial_t\nu\|_{L^\infty(\Omega\times[0,T])}$, gives

$$\begin{aligned}&\Delta t\gamma\|\delta_t u_{\Delta t}\|^2_{\ell^\infty(\mathbf{L}^2(\Omega))}+\check{\rho}\|\delta_t u_{\Delta t}\|^2_{\ell^2(\mathbf{L}^2(\Omega))}+\\&(2\sigma-\hat{\nu})\,\Delta t\|\nabla\delta_t u_{\Delta t}\|^2_{l^2(\mathbf{L}^2(\Omega))}+\|\sqrt{\nu}\nabla u_{\Delta t}\|^2_{\ell^\infty(\mathbf{L}^2(\Omega))}<\\&\Delta t\,(\gamma-\check{\rho})\,\|\delta_t u^0\|^2_{\mathbf{L}^2(\Omega)}+\|\sqrt{\nu^0}\nabla u^0\|^2_{\mathbf{L}^2(\Omega)}+\Delta t\hat{\nu}_t\sum_{k=0}^{K-1}\|\nabla u^k\|^2_{\mathbf{L}^2(\Omega)},\end{aligned}\tag{28}$$

which concludes the proof.

Note that this scheme is useful for equations with time-dependent coefficients since it avoids recomputing the stiffness and mass matrices at each time step.

3.2 Direction-Splitting Algorithms

If Ω is a rectangle in two space dimensions or a parallelepiped in three space dimensions, the above algorithm is suitable for further direction splitting. The factorized form of the direction-splitting algorithm for the three-dimensional version of (27) is given by

$$\left(I-\frac{\sigma}{\gamma}\Delta t\partial_{xx}\right)\left(I-\frac{\sigma}{\gamma}\Delta t\partial_{yy}\right)\left(I-\frac{\sigma}{\gamma}\Delta t\partial_{zz}\right)\delta_t u^{k+1} = \left(1-\frac{\rho^k}{\gamma}\right)\delta_t u^k + \frac{1}{\gamma}\nabla\cdot\left(\nu^{k+1}\nabla u^k\right), \tag{29}$$

where I is the identity operator. The factorized form in two space dimensions is obtained by truncating the operator product in the left- hand side. This direction-splitting scheme is also unconditionally stable and satisfies stability estimates that are similar to those stated Theorem 3.1:

Theorem 3.2. *Under the assumptions of Theorem 3.1, the following stability estimate holds for the solution of* (29) *in three space dimensions:*

$$\begin{aligned}
&\Delta t\gamma\|\delta_t u_{\Delta t}\|^2_{\ell^\infty(\mathbf{L}^2(\Omega))} + \check{\rho}\|\delta_t u_{\Delta t}\|^2_{\ell^2(\mathbf{L}^2(\Omega))} + \\
&(2\sigma-\hat{\nu})\Delta t\|\nabla\delta_t u_{\Delta t}\|^2_{l^2(\mathbf{L}^2(\Omega))} + \|\sqrt{\nu}\nabla u_{\Delta t}\|^2_{\ell^\infty(\mathbf{L}^2(\Omega))} \le \\
&\Delta t(\gamma-\check{\rho})\|\delta_t u^0\|^2_{\mathbf{L}^2(\Omega)} + \|\sqrt{\nu^0}\nabla u^0\|^2_{\mathbf{L}^2(\Omega)} + \Delta t\hat{\nu}_t\sum_{k=0}^{K-1}\|\nabla u^k\|^2_{\mathbf{L}^2(\Omega)}.
\end{aligned}$$

Proof. We first note that (29) can be rewritten as follows:

$$\gamma\delta_t u^{k+1} - \sigma\Delta t\nabla^2\delta_t u^{k+1} + B\delta_t u^{k+1} = (\gamma-\rho)\,\delta_t u^k + \nabla\cdot\nu^{k+1}\nabla u^k,$$

where the operator B, with domain $H^1_0(\Omega)\cap H^3(\Omega)$, is defined as follows:

$$Bv = \frac{(\sigma\Delta t)^2}{\gamma}\partial_{xx}\partial_{yy}v + \frac{(\sigma\Delta t)^2}{\gamma}\partial_{yy}\partial_{zz}v + \frac{(\sigma\Delta t)^2}{\gamma}\partial_{xx}\partial_{zz}v - \frac{(\sigma\Delta t)^3}{\gamma^2}\partial_{xx}\partial_{yy}\partial_{zz}v.$$

It is remarkable that all the mixed derivatives can be integrated by parts if the domain has a simple shape as shown in Guermond et al. [16], i.e., the operator B is nonnegative. Then we can apply the same arguments as in Theorem 3.1 to obtain the desired result.

3.3 Variable Density Navier–Stokes Equations

In the case of Navier–Stokes equations with variable density, the perturbation of the incompressibility constraint (19) needs some modification too. The usual pressure Poisson equation associated with the projection schemes in such situation is

$$\nabla\cdot\left(\frac{1}{\rho}\nabla\phi^{k+1}\right)=\frac{\beta}{\Delta t}\nabla\cdot u^{k+1},\quad \partial_n\phi^{k+1}|_{\partial\Omega}=0, \tag{30}$$

with β being a coefficient depending on the discretization of the velocity time derivative. This formulation is inconvenient for a direction splitting. Besides, in case of large density variations, the resulting linear system is hard to solve. To avoid this difficulty, Guermond and Salgado [10] proposed to use the following perturbation of the incompressibility:

$$\Delta\phi^{k+1}=\frac{\beta\check{\rho}}{\Delta t}\nabla\cdot u^{k+1},\quad \partial_n\phi^{k+1}|_{\partial\Omega}=0, \tag{31}$$

with $\check{\rho}$ being a positive constant such that $\check{\rho}\le\rho(x,t),\forall x,t$. Then, the overall first-order approximation to the time-dependent Stokes problem is given by

$$\begin{cases}\gamma\delta_t u^{k+1}-\sigma\Delta u^{k+1}=\left(\gamma-\rho^k\right)\delta_t u^k+\nabla\cdot\left((v^{k+1}-\sigma)\nabla u^k\right)-\nabla p^k+f^{k+1},\\ \Delta p^{k+1}=\dfrac{\check{\rho}}{\Delta t}\nabla\cdot u^{k+1},\quad \partial_n p^{k+1}|_{\partial\Omega}=0.\end{cases} \tag{32}$$

The unconditional stability and optimal convergence of this scheme can be proven along the same lines as the analysis in Guermond and Salgado [10]; however, the proof is very technical and is beyond the scope of this paper.

Similarly, this idea can be combined with the direction-splitting perturbation into the equation

$$Ap^{k+1}=\frac{\check{\rho}}{\Delta t}\nabla\cdot u^{k+1},\quad \partial_n p^{k+1}|_{\partial\Omega}=0, \tag{33}$$

which can in turn be combined with the following discretization of the momentum equation

$$\begin{aligned}&\left(I-\frac{\sigma}{\gamma}\Delta t\partial_{xx}\right)\left(I-\frac{\sigma}{\gamma}\Delta t\partial_{yy}\right)\left(I-\frac{\sigma}{\gamma}\Delta t\partial_{zz}\right)\delta_t u^{k+1}=\\ &\left(1-\frac{\rho^k}{\gamma}\right)\delta_t u^k+\frac{1}{\gamma}\nabla\cdot\left(v^{k+1}\nabla u^k\right)-\nabla p^k+f^{k+1},\end{aligned} \tag{34}$$

to yield a first-order direction-splitting scheme. The analysis of this scheme would be significantly more complicated than the analysis of (32).

Both schemes in this section can be extended to second order of accuracy.

4 Schemes for Equations Involving Mixed Derivatives

In some cases the momentum equations of the (unsteady) Stokes system may need to be used in the following equivalent form:

$$\partial_t \mathrm{u} - \nu\nabla\cdot\left(\nabla\mathrm{u} + \nabla\mathrm{u}^T\right) - \lambda\nabla\nabla\cdot\mathrm{u} + \nabla\mathrm{p} = 0, \qquad \nabla\cdot\mathrm{u} = 0, \tag{35}$$

where we assume for simplicity that ν is constant. Examples of such situations are flows involving fluid-structure interaction or floating rigid particles. Using the relation $\nabla\cdot\nabla\mathrm{u}^T = \nabla\nabla\cdot\mathrm{u}$, the above equation can be reformulated as follows:

$$\partial_t \mathrm{u} - \nu\Delta\mathrm{u} - (\nu+\lambda)\nabla\nabla\cdot\mathrm{u} + \nabla\mathrm{p} = 0, \qquad \nabla\cdot\mathrm{u} = 0, \tag{36}$$

Sometimes, an additional term $\nabla\nabla\cdot\mathrm{u}$ is added to the momentum equation for a better control of the divergence of the velocity field. This term couples the different Cartesian components of the velocity and makes the overall solution procedure clumsy. This additional coupling can be avoided if we use the following scheme. For simplicity we present the scheme in two space dimensions (its extension to three space dimensions is evident) and ignore the pressure because it can be handled by any of the splitting approaches described above. The first-order version of the scheme is as follows:

$$\begin{cases} \frac{1}{\Delta t}(u_1^{k+1} - u_1^k) & -\nu\nabla^2 u_1^{k+1} - \gamma\partial_{xx}u_1^{k+1} - \gamma\alpha\partial_{xx}(u_1^{k+1} - u_1^k) = \nu\partial_{xy}u_2^k \\ \frac{1}{\Delta t}(u_2^{k+1} - u_2^k) & -\nu\nabla^2 u_2^{k+1} - \gamma\partial_{yy}u_2^{k+1} - \gamma\alpha\partial_{yy}(u_2^{k+1} - u_2^k) = \nu\partial_{xy}u_1^k, \end{cases} \tag{37}$$

where $\alpha \geq 1$, u_j^k is the j-th Cartesian component of the velocity vector u^k, and we set $\gamma := \nu + \lambda$. The stability of this algorithm is established in the following theorem:

Theorem 4.3. *Under suitable initialization and smoothness assumptions and assuming that* $\alpha \geq 1$, *the algorithm* (37) *is unconditionally stable, i.e.,*

$$\begin{aligned} &\|u_{\Delta t}\|^2_{\ell^\infty(\mathbf{L}^2(\Omega))} + \Delta t\|\delta_t u\|^2_{\ell^2(\mathbf{L}^2(\Omega))} + \nu\|\nabla u\|^2_{\ell^2(\mathbf{L}^2(\Omega))} + \gamma\|\nabla\cdot u_{\Delta t}\|^2_{\ell^2(L^2(\Omega))} \\ &(\alpha+1)\gamma\Delta t\Big(\|\partial_x u^{k+1}_{1\Delta t}\|^2_{\ell^\infty(L^2(\Omega))} + \|\partial_y u_{2\Delta t}\|^2_{\ell^\infty(L^2(\Omega))}\Big) \leq \\ &\|u^0\|^2_{\mathbf{L}^2(\Omega)} + (\alpha+1)\gamma\Delta t\left(\|\partial_x u_1^0\|^2_{\mathbf{L}^2(\Omega)} + \|\partial_y u_2^0\|^2_{\mathbf{L}^2(\Omega)}\right) \end{aligned} \tag{38}$$

Proof. Multiplying (37) by $2\Delta t u^{k+1}$ and using the identity $2(a-b,a) = \|a\|^2 + \|a-b\|^2 - \|b\|^2$, we obtain

$$\begin{aligned} &\|u_1^{k+1}\|^2_{L^2(\Omega)} + \Delta t^2\|\delta_t u_1^{k+1}\|^2_{L^2(\Omega)} + \nu\Delta t\|\nabla u_1^{k+1}\|^2_{\mathbf{L}^2(\Omega)} + \\ &2\gamma\Delta t\Big(\|\partial_x u_1^{k+1}\|^2_{L^2(\Omega)} + (\partial_x u_1^{k+1}, \partial_y u_2^{k+1}) + (\partial_x u_1^{k+1}, \partial_y(u_2^k - u_2^{k+1}))\Big) + \end{aligned}$$

$$2\alpha\gamma\Delta t(\partial_x u_1^{k+1}, \partial_x(u_1^{k+1} - u_1^k)) = \|u_1^k\|^2_{L^2(\Omega)}$$

$$\|u_2^{k+1}\|^2_{L^2(\Omega)} + \Delta t^2\|\delta_t u_2^{k+1}\|^2_{L^2(\Omega)} + \nu\Delta t\|\nabla u_2^{k+1}\|^2_{\mathbf{L}^2(\Omega)} +$$

$$2\gamma\Delta t\Big(\|\partial_y u_2^{k+1}\|^2_{L^2(\Omega)} + (\partial_y u_2^{k+1}, \partial_x u_1^{k+1}) + (\partial_y u_2^{k+1}, \partial_x(u_1^k - u_1^{k+1}))\Big) +$$

$$2\alpha\gamma\Delta t(\partial_y u_2^{k+1}, \partial_y(u_2^{k+1} - u_2^k)) = \|u_2^k\|^2_{L^2(\Omega)}$$

Summing the two equations gives

$$\|u^{k+1}\|^2_{\mathbf{L}^2(\Omega)} + \Delta t^2\|\delta_t u^{k+1}\|^2_{\mathbf{L}^2(\Omega)} + \nu\Delta t\|\nabla u^{k+1}\|^2_{\mathbf{L}^2(\Omega)} +$$

$$2\gamma\Delta t\Big(\|\nabla\cdot u^{k+1}\|^2_{L^2(\Omega)} + (\nabla\cdot u^{k+1}, \partial_x(u_1^k - u_1^{k+1}) + \partial_y(u_2^k - u_2^{k+1}))\Big)$$

$$+ 2(\alpha+1)\gamma\Delta t\Big((\partial_x u_1^{k+1}, \partial_x(u_1^{k+1} - u_1^k)) + (\partial_y u_2^{k+1}, \partial_y(u_2^{k+1} - u_2^k)\Big) = \|u^k\|^2_{\mathbf{L}^2(\Omega)}.$$

Using the inequality $|ab| \leq \frac{1}{4}a^2 + b^2$, we obtain

$$\|u^{k+1}\|^2_{\mathbf{L}^2(\Omega)} + \Delta t^2\|\delta_t u^{k+1}\|^2_{\mathbf{L}^2(\Omega)} + \nu\Delta t\|\nabla u^{k+1}\|^2_{\mathbf{L}^2(\Omega)} + 2\gamma\Delta t\Big(\frac{1}{2}\|\nabla\cdot u^{k+1}\|^2_{L^2(\Omega)} -$$

$$\|\partial_x(u_1^k - u_1^{k+1}\|^2_{L^2(\Omega)} - \|\partial_y(u_2^k - u_2^{k+1}\|^2_{L^2(\Omega)}\Big) + (\alpha+1)\gamma\Delta t\Big(\|\partial_x u_1^{k+1}\|^2_{L^2(\Omega)} +$$

$$\|\partial_x(u_1^{k+1} - u_1^k))\|^2_{L^2(\Omega)} + \|\partial_y u_2^{k+1}\|^2_{L^2(\Omega)} + \|\partial_y(u_2^{k+1} - u_2^k))\|^2_{L^2(\Omega)}\Big)$$

$$\leq \|u^k\|^2_{\mathbf{L}^2(\Omega)} + (\alpha+1)\gamma\Delta t(\|\partial_x u_1^k\|^2_{L^2(\Omega)} + \|\partial_y u_2^k\|^2_{L^2(\Omega)}).$$

Using the assumption that $\alpha \geq 1$, we finally derive

$$\|u^{k+1}\|^2_{\mathbf{L}^2(\Omega)} + \Delta t^2\|\delta_t u^{k+1}\|^2_{\mathbf{L}^2(\Omega)} + \nu\Delta t\|\nabla u^{k+1}\|^2_{\mathbf{L}^2(\Omega)} + \gamma\Delta t\|\nabla\cdot u^{k+1}\|^2_{L^2(\Omega)}$$

$$+ (\alpha+1)\gamma\Delta t\Big(\|\partial_x u_1^{k+1}\|^2_{L^2(\Omega)} + \|\partial_y u_2^{k+1}\|^2_{L^2(\Omega)}\Big)$$

$$\leq \|u^k\|^2_{\mathbf{L}^2(\Omega)} + (\alpha+1)\gamma\Delta t(\|\partial_x u_1^k\|^2_{L^2(\Omega)} + \|\partial_y u_2^k\|^2_{L^2(\Omega)}).$$

Summing for $k = 1, \ldots, K-1$ yields the desired result.

This scheme is also suitable for a further direction splitting.

5 Conclusions

From the discussions above we can draw the following conclusions: (a) The most efficient schemes for the unsteady Navier–Stokes equations are based on some sort of decoupling of pressure and velocity at each time step. A very efficient algorithm for parallel clusters is provided by (14)–(20). (b) There exist direction-splitting algorithms that are unconditionally stable for the momentum equation in complex-shaped domains. The pressure equation, however, must be extended to a simple-shaped domain if the perturbation of the incompressibility constraint is in the form of (19). It is an open question whether this type of pressure equation can "fit" the boundary of a complex domain with appropriate boundary conditions. (c) In case of variable density/viscosity flows, which include, for instance, multicomponent flows, there is no need for recomputation of the implicit discrete operator at each time step. The perturbation of the momentum equation in the form of (27) is unconditionally stable (the advection is not taken into account in this statement). The example provided in this paper is first-order accurate in time, but it can easily be extended to higher order.

Acknowledgements This material is based upon work supported by the National Science Foundation grants DMS-0713829, by the Air Force Office of Scientific Research, USAF, under grant/contract number FA9550-09-1-0424, and a discovery grant of the National Science and Engineering Research Council of Canada. This publication is also partially based on work supported by Award No. KUS-C1-016-04, made by King Abdullah University of Science and Technology (KAUST).

References

1. Angot, P.: Analysis of singular perturbations on the Brinkman problem for fictitious domain models of viscous flows. Math. Methods Appl. Sci. **22**(16), 1395–1412 (1999)
2. Angot, P., Keating, J., Minev, P.: A direction splitting algorithm for incompressible flow in complex geometries. Comput. Methods Appl. Mech. Eng. **117**, 111–120 (2012)
3. Bramble, J., Zhang, X.: The analysis of multigrid methods. In: Handbook of Numerical Analysis, vol. VII, pp. 173–415. North-Holland, Amsterdam (2000)
4. Bramble, J., Pasciak, J., Xu, J.: Parallel multilevel preconditioners. Math. Comp. **55**(191), 1–22 (1990)
5. Chorin, A.: Numerical solution of the Navier-Stokes equations. Math. Comp. **22**, 745–762 (1968)
6. Douglas, J., Jr.: Alternating direction methods for three space variables. Numer. Math. **4**, 41–63 (1962)
7. Guermond, J.-L.: Some practical implementations of projection methods for Navier-Stokes equations. Modél. Math. Anal. Num. **30**, 637–667 (1996)
8. Guermond, J.-L., Minev, P.: A new class of massively parallel direction splitting schemes for the incompressible Navier-Stokes equations. Comp. Methods Appl. Mech. Eng. **200**: 2083–2093 (2011)
9. Guermond, J.-L., Minev, P.: Start-up flow in a three-dimensional lid-driven cavity by means of a massively parallel direction splitting algorithm. Int. J. Numer. Meth. Fluids. **68**, 856–871 (2012)
10. Guermond, J.-L., Salgado, A.: A splitting method for incompressible flows with variable density based on a pressure Poisson equation. J. Comput. Phys. **228**(8), 2834–2846 (2009)
11. Guermond, J., Shen, J.: Quelques résultats nouveaux sur les méthodes de projection. C. R. Acad. Sci. Paris Sér. I **333**, 1111–1116 (2001)
12. Guermond, J., Shen, J.: Velocity-correction projection methods for incompressible flows. SIAM J. Numer. Anal. **41**(1), 112–134 (2003)

13. Guermond, J.L., Shen, J.: On the error estimates for the rotational pressure-correction projection methods. Math. Comp. **73**(248), 1719–1737 (electronic) (2004)
14. Guermond, J.-L., Quartapelle, L.: On stability and convergence of projection methods based on pressure Poisson equation. Int. J. Numer. Methods Fluids **26**(9), 1039–1053 (1998)
15. Guermond, J.-L., Minev, P., Shen, J.: An overview of projection methods for incompressible flows. Comput. Methods Appl. Mech. Eng. **195**, 6011–6054 (2006)
16. Guermond, J.-L., Minev, P., Salgado, A.: Convergence analysis of a class of massively parallel direction splitting algorithms for the Navier-Stokes equations simple domains. Math. Comp. **81**, 1951–1977 (2012)
17. Karniadakis, G.E., Israeli, M., Orszag, S.A.: High-order splitting methods for the incompressible Navier-Stokes equations. J. Comput. Phys. **97**, 414–443 (1991)
18. Kim, J., Moin, P.: Application of a fractional-step method to incompressible Navier-Stokes equations. J. Comput. Phys. **59**(2), 308–323 (1985)
19. Korobytsina, Z.L.: Fictitious domain method for linear parabolic equations (in Russian). Differ. Equ. **21**(5), 854–862 (1985)
20. Kuttykozhaeva, S., Zhumagulov, B., Smagulov, S.: An unimprovable estimate of the convergence rate in the fictitious domain method for the Navier-Stokes equations. Dokl. Math. **67**(3), 382–385 (2003)
21. Orszag, S.A., Israeli, M., Deville, M.: Boundary conditions for incompressible flows. J. Sci. Comput. **1**, 75–111 (1986)
22. Rannacher, R.: On Chorin's projection method for the incompressible Navier-Stokes equations. In: Lecture Notes in Mathematics, Springer, vol. 1530 (1991)
23. Samarskii, A.: Regularization of difference schemes. USSR Comput. Math. Math. Phys. **7**, 79–120 (1967)
24. Samarskii, A., Vabishchevich, P.: Additive Schemes for Problems in Mathematical Physics. Nauka, Moskva (1999) (in Russian)
25. Shen, J.: On error estimates of the projection methods for the Navier-Stokes equations: first-order schemes. SIAM J. Numer. Anal. **29**, 57–77 (1992)
26. Shen, J.: On error estimates of projection methods for the Navier-Stokes equations: second-order schemes. Math. Comp. **65**(215), 1039–1065 (1996)
27. Temam, R.: Sur l'approximation de la solution des équations de Navier-Stokes par la méthode des pas fractionnaires ii. Arch. Ration. Mech. Anal. **33**, 377–385 (1969)
28. Timmermans, L., Minev, P., Vosse, F.V.D.: An approximate projection scheme for incompressible flow using spectral elements. Int. J. Numer. Methods Fluids **22**, 673–688 (1996)
29. Weinan, E., Liu, J.: Gauge method for viscous incompressible flows. Commun. Math. Sci. **1**(2), 317–332 (2003)

Efficient Solvers for Some Classes of Time-Periodic Eddy Current Optimal Control Problems

Michael Kolmbauer and Ulrich Langer

Abstract In this paper, we present and discuss the results of our numerical studies of preconditioned MinRes methods for solving the optimality systems arising from the multiharmonic finite element approximations to time-periodic eddy current optimal control problems in different settings including different observation and control regions, different tracking terms, as well as box constraints for the Fourier coefficients of the state and the control. These numerical studies confirm the theoretical results published by the first author in a recent paper.

Keywords Time-periodic eddy current optimal control problems • Multiharmonic finite element discretization • MinRes solver • Preconditioners

Mathematics Subject Classification (2010): 49J20, 65T40, 65M60, 65F08

1 Introduction

This work is devoted to the study of efficient solution procedures for the following time-periodic eddy current optimal control problem: Minimize the functional

M. Kolmbauer
DK Computational Mathematics, Johannes Kepler University Linz,
Altenberger Str. 69, 4040 Linz, Austria
e-mail: kolmbauer@numa.uni-linz.ac.at

U. Langer (✉)
Institute of Computational Mathematics, Johannes Kepler University Linz,
Altenberger Str. 69, 4040 Linz, Austria
e-mail: ulanger@numa.uni-linz.ac.at

O.P. Iliev et al. (eds.), *Numerical Solution of Partial Differential Equations: Theory, Algorithms, and Their Applications*, Springer Proceedings in Mathematics & Statistics 45,
DOI 10.1007/978-1-4614-7172-1_11,

$$J(\mathrm{y},\mathrm{u}) = \frac{\alpha}{2}\int_{\Omega_1\times(0,T)} |\mathrm{y}-\mathrm{y_d}|^2 \mathrm{dx\,dt} + \frac{\beta}{2}\int_{\Omega_1\times(0,T)} |\mathrm{curl\,y}-\mathrm{y_c}|^2 \mathrm{dx\,dt} \\ + \frac{\lambda}{2}\int_{\Omega_2\times(0,T)} |\mathrm{u}|^2 \mathrm{dx\,dt}, \tag{1}$$

subject to the state equations

$$\begin{cases} \sigma\dfrac{\partial \mathrm{y}}{\partial t} + \mathrm{curl}(\nu\,\mathrm{curl\,y}) = \mathrm{u}, & \text{in } \Omega\times(0,T), \\ \mathrm{div}(\sigma \mathrm{y}) = 0, & \text{in } \Omega\times(0,T), \\ \mathrm{y}\times\mathrm{n} = 0, & \text{on } \partial\Omega\times(0,T), \\ \mathrm{y}(0) = \mathrm{y}(T), & \text{in } \Omega, \end{cases} \tag{2}$$

where Ω is a bounded, simply connected Lipschitz domain with the boundary $\partial\Omega$. The domains Ω_1 and Ω_2 are nonempty Lipschitz subdomains of Ω, i.e., $\Omega_1, \Omega_2 \subset \Omega \subset \mathbb{R}^3$. The reluctivity $\nu \in L^\infty(\Omega)$ and the conductivity $\sigma \in L^\infty(\Omega)$ are supposed to be uniformly positive, i.e.,

$$0 < \nu_{\min} \leq \nu(\mathrm{x}) \leq \nu_{\max}, \quad \text{and} \quad 0 < \sigma_{\min} \leq \sigma(\mathrm{x}) \leq \sigma_{\max}, \qquad \mathrm{x} \in \Omega.$$

We mention that the electric conductivity σ vanishes in regions consisting of nonconducting materials. In order to fulfill the assumption made above on the uniform positivity of σ, one can replace $\sigma(x)$ by $\max\{\varepsilon, \sigma(x)\}$ with some suitably chosen positive ε; see, e.g., [10, 12] for more details. We here assume that the reluctivity ν is independent of $|\mathrm{curl\,y}|$, i.e., we only consider linear eddy current problems. The regularization parameter λ also representing a weight for the cost of the control is assumed to be a suitably chosen positive real number. The weight parameters α and β are nonnegative. In fact, we only study the cases $(\alpha = 1, \beta = 0)$ and $(\alpha = 0, \beta = 1)$. The functions $\mathrm{y_d}$ and $\mathrm{y_c}$ from $L_2((0,T), \mathrm{L}_2(\Omega))$ are the given desired state and the desired curl of the state, respectively.

The problem setting (1)–(2) has been analyzed in [11, 12], wherein, due to the time-periodic structure, a time discretization in terms of a truncated Fourier series, also called multiharmonic approach, is used. In [12], we consider the special case of a fully distributed optimal control problem for tracking some $\mathrm{y_d}$ in the complete computational domain, i.e., $\Omega_1 = \Omega_2 = \Omega$ and $\beta = 0$ in (1), whereas [11] is devoted to the various other settings including different observation and control regions, different tracking terms, as well as box constraints for the Fourier coefficients of the state and the control. Similar optimal control problems for time-periodic parabolic equations and their numerical treatment by means of the multiharmonic finite element method (FEM) have recently been considered in [9] and [8]. Other approaches to time-periodic parabolic optimal control problems have been discussed in [1]. There are many publications on optimal control problems with PDE constraints given by initial-boundary value problems for parabolic equations; see, e.g., [14] for a comprehensive presentation. There are less publications on optimal control problems where initial-boundary value problems for eddy current equations are considered as PDE constraints; see, e.g., [15,16], where one can also find interesting applications. The multiharmonic approach allows us to switch from the time domain to the frequency domain and, therefore, to replace a time-dependent problem by a system of time-independent problems for the Fourier coefficients. Since we are here interested in studying robust solvers, this special time discretization technique justifies the following assumption: Let us assume that the desired states $\mathrm{y_d}$ and $\mathrm{y_c}$ are multiharmonic, i.e., $\mathrm{y_d}$ and $\mathrm{y_c}$ have the form of a truncated Fourier series:

$$\begin{aligned} \mathrm{y_d} &= \sum_{k=0}^{N} \mathrm{y_{d,k}^c} \cos(k\omega t) + \mathrm{y_{d,k}^s} \sin(k\omega t), \\ \mathrm{y_c} &= \sum_{k=0}^{N} \mathrm{y_{c,k}^c} \cos(k\omega t) + \mathrm{y_{c,k}^s} \sin(k\omega t). \end{aligned} \tag{3}$$

Consequently, the state y and the control u are multiharmonic as well and, therefore, have a representation in terms of a truncated Fourier series with the same number of modes N, i.e.,

$$\begin{aligned} \mathrm{y} &= \sum_{k=0}^{N} \mathrm{y_k^c} \cos(k\omega t) + \mathrm{y_k^s} \sin(k\omega t), \\ \mathrm{u} &= \sum_{k=0}^{N} \mathrm{u_k^c} \cos(k\omega t) + \mathrm{u_k^s} \sin(k\omega t). \end{aligned} \tag{4}$$

Using the multiharmonic representation of $\mathrm{y_d}$, $\mathrm{y_c}$, y, and u, the minimization problem (1)–(2) can be stated in the frequency domain: Minimize the functional

$$\begin{aligned} J_N = \frac{1}{2}\sum_{k=0}^{N}\Big[\sum_{j\in\{c,s\}}\Big[\alpha\int_{\Omega_1}|\mathrm{y_k^j}-\mathrm{y_{d,k}^j}|^2\mathrm{dx}+\beta\int_{\Omega_1}|\mathrm{curl}\,\mathrm{y_k^j}-\mathrm{y_{c,k}^j}|^2\mathrm{dx} \\ +\lambda\sum_{j\in\{c,s\}}\int_{\Omega_2}|\mathrm{u_k^j}|^2\mathrm{dx}\Big]\Big], \end{aligned} \tag{5a}$$

subject to the state equation

$$\begin{cases} k\omega\,\sigma\mathrm{y_k^s}+\mathrm{curl}(\nu\;\mathrm{curl}\,\mathrm{y_k^c})=\mathrm{u_k^c}, & \text{in } \Omega, k=1,\ldots,N, \\ -k\omega\,\sigma\mathrm{y_k^c}+\mathrm{curl}(\nu\;\mathrm{curl}\,\mathrm{y_k^s})=\mathrm{u_k^s}, & \text{in } \Omega, k=1,\ldots,N, \\ \mathrm{curl}(\nu\;\mathrm{curl}\,\mathrm{y_0^c})=\mathrm{u_0^c}, & \text{in } \Omega, \\ \mathrm{y_k^c}\times\mathrm{n}=\mathrm{y_k^s}\times\mathrm{n}=0, & \text{on } \partial\Omega, k=1,\ldots,N, \\ \mathrm{y_k^0}\times\mathrm{n}=0, & \text{on } \partial\Omega, \end{cases} \tag{5b}$$

completed by the divergence constraints

$$\begin{cases} k\omega\,\mathrm{div}(\sigma \mathrm{y}_k^c) = 0, & \text{in } \Omega, k = 1,\ldots,N, \\ k\omega\,\mathrm{div}(\sigma \mathrm{y}_k^s) = 0, & \text{in } \Omega, k = 1,\ldots,N, \\ \quad \mathrm{div}(\sigma \mathrm{y}_0^c) = 0, & \text{in } \Omega. \end{cases} \tag{5c}$$

Additionally, we add control constraints associated to the Fourier coefficients of the control u, i.e.,

$$\begin{aligned} &\underline{\mathrm{u}}_k^c \le \mathrm{u}_k^c \le \overline{\mathrm{u}}_k^c, \quad \text{a.e. in } \Omega, k = 0,1,\ldots,N, \\ &\underline{\mathrm{u}}_k^s \le \mathrm{u}_k^s \le \overline{\mathrm{u}}_k^s, \quad \text{a.e. in } \Omega, k = 1,\ldots,N, \end{aligned} \tag{5d}$$

and state constraints associated to the Fourier coefficients of the state y, i.e.,

$$\begin{aligned} &\underline{\mathrm{y}}_k^c \le \mathrm{y}_k^c \le \overline{\mathrm{y}}_k^c, \quad \text{a.e. in } \Omega, k = 0,1,\ldots,N, \\ &\underline{\mathrm{y}}_k^s \le \mathrm{y}_k^s \le \overline{\mathrm{y}}_k^s, \quad \text{a.e. in } \Omega, k = 1,\ldots,N. \end{aligned} \tag{5e}$$

This minimization problem is typically solved by deriving the corresponding optimality system, which fortunately decouples in terms of the mode k. The decoupled systems are then discretized in space by means of the FEM. Since even the simple box constraints (5d)–(5e) give rise to nonlinear optimality systems, we apply a primal–dual active set strategy (semi-smooth Newton) approach for their solution [5]. The resulting procedure is summarized in Algorithm 1.

Algorithm 1: Primal–dual active set strategy

```
Input: number of modes N, initial guesses x^(k,0) ∈ R^n (k = 0,...,N) .
Output: approximate solution x^(k,l) ∈ R^n (k = 0,...,N).
for k ← 0 to N do
    Determine the active sets E^c_{k,0} and E^s_{k,0};
end
Set l := 0;
while not converged do
    for k ← 0 to N do
        Compute b_E^(k,l+1), A_E^(k,l+1);
        Solve A_E^(k,l+1) x^(k,l+1) = b_E^(k,l+1);
        Determine the active sets E^c_{k,l+1} and E^s_{k,l+1};
    end
    Set l := l + 1;
end
```

The specific structure of the Jacobi matrix $\mathcal{A}_{\mathcal{E}}^{(k,l+1)}$ depends on the actual computational setting. In our applications, the matrix $\mathcal{A}_{\mathcal{E}}^{(k,l+1)}$ has either the form $\mathcal{A}_1$ (cf. (6a)) or the form $\mathcal{A}_2$, cf. (6b). It is clear that the efficient and parameter-robust solution of the $(N+1)$ linear systems of equations at each semi-smooth Newton step is essential for the efficiency of the proposed method. For further details we refer to [11].

2 Parameter-Robust and Efficient Solution Procedures

In order to discretize the problems in space, we use the edge (Nédélec) finite element space $\mathcal{ND}_0^0(\mathcal{T}_h)$, that is a conforming finite element subspace of $\mathrm{H}_0(\mathrm{curl},\Omega)$, and the nodal (Lagrange) finite element space $\mathcal{S}_0^1(\mathcal{T}_h)$, that is a conforming finite element subspace of $H_0^1(\Omega)$. Let $\{\varphi_i\}_{i=1,N_h}$ and $\{\psi_i\}_{i=1,M_h}$ denote the usual edge basis of $\mathcal{ND}_0^0(\mathcal{T}_h)$ and the usual nodal basis of $\mathcal{S}_0^1(\mathcal{T}_h)$, respectively. We are now in the position to define the following FEM matrices:

$$
\begin{aligned}
(\mathrm{K}_\nu)_{ij} &= (\nu \,\mathrm{curl}\,\varphi_i, \mathrm{curl}\,\varphi_j)_{0,\Omega},\\
(\mathrm{M}_{\sigma,k\omega})_{ij} &= k\omega(\sigma\varphi_i, \varphi_j)_{0,\Omega},\\
(\mathrm{M})_{ij} &= (\varphi_i, \varphi_j)_{0,\Omega},\\
(\mathrm{D}_{\sigma,k\omega})_{ij} &= k\omega(\sigma\varphi_i, \nabla\psi_j)_{0,\Omega},
\end{aligned}
$$

where $(\cdot,\cdot)_{0,\Omega}$ denotes the inner product in $\mathrm{L}_2(\Omega)$. Throughout this paper we are repeatedly faced with the following two types of system matrices:

$$
\mathcal{A}_1 = \begin{pmatrix}
* & 0 & \mathrm{K}_\nu & -\mathrm{M}_{\sigma,k\omega}\\
0 & * & \mathrm{M}_{\sigma,k\omega} & \mathrm{K}_\nu\\
\mathrm{K}_\nu & \mathrm{M}_{\sigma,k\omega} & -\lambda^{-1}* & 0\\
-\mathrm{M}_{\sigma,k\omega} & \mathrm{K}_\nu & 0 & -\lambda^{-1}*
\end{pmatrix} \tag{6a}
$$

$$
\mathcal{A}_2 = \begin{pmatrix}
* & 0 & \mathrm{K}_\nu & -\mathrm{M}_{\sigma,k\omega} & 0 & 0 & {\mathrm{D}_{\sigma,k\omega}}^T & 0\\
0 & * & \mathrm{M}_{\sigma,k\omega} & \mathrm{K}_\nu & 0 & 0 & 0 & {\mathrm{D}_{\sigma,k\omega}}^T\\
\mathrm{K}_\nu & \mathrm{M}_{\sigma,k\omega} & -\lambda^{-1}* & 0 & {\mathrm{D}_{\sigma,k\omega}}^T & 0 & 0 & 0\\
-\mathrm{M}_{\sigma,k\omega} & \mathrm{K}_\nu & 0 & -\lambda^{-1}* & 0 & {\mathrm{D}_{\sigma,k\omega}}^T & 0 & 0\\
0 & 0 & \mathrm{D}_{\sigma,k\omega} & 0 & 0 & 0 & 0 & 0\\
0 & 0 & 0 & \mathrm{D}_{\sigma,k\omega} & 0 & 0 & 0 & 0\\
\mathrm{D}_{\sigma,k\omega} & 0 & 0 & 0 & 0 & 0 & 0 & 0\\
0 & \mathrm{D}_{\sigma,k\omega} & 0 & 0 & 0 & 0 & 0 & 0
\end{pmatrix}. \tag{6b}
$$

Therein, the placeholder $*$ stands for a symmetric and positive semi-definite matrix, that actually depends on the considered setting (cf. Table 1). We refer to problems described by matrices of the types $\mathcal{A}_1$ and $\mathcal{A}_2$ as *Formulation OC-FEM 1* and *Formulation OC-FEM 2*, respectively. In fact, the system matrices $\mathcal{A}_1$

and $\mathcal{A}_2$ are symmetric and indefinite and have a two- or threefold saddle point structure, respectively. Since $\mathcal{A}_1$ and $\mathcal{A}_2$ are symmetric, the corresponding systems can be solved by a preconditioned minimal residual (MinRes) method (cf. [13]). Typically, the convergence rate of any iterative Krylov subspace method applied to the unpreconditioned system deteriorates, with respect to the mesh size h, the parameters $k = 0, 1, \ldots, N$ and ω involved in the spectral time discretization and the problem parameters ν, σ, and λ (cf. also Tables 2 and 3). Therefore, preconditioning is an important issue.

The proper choice of parameter-robust and efficient preconditioners has been addressed by the authors in [11, 12]. While for equations with system matrices of type (6a), we propose to use the preconditioner

$$\mathcal{C} := \text{diag}\left(\sqrt{\lambda}\mathrm{F}, \sqrt{\lambda}\mathrm{F}, \frac{1}{\sqrt{\lambda}}\mathrm{F}, \frac{1}{\sqrt{\lambda}}\mathrm{F}\right), \tag{7}$$

with the block $\mathrm{F} = \mathrm{K}_\nu + \mathrm{M}_{\sigma,\mathrm{k}\omega} + 1/\sqrt{\lambda}\mathrm{M}$; for equations with system matrices of type (6b), we advise to use the preconditioner

$$\mathcal{C}_M = \text{diag}\left(\sqrt{\lambda}\mathrm{F}, \sqrt{\lambda}\mathrm{F}, \frac{1}{\sqrt{\lambda}}\mathrm{F}, \frac{1}{\sqrt{\lambda}}\mathrm{F}, \frac{1}{\sqrt{\lambda}}\mathrm{S_J}, \frac{1}{\sqrt{\lambda}}\mathrm{S_J}, \sqrt{\lambda}\mathrm{S_J}, \sqrt{\lambda}\mathrm{S_J}\right), \tag{8}$$

where $\mathrm{S_J} = \mathrm{D}_{\sigma,\mathrm{k}\omega}{}^T \mathrm{F}^{-1} \mathrm{D}_{\sigma,\mathrm{k}\omega}$. In a MinRes setting, the quality of the preconditioners $\mathcal{C}$ and $\mathcal{C}_M$, used for the system matrices $\mathcal{A}_1$ and $\mathcal{A}_2$, respectively, is in general determined by the condition number κ_1 or κ_2 of the preconditioned system, defined as follows:

$$\kappa_1 := \|\mathcal{C}^{-1}\mathcal{A}_1\|_{\mathcal{C}} \|\mathcal{A}_1^{-1}\mathcal{C}\|_{\mathcal{C}} \quad \text{and} \quad \kappa_2 := \|\mathcal{C}_M^{-1}\mathcal{A}_2\|_{\mathcal{C}_M} \|\mathcal{A}_2^{-1}\mathcal{C}_M\|_{\mathcal{C}_M}. \tag{9}$$

In Table 1, we list the theoretical results that have been derived for different settings of (5) in [11, 12]. We especially want to point out that the bounds for the condition numbers are at least uniform in the space discretization parameter h as well as the time discretization parameters ω and N. This has the important consequence that the proposed preconditioned MinRes method converges within a few iterations, independent of the discretization parameters that are directly related to the size of the system matrices.

3 Numerical Validation

The main aim of this paper is to verify the theoretical proven convergence rates by numerical experiments. We consider an academic test problem of the form (1)–(2) or rather (5) in the unit cube $\Omega = (0, 1)^3$ and report on various numerical test for various computational settings and varying parameters. Since we are here only

Table 1 Condition number estimates for different settings. Here (σ) denotes robustness with respect to $\sigma \in \mathbb{R}^+$

Test case	α	β	Domains	Equations	Condition number estimate
I	1	0	$\Omega_1 = \Omega_2$	(5a)–(5b)	$\kappa_1 \leq \sqrt{3} \neq c(h,\omega,N,\sigma,\nu,\lambda)$
II	1	0	$\Omega_1 = \Omega_2$	(5a)–(5c)	$\kappa_2 \leq \sqrt{3}(1+\sqrt{5}) \neq c(h,\omega,N,\sigma,\nu,\lambda)$
III	0	1	$\Omega_1 = \Omega_2$	(5a)–(5c)	$\kappa_2 \leq c \neq c(h,\omega,N,(\sigma))$
IV	1	0	$\Omega_1 \neq \Omega_2$	(5a)–(5c)	$\kappa_2 \leq c \neq c(h,\omega,N,(\sigma),\Omega_1,\Omega_2)$
V	1	0	$\Omega_1 = \Omega_2$	(5a)–(5d)	$\kappa_2 \leq c \neq c(h,\omega,N,(\sigma),\text{index sets})$
VI	1	0	$\Omega_1 = \Omega_2$	(5a)–(5b) + (5e)	$\kappa_1 \leq c \neq c(h,\omega,N,\sigma,\nu,\lambda,\text{index sets})$

interested in the study of the robustness of the solver, it is obviously sufficient to consider the solution of the system corresponding to the block of the mode $k = 1$. The numerical results presented in this section were attained using ParMax.[1] We demonstrate the robustness of the block-diagonal preconditioners with respect to the involved parameters. Therefore, for the solution of the preconditioning equations arising from the diagonal blocks F, we use the sparse direct solver UMFPACK,[2] that is very efficient for several thousand unknowns in the case of three-dimensional problems [2–4]. For numerical tests, where the diagonal blocks are replaced by an auxiliary space preconditioner [6, 7], we refer the reader to [10] and [12].

3.1 *Test Case I*

Tables 2–5 provide the number of MinRes iterations needed for reducing the initial residual by a factor of 10^{-8}. These experiments demonstrate the independence of the MinRes convergence rate of the parameters ω, σ, λ and the mesh size h for all computed constellations. Indeed, the number of iterations is bounded by 28, that is very close to the theoretical bound 30 given by the condition number estimate $\sqrt{3}$. We mention that varying ω also covers the variation of $k\omega$ in terms of k. Furthermore, in Tables 2 and 3, we also report the number of unpreconditioned MinRes iterations, that are necessary for reducing the initial residual by a factor of 10^{-8}. The large number of iterations in the unpreconditioned case underlines the importance of appropriate preconditioning.

3.2 *Test Case II*

Table 6 provides the number of MinRes iterations needed for reducing the initial residual by a factor 10^{-8}. These experiments demonstrate the independence of the

[1] http://www.numa.uni-linz.ac.at/P19255/software.shtml.

[2] http://www.cise.ufl.edu/research/sparse/umfpack/.

Table 2 Formulation OC-FEM 1 for test case I. Number of MinRes iterations for $DOF = 2,416$, $\nu = \sigma = 1$, and different values of λ and ω. $[\cdot]$ denotes the number of MinRes iterations without preconditioner

$\lambda \setminus \omega$	10^{-10}	10^{-8}	10^{-6}	10^{-4}	10^{-2}	1	10^{2}	10^{4}	10^{6}	10^{8}	10^{10}
10^{-10}	7	7	7	7	7	7	7	7	7	6	4
	[587]	[587]	[586]	[587]	[587]	[587]	[587]	[591]	[485]	[263]	[116]
10^{-6}	21	21	21	21	21	21	20	12	6	4	4
	[373]	[373]	[373]	[373]	[373]	[373]	[373]	[263]	[116]	[114]	[114]
10^{-2}	20	20	20	20	20	20	20	12	6	4	4
	[1,134]	[1,134]	[1,134]	[1,136]	[1,135]	[1,134]	[227]	[114]	[114]	[114]	[114]
1	10	10	10	10	10	14	20	12	6	4	4
	[2,349]	[2,351]	[2,349]	[2,350]	[2,350]	[2,274]	[222]	[114]	[114]	[114]	[114]
10^{2}	6	6	6	6	8	10	20	12	6	4	4
	[2,688]	[2,681]	[2,696]	[2,667]	[3,291]	[2,494]	[224]	[114]	[114]	[114]	[114]
10^{6}	4	4	4	6	6	10	20	12	6	4	4
	[1,152]	[1,159]	[3,434]	[4,697]	[4,867]	[2,493]	[222]	[114]	[114]	[114]	[114]
10^{10}	2	4	4	4	4	10	20	12	6	4	4
	[1,157]	[1,163]	[4,937]	[5,881]	[4,791]	[2,501]	[224]	[114]	[114]	[114]	[114]

Table 3 Formulation OC-FEM 1 for test case I. Number of MinRes iterations for $DOF = 16,736$, $\nu = \sigma = 1$, and different values of λ and ω. $[\cdot]$ denotes the number of MinRes iterations without preconditioner. [-] indicates that MinRes did not converge within 10,000 iterations

$\lambda \setminus \omega$	10^{-10}	10^{-8}	10^{-6}	10^{-4}	10^{-2}	1	10^{2}	10^{4}	10^{6}	10^{8}	10^{10}
10^{-10}	9	9	9	9	9	9	9	10	6	4	4
	[708]	[708]	[708]	[708]	[708]	[708]	[708]	[711]	[578]	[308]	[134]
10^{-6}	21	21	21	21	21	21	20	18	6	4	4
	[825]	[824]	[825]	[825]	[825]	[825]	[824]	[307]	[134]	[132]	[132]
10^{-2}	18	18	18	18	18	20	22	20	6	4	4
	[6,698]	[6,669]	[6,696]	[6,698]	[6,690]	[6,676]	[1,095]	[132]	[132]	[132]	[132]
1	10	10	10	10	10	14	22	20	6	4	4
	[-]	[-]	[-]	[-]	[-]	[-]	[1,094]	[132]	[132]	[132]	[132]
10^{2}	6	6	6	6	8	10	22	20	6	4	4
	[-]	[-]	[-]	[-]	[-]	[-]	[1,094]	[132]	[132]	[132]	[132]
10^{6}	4	4	4	6	6	10	22	20	6	4	4
	[7,365]	[7,547]	[-]	[-]	[-]	[-]	[1,094]	[132]	[132]	[132]	[132]
10^{10}	2	4	4	4	4	10	22	20	6	4	4
	[7,381]	[1,545]	[-]	[-]	[-]	[-]	[1,094]	[132]	[132]	[132]	[132]

MinRes convergence rate of the parameters ω, σ, λ and the mesh size h since the number of iterations is bounded by 88 for all computed constellations. The condition number estimate from Table 1 yields 106 as a bound for the maximal number of iterations.

Table 4 Formulation OC-FEM 1 for test case I. Number of MinRes iterations for $DOF = 124{,}096$, $\nu = \sigma = 1$, and different values of λ and ω

$\lambda \setminus \omega$	10^{-10}	10^{-8}	10^{-6}	10^{-4}	10^{-2}	1	10^{2}	10^{4}	10^{6}	10^{8}	10^{10}
10^{-10}	13	13	13	13	13	13	13	13	8	4	4
10^{-8}	21	21	21	21	21	21	21	17	8	4	4
10^{-6}	21	21	21	21	21	21	21	20	8	4	4
10^{-4}	20	20	20	20	20	20	28	22	8	4	4
10^{-2}	16	16	16	16	16	18	22	22	8	4	4
1	10	10	10	10	10	12	20	22	8	4	4
10^{2}	6	6	6	6	8	10	20	22	8	4	4
10^{4}	4	4	4	6	6	10	20	22	8	4	4
10^{6}	4	4	4	4	6	10	20	22	8	4	4
10^{8}	2	4	4	4	6	10	20	22	8	4	4
10^{10}	3	4	4	4	4	10	20	22	8	4	4

Table 5 Formulation OC-FEM 1 for test case I. Number of MinRes iterations for $DOF = 124{,}096$, $\omega = \sigma = 1$, and different values of λ and ν

$\lambda \setminus \nu$	10^{-10}	10^{-8}	10^{-6}	10^{-4}	10^{-2}	1	10^{2}	10^{4}	10^{6}	10^{8}	10^{10}
10^{-10}	2	2	3	3	5	13	21	16	6	4	3
10^{-8}	2	2	3	4	7	21	20	10	4	4	3
10^{-6}	2	3	3	5	13	21	16	6	4	4	4
10^{-4}	2	3	4	7	21	20	10	6	4	4	4
10^{-2}	3	4	6	13	21	18	8	4	4	6	6
1	4	4	8	17	28	12	6	4	6	6	9
10^{2}	4	4	8	20	22	10	6	4	6	6	8
10^{4}	4	4	8	22	20	10	6	4	4	4	8
10^{6}	4	4	8	22	20	10	4	4	4	4	8
10^{8}	4	4	8	22	20	10	4	4	4	4	8
10^{10}	4	4	8	22	20	10	4	2	4	4	8

Table 6 Formulation OC-FEM 2 for test case II. Number of MinRes iterations for $\nu = \sigma = 1$, different values of λ and ω, and $DOF = 19{,}652$ / **143,748**

$\lambda \setminus \omega$	10^{-10}	10^{-6}	10^{-2}	1	10^{2}	10^{6}	10^{10}
10^{-10}	21 / **27**	19 / **25**	17 / **25**	17 / **25**	17 / **25**	12 / **16**	10 / **10**
10^{-6}	33 / **32**	33 / **32**	33 / **32**	33 / **32**	29 / **33**	10 / **14**	8 / **8**
10^{-2}	22 / **20**	22 / **20**	26 / **23**	31 / **29**	34 / **35**	14 / **16**	10 / **10**
1	12 / **12**	14 / **14**	14 / **14**	14 / **14**	24 / **24**	10 / **12**	8 / **8**
10^{2}	11 / **11**	13 / **13**	13 / **13**	18 / **18**	34 / **34**	14 / **16**	10 / **10**
10^{6}	13 / **13**	13 / **15**	21 / **21**	28 / **30**	56 / **58**	22 / **24**	14 / **14**
10^{10}	31 / **46**	34 / **65**	33 / **33**	42 / **42**	80 / **88**	30 / **38**	16 / **16**

3.3 Test Case III

Numerical results for the observation of the magnetic flux density are reported in Tables 7–9. The robustness with respect to the space and time discretization

Table 7 Observation of the magnetic flux density B in Formulation OC-FEM 2 for test case *III*. Number of MinRes iterations for $\nu = \sigma = \lambda = 1$ and for different values of ω and various *DOF*

	ω										
DOF	10^{-10}	10^{-8}	10^{-6}	10^{-4}	10^{-2}	1	10^{2}	10^{4}	10^{6}	10^{8}	10^{10}
500	13	13	14	14	14	16	23	12	9	8	7
2,916	11	12	13	13	13	15	29	16	10	8	8
19,652	11	11	12	12	12	14	30	21	11	8	8
143,748	11	11	12	12	12	14	28	27	13	8	8

Table 8 Observation of the magnetic flux density B in Formulation OC-FEM 2 for test case *III*. Number of MinRes iterations for $\sigma = \omega = 1$, different values of λ and ν, and $DOF =$ 19,652/**143,748**. [-] indicates that MinRes did not converge within 10,000 iterations

$\lambda \setminus \nu$	10^{-10}	10^{-6}	10^{-2}	1	10^{2}	10^{6}	10^{10}
10^{-10}	174 / **325**	175 / **326**	175 / **327**	213 / **411**	290 / **505**	14 / **14**	8 / **8**
10^{-6}	146 / **289**	146 / **289**	177 / **359**	215 / **392**	58 / **53**	8 / **10**	8 / **8**
10^{-2}	272 / **543**	272 / **543**	306 / **523**	55 / **52**	13 / **13**	9 / **8**	13 / **15**
1	290 / **543**	290 / **541**	240 / **325**	14 / **14**	8 / **8**	8 / **8**	12 / **14**
10^{2}	475 / **948**	479 / **941**	83 / **79**	18 / **18**	12 / **12**	14 / **14**	26 / **36**
10^{6}	193 / **688**	195 / **680**	55 / **55**	28 / **30**	18 / **18**	24 / **26**	360 / **[-]**
10^{10}	36 / **56**	39 / **55**	84 / **88**	42 / **42**	26 / **26**	50 / **54**	[-] / **[-]**

Table 9 Observation of the magnetic flux density B in Formulation OC-FEM 2 for test case *III*. Number of MinRes iterations for $\nu = \sigma = \omega = 1$ and for different values of λ and various *DOF*

	λ										
DOF	10^{-10}	10^{-8}	10^{-6}	10^{-4}	10^{-2}	1	10^{2}	10^{4}	10^{6}	10^{8}	10^{10}
500	36	36	37	39	40	16	19	26	30	36	44
2,916	115	113	121	121	55	15	18	24	28	38	44
19,652	213	214	215	195	55	14	18	24	28	36	42
143,748	411	402	392	265	52	14	18	24	30	36	42

parameters h and ω is demonstrated in Table 7. Table 8 describes the non-robust behavior with respect to the parameters λ and ν. In Table 9 we observe that for large mesh sizes, good iteration numbers are observed even for small λ. Nevertheless, for fixed λ, the iteration numbers are growing with respect to the involved degrees of freedom.

The next experiment demonstrates that robustness with respect to the time discretization parameter ω cannot be achieved by using the preconditioner $\mathcal{C}$ in *Formulation OC-FEM 1*. In Table 10 the number of MinRes iteration needed for reducing the initial residual by a factor of 10^{-8} is displayed. In Table 11, the same experiment as in Table 8 is performed, but using *Formulation OC-FEM 1* instead of *Formulation OC-FEM 2*. Indeed, comparing Table 7 with Table 10 and Table 8 with Table 11 clearly shows that it is essential to work with *Formulation OC-FEM 2*. Besides the robustness with respect to the frequency ω, that is related to the time discretization parameters, we additionally observe better iteration numbers with respect to the regularization parameter λ in the interesting region $0 < \lambda < 1$.

Table 10 Observation of the magnetic flux density B in Formulation OC-FEM 1 for test case *III*. Number of MinRes iterations for $\nu = \sigma = \lambda = 1$ and for different values of ω and various *DOF*. [-] indicates that MinRes did not converge within 10,000 iterations

	ω										
DOF	10^{-10}	10^{-8}	10^{-6}	10^{-4}	10^{-2}	1	10^{2}	10^{4}	10^{6}	10^{8}	10^{10}
392	4,133	[-]	46	20	16	15	21	9	5	4	3
2,416	[-]	[-]	64	29	15	13	27	12	6	4	4
16,736	[-]	[-]	102	28	15	13	26	18	7	4	4
124,096	[-]	[-]	28	13	12	26	24	9	5	4	4

Table 11 Observation of the magnetic flux density B in Formulation OC-FEM 1 for test case *III*. Number of MinRes iterations for $DOF = 16,736$, $\sigma = \omega = 1$, and different values of λ and ν. [-] indicates that MinRes did not converge within 10,000 iterations

$\lambda \setminus \nu$	10^{-10}	10^{-8}	10^{-6}	10^{-4}	10^{-2}	1	10^{2}	10^{4}	10^{6}	10^{8}	10^{10}
10^{-10}	739	901	1,073	1,140	1,462	1,153	1,548	182	32	19	[-]
10^{-6}	357	361	357	385	478	607	96	17	10	9	18
10^{-2}	234	234	234	253	279	50	9	6	7	6	9
1	260	260	260	259	214	13	7	5	6	6	8
10^{2}	462	462	469	440	76	11	6	4	6	6	7
10^{6}	79	79	79	73	21	10	4	4	4	4	6
10^{10}	10	10	9	19	22	10	4	3	4	4	6

Table 12 Different control and observation domains in Formulation OC-FEM 2 / **OC-FEM 1** for test case IV. Number of MinRes iterations for $\nu = \sigma = \lambda = 1$ and for different values of ω and various *DOF*

	ω						
DOF	10^{-10}	10^{-6}	10^{-2}	1	10^{2}	10^{6}	10^{10}
2,916	19 / **34**	20 / **67**	23 / **52**	30 / **30**	30 / **22**	12 / **6**	8 / **4**
19,652	19 / **32**	20 / **82**	24 / **51**	30 / **30**	32 / **22**	12 / **6**	8 / **4**
143,748	19 / **29**	19 / **83**	23 / **48**	29 / **30**	32 / **20**	14 / **8**	8 / **4**

3.4 *Test Case VI*

In this subsection we consider a numerical example with different observation and control domains Ω_1 and Ω_2, i.e., $\Omega_1 = \Omega = (0,1)^3$ and $\Omega_2 = (0.25, 0.75)^3$. Let us mention that we have to ensure that Ω_1 and Ω_2 are resolved by the mesh. The corresponding numerical results are documented in Tables 12–14. Robustness with respect to the space and time discretization parameters h and ω is demonstrated in Table 12. Table 13 describes the non-robust behavior with respect to the parameters λ and ν. Table 12 in combination with Table 14 indicates that, for the *Formulation OC-FEM 1* in combination with the preconditioner $\mathcal{C}$, robustness with respect to the frequency ω, that is related to the time discretization parameters, cannot be obtained. Here, we want to mention that the good iteration numbers observed in Table 12 are caused by the special choice of $\lambda = 1$.

Table 13 Different control and observation domains in Formulation OC-FEM 2 / **OC-FEM 1** for test case IV. Number of MinRes iterations for $DOF = 19{,}652$ / $\mathbf{16{,}736}$, $\sigma = \omega = 1$, and different values of λ and ν. [-] indicates that MinRes did not converge within 10,000 iterations

$\lambda \setminus \nu$	10^{-10}	10^{-6}	10^{-2}	1	10^2	10^6	10^{10}
10^{-10}	1,038 / **34**	661 / **36**	[-] / **2,701**	[-] / **[-]**	[-] / **983**	49 / **60**	9 / **[-]**
10^{-6}	342 / **31**	363 / **32**	6,843 / **2,630**	7,142 / **828**	619 / **81**	26 / **41**	8 / **73**
10^{-2}	188 / **29**	209 / **37**	607 / **169**	204 / **61**	114 / **43**	79 / **37**	106 / **47**
1	40 / **19**	41 / **22**	52 / **39**	30 / **30**	26 / **25**	26 / **22**	26 / **24**
10^2	41 / **10**	42 / **11**	70 / **22**	40 / **13**	26 / **12**	22 / **11**	28 / **10**
10^6	24 / **6**	30 / **6**	76 / **22**	38 / **10**	24 / **6**	26 / **6**	414 / **6**
10^{10}	22 / **4**	34 / **6**	148 / **22**	46 / **10**	44 / **4**	68 / **4**	[-] / **6**

Table 14 Different control and observation domains in Formulation OC-FEM 1 for test case IV. Number of MinRes iterations for $DOF = 16{,}736$, $\sigma = \nu = 1$, and different values of λ and ω. [-] indicates that MinRes did not converge within 10,000 iterations

$\lambda \setminus \omega$	10^{-10}	10^{-8}	10^{-6}	10^{-4}	10^{-2}	1	10^2	10^4	10^6	10^8	10^{10}
10^{-10}	9,338	9,347	9,346	9,340	[-]	[-]	2,630	66	11	6	4
10^{-6}	571	571	571	1,075	983	828	169	20	6	4	4
10^{-2}	49	49	122	103	81	61	22	20	6	4	4
1	32	33	82	67	51	30	22	20	6	4	4
10^2	23	112	60	46	43	13	22	20	6	4	4
10^6	[-]	46	41	39	12	10	22	20	6	4	4
10^{10}	[-]	58	37	12	6	10	22	20	6	4	4

3.5 Test Case V

Numerical results for the case of state constraints imposed on the Fourier coefficients are presented in Tables 15, 16. Here we choose 15,512 random points as the active sets $\mathcal{E}^c$ and $\mathcal{E}^s$ and solve the resulting Jacobi system. The dependence of the MinRes convergence rate on the Moreau–Yosida regularization parameter ε is demonstrated in Table 15. Table 16 clearly demonstrates the robustness with respect to the parameters λ and ω. We refer the reader to [11] for a detailed description of the treatment of state constraints via the Moreau–Yosida regularization. Furthermore, we mention that the presence of constrains imposed on the control Fourier coefficients finally results in (linearized) systems with system matrices having the same structure as the system matrix arising from the case of different observation and control domains.

4 Summary and Conclusion

We demonstrated in many numerical experiments that the preconditioners derived and analyzed in [12] and [11] lead to parameter-robust and efficient solvers in many

Table 15 State constraints in Formulation OC-FEM 1 for test case VI. Number of MinRes iterations for $\nu = \sigma = \omega = 1$, different values of λ and ε, and $DOF = 16{,}736$ / **124,096**. [-] indicates that MinRes did not converge within 10,000 iterations

$\lambda \setminus \varepsilon$	10^{-10}	10^{-6}	10^{-2}	1	10^{2}	10^{6}	10^{10}
10^{-10}	88 / **142**	59 / **94**	31 / **46**	17 / **22**	9 / **13**	9 / **13**	9 / **13**
10^{-6}	992 / **3,275**	612 / **1,930**	220 / **372**	36 / **35**	21 / **21**	21 / **21**	21 / **21**
10^{-2}	[-] / **[-]**	[-] / **[-]**	351 / **383**	29 / **29**	20 / **18**	20 / **18**	20 / **18**
1	[-] / **[-]**	[-] / **[-]**	191 / **206**	24 / **24**	16 / **16**	14 / **13**	14 / **12**
10^{2}	[-] / **[-]**	[-] / **[-]**	120 / **124**	13 / **13**	12 / **12**	10 / **10**	10 / **10**
10^{6}	[-] / **[-]**	5,882 / **6,619**	12 / **11**	10 / **10**	10 / **10**	10 / **10**	10 / **10**
10^{10}	[-] / **[-]**	162 / **167**	10 / **10**	10 / **10**	10 / **10**	10 / **10**	10 / **10**

Table 16 State constraints in Formulation OC-FEM 1 for test case VI. Number of MinRes iterations for $DOF = 124{,}096$, $\nu = \sigma = \varepsilon = 1$, and different values of λ and ω

$\lambda \setminus \omega$	10^{-10}	10^{-8}	10^{-6}	10^{-4}	10^{-2}	1	10^{2}	10^{4}	10^{6}	10^{8}	10^{10}
10^{-10}	22	22	22	22	22	22	22	22	12	6	4
10^{-6}	35	35	35	35	35	35	35	22	8	4	4
10^{-2}	30	30	30	30	30	29	22	22	8	4	4
1	20	20	20	20	20	24	20	22	8	4	4
10^{2}	16	16	16	16	18	13	20	22	8	4	4
10^{6}	13	13	14	18	12	10	20	22	8	4	4
10^{10}	13	13	16	12	6	10	20	22	8	4	4

practically important cases. Therefore, we reported on a broad range of numerical experiments, that confirm the theoretical convergence rates. Consequently, the multiharmonic finite element discretization technique in combination with efficient and parameter-robust solvers leads to a very competitive method. Furthermore, we want to mention that due to the decoupling nature of the frequency domain equations with respect to the individual modes, a parallelization of the proposed method is straightforward (cf. Algorithm 1).

Acknowledgements The authors gratefully acknowledge the financial support by the Austrian Science Fund (FWF) under the grants P19255 and W1214 (project DK04). The authors also thank the Austria Center of Competence in Mechatronics (ACCM), which is a part of the COMET K2 program of the Austrian Government, for supporting their work on eddy current problems.

References

1. Abbeloos, D., Diehl, M., Hinze, M., Vandewalle, S.: Nested multigrid methods for time-periodic, parabolic optimal control problems. Comput. Visual. Sci. **14**(1), 27–38 (2011)
2. Davis, T.A.: Algorithm 832: Umfpack v4.3—an unsymmetric-pattern multifrontal method. ACM Trans. Math. Softw. **30**, 196–199 (2004)
3. Davis, T.A.: A column pre-ordering strategy for the unsymmetric-pattern multifrontal method. ACM Trans. Math. Softw. **30**, 165–195 (2004)

4. Davis, T.A., Duff, I.S.: A combined unifrontal/multifrontal method for unsymmetric sparse matrices. ACM Trans. Math. Softw. **25**, 1–20 (1999)
5. Hintermüller, M., Ito, K., Kunisch, K.: The primal-dual active set strategy as a semismooth Newton method. SIAM J. Optim. **13**(3), 865–888 (2002)
6. Hiptmair, R., Xu, J.: Nodal auxiliary space preconditioning in $\mathbf{H}(\mathbf{curl})$ and $\mathbf{H}(\mathrm{div})$ spaces. SIAM J. Numer. Anal. **45**(6), 2483–2509 (2007)
7. Kolev, T.V., Vassilevski, P.S.: Parallel auxiliary space AMG for $H(\mathrm{curl})$ problems. J. Comput. Math. **27**(5), 604–623 (2009)
8. Kollmann, M., Kolmbauer, M.: A preconditioned MinRes solver for time-periodic parabolic optimal control problems. Numer. Lin. Algebra Appl. (2012). doi: 10.1002/nla.1842
9. Kollmann, M., Kolmbauer, M., Langer, U., Wolfmayr, M., Zulehner, W.: A finite element solver for a multiharmonic parabolic optimal control problem. Comput. Math. Appl. **65**(3), 469–486 (2013)
10. Kolmbauer, M.: The multiharmonic finite element and boundary element method for simulation and control of eddy current problems. Ph.D. thesis, Johannes Kepler University, Institute of Computational Mathematics, Linz, Austria (2012)
11. Kolmbauer, M.: Efficient solvers for multiharmonic eddy current optimal control problems with various constraints and their analysis. IMA J. Numer. Anal. (2012). doi: 10.1093/imanum/drs025
12. Kolmbauer, M., Langer, U.: A robust preconditioned MinRes solver for distributed time-periodic eddy current optimal control problems. SIAM J. Sci. Comput. **34**(6), B785–B809 (2012)
13. Paige, C.C., Saunders, M.A.: Solutions of sparse indefinite systems of linear equations. SIAM J. Numer. Anal. **12**(4), 617–629 (1975)
14. Tröltzsch, F.: Optimal Control of Partial Differential Equations. Theory, Methods and Applications. Graduate Studies in Mathematics, vol. 112. AMS, Providence (2010)
15. Tröltzsch, F., Yousept, I.: PDE-constrained optimization of time-dependent 3D electromagnetic induction heating by alternating voltages. ESAIM: M2AN **46**, 709–729 (2012)
16. Yousept, I.: Optimal control of Maxwell's equations with regularized state constraints. Comput. Optim. Appl. **52**(2), 559–581 (2012)

Robust Algebraic Multilevel Preconditioners for Anisotropic Problems

J. Kraus, M. Lymbery, and S. Margenov

Abstract We present an overview on the state of the art of robust AMLI preconditioners for anisotropic elliptic problems. The included theoretical results summarize the convergence analysis of both linear and nonlinear AMLI methods for finite element discretizations by conforming and nonconforming linear elements and by conforming quadratic elements. The initially proposed hierarchical basis approach leads to robust multilevel algorithms for linear but not for quadratic elements for which an alternative AMLI method based on additive Schur complement approximation (ASCA) has been developed by the authors just recently. The presented new numerical results are focused on cases beyond the limitations of the rigorous AMLI theory. They reveal the potential and prospects of the ASCA approach to enhance the robustness of the resulting AMLI methods especially in situations when the matrix-valued coefficient function is not resolved on the coarsest mesh in the multilevel hierarchy.

Keywords Heterogeneous anisotropic problems • AMLI • Robust Preconditioning

Mathematics Subject Classification (2010): 65N20, 65M25

J. Kraus
Johann Radon Institute for Computational and Applied Mathematics,
Austrian Academy of Sciences, Altenberger Str. 69, A-4040 Linz, Austria
e-mail: johannes.kraus@oeaw.ac.at

M. Lymbery • S. Margenov (✉)
Institute of Information and Communication Technologies, Bulgarian Academy
of Sciences, Acad. G. Bonchev Str., Bl. 25A, 1113 Sofia, Bulgaria
e-mail: mariq@parallel.bas.bg; margenov@parallel.bas.bg

O.P. Iliev et al. (eds.), *Numerical Solution of Partial Differential Equations: Theory, Algorithms, and Their Applications*, Springer Proceedings in Mathematics & Statistics 45,
DOI 10.1007/978-1-4614-7172-1_12,

1 Introduction

Anisotropy arises in many applications such as heat transfer, electrostatics, magnetostatics, flow in porous media (see, e.g., [12]), and many other areas in science and engineering. For instance, in porous media a strong anisotropy of conductivity can be due to fractures, where the direction of dominating anisotropy is determined by the orientation of the fractures. The presence of fracture corridors can form long and tiny highly anisotropic channels. The network of channels is resolved at the finest mesh. The ratio of anisotropy in the channels can be of 5–6 orders of magnitude. Such kind of high-contrast and high-frequency anisotropic problems are still beyond the limits of robust algebraic multilevel preconditioning. At the end of the paper we experimentally study the robustness of algebraic multilevel iteration (AMLI) methods on model problems with channels.

In this paper we consider the elliptic boundary value problem

$$\begin{aligned} Lu \equiv -\nabla \cdot (\mathbf{a}(x)\nabla u(x)) = f(x) &\quad \text{in} \quad \Omega, \\ u = 0 &\quad \text{on} \quad \Gamma_D, \\ (\mathbf{a}(x)\nabla u(x)) \cdot \mathbf{n} = 0 &\quad \text{on} \quad \Gamma_N, \end{aligned} \tag{1}$$

where Ω is a polygonal domain in $\mathbb{R}^2$, $f(x)$ is a given function in $L^2(\Omega)$, the coefficient matrix $\mathbf{a}(x)$ is symmetric positive definite and uniformly bounded in Ω, and $\mathbf{n}$ is the outward unit vector normal to the boundary $\Gamma = \partial\Omega$, where $\Gamma = \bar{\Gamma}_D \cup \bar{\Gamma}_N$. We assume also that the elements of the diffusion coefficient matrix $\mathbf{a}(x)$ are piecewise smooth functions on $\bar{\Omega}$.

The weak formulation of the problem reads as follows: Given $f \in L^2(\Omega)$, find $u \in \mathcal{V} \equiv H_D^1(\Omega) = \{v \in H^1(\Omega) : v = 0 \text{ on } \Gamma_D\}$, satisfying

$$\begin{aligned} \mathcal{A}(u,v) &= (f,v) := \int_\Omega f(x)v(x)dx \quad \forall v \in H_D^1(\Omega), \quad \text{where} \\ \mathcal{A}(u,v) &:= \int_\Omega \mathbf{a}(x)\nabla u(x) \cdot \nabla v(x)dx. \end{aligned} \tag{2}$$

We assume that the domain Ω is discretized by the triangulation $\mathcal{T}_0$ which is obtained by a proper number of ℓ uniform refinement steps of a given coarser triangulation $\mathcal{T}_\ell$. We suppose also that $\mathcal{T}_\ell$ is aligned with the discontinuities of $\mathbf{a}(x)$ so that over each element $T \in \mathcal{T}_\ell$, the entries of the coefficient matrix (diffusion tensor) $\mathbf{a}(x)$ are smooth functions. This assumption is mainly needed for theoretical considerations and is disregarded in the computational examples presented in Sect. 5.

The variational problem (2) is discretized using the finite element method (FEM), i.e., the continuous space $\mathcal{V}$ is replaced by a finite-dimensional space $\mathcal{V}_h$. Then the finite element formulation is the following: find $u_h \in \mathcal{V}_h$, satisfying

$$\begin{aligned} \mathcal{A}_h(u_h,v_h) &= (f,v_h) \quad \forall v_h \in \mathcal{V}_h, \quad \text{where} \\ \mathcal{A}_h(u_h,v_h) &:= \textstyle\sum_{e \in \mathcal{T}_h} \int_e \mathbf{a}(e)\nabla u_h \cdot \nabla v_h dx. \end{aligned} \tag{3}$$

We note that the element-by-element additive setting of $\mathcal{A}_h(u_h, v_h)$ is applicable to both conforming and nonconforming FEM discretizations.

Here $\mathbf{a}(e)$ is a piecewise constant symmetric positive definite matrix, defined by the integral averaged values of $\mathbf{a}(x)$ over each element from the coarsest triangulation $\mathcal{T}_\ell$, i.e.,

$$\mathbf{a}(e) = \frac{1}{|e|}\int_e \mathbf{a}(x)dx, \qquad \forall e \in \mathcal{T}_\ell.$$

In this way strong coefficient jumps across the boundaries between adjacent finite elements from $\mathcal{T}_\ell$ are allowed.

The resulting FEM linear system of equations reads as

$$A_h\mathbf{u}_h = \mathbf{f}_h, \tag{4}$$

with A_h and $\mathbf{f}_h$ being the corresponding global stiffness matrix and global right-hand side and h being the discretization (mesh size) parameter for the underlying triangulation $\mathcal{T}_0 = \mathcal{T}_h$ of Ω.

The stiffness matrix is symmetric, positive definite, and sparse. The sparsity property means that the number of nonzero entries in each row/column is uniformly bounded with respect to the number of the unknowns $N = O(h^{-2})$.

In the case of advanced real-life applications (and in the context of this paper), A_h could be very large, that is, N is of order 10^6 up to 10^9. For such problems, the advantages of the iterative solution methods increase quickly with the size of the problem. The conjugate gradient (CG) method invented 60 years ago by Hestenes and Stiefel [15] is the fastest basic iterative scheme for such kind of problems. It provides a sequence of best approximations to the exact solution in the Krylov subspaces generated by the stiffness matrix. The number of CG iterations n_{it}^{CG} depends on the spectral condition number of the matrix $\kappa(A_h)$. In the case of two-dimensional FEM elliptic systems, $\kappa(A_h) = O(N)$ and $n_{it}^{CG} = O(\sqrt{\kappa(A_h)}) = O(N^{1/2})$. The aim of the preconditioning is to relax the mesh-size dependency of the iterations' count. The following estimate characterizes the preconditioned conjugate gradient (PCG) method:

$$n_{it}^{PCG} \leq \frac{1}{2}\sqrt{\kappa(B^{-1}A_h)}\ln\left(\frac{2}{\varepsilon}\right) + 1, \tag{5}$$

where B is a symmetric and positive definite preconditioning matrix (also called preconditioner) and n_{it}^{PCG} is the related number of PCG iterations sufficient to get a prescribed relative accuracy of $\varepsilon > 0$. The general strategy for efficient preconditioning simply follows from the estimate (5). It reads as follows: (i) The condition number of the preconditioned matrix is much less than the original one, i.e., $\kappa\left(B^{-1}A\right) << \kappa(A)$; (ii) The computational complexity to solve the preconditioned system is much smaller than the complexity to solve the original problem, i.e.,

$\mathcal{N}(B^{-1}\mathbf{v}) << \mathcal{N}(A^{-1}\mathbf{v})$. One could say that these conditions are contradictory. Indeed, when $\kappa(B^{-1}A)$ tends to its minimal value, the preconditioner should tend to A and $\mathcal{N}(B^{-1}\mathbf{v}) \to \mathcal{N}(A^{-1}\mathbf{v})$. Fortunately, such kind of reasonings are too pessimistic according to the recent state of the art of the preconditioning algorithms.

Definition 1. The preconditioner is called optimal if the related PCG algorithm has optimal order of computational complexity $\mathcal{N}^{PCG} = O(N)$, that is, if $\kappa(B^{-1}A) = O(1)$ and $\mathcal{N}(B^{-1}\mathbf{v}) = O(N)$.

The existence of optimal iterative solution methods has been an open question before the early 1960s. Now, the optimal order multigrid and multilevel methods are well known in the community of researchers and engineers dealing with large-scale scientific computations and their advanced applications.

This paper is devoted to some recent achievements in the development of robust preconditioners for FEM elliptic systems belonging to the class of multilevel block factorization methods of the AMLI type. It provides a survey on robust AMLI methods for anisotropic elliptic problems, covering a significantly enriched state of the art in this field as compared to the related earlier paper [21].

Based on a sequence of nested finite element meshes, the AMLI methods were originally introduced by Axelsson and Vassilevski in [6] for the case of isotropic elliptic problems discretized by conforming linear finite elements. They are optimal with respect to the mesh parameter (problem size) and can handle straightforwardly arbitrary coefficient jumps on the coarsest mesh. The originally introduced AMLI methods are based on a hierarchical basis (HB) splitting of the stiffness matrix and a recursive application of HB two-level preconditioning. Since then the AMLI theory has evolved beyond the HB framework; see, e.g., [2, 16, 17, 25]. The construction of AMLI is always based on a recursive approximate (two-by-two) block factorization. Under rather general assumptions, the HB AMLI methods are robust in the case of linear (conforming and nonconforming) elements which does not hold for higher-order FEM. Here we present complimentary some very recent results for quadratic elements where the approximate block factorization on each level exploits an additive Schur complement approximation (ASCA), thereby avoiding the HB splitting; see [18, 20] for further details. The resulting (nonlinear) AMLI is very robust with respect to anisotropy that does not have to be aligned with the grid if it is complemented by a proper block-relaxation process. The efficiency of the interplay between these two components can be enhanced if one applies the ASCA and the block smoother on specific, augmented coarse grids (cf. Sect. 4 and [20]).

In our presentation we follow the mathematical concept of high anisotropy or orthotropy introduced in [12]. For any $x \in \Omega$, we denote the eigenvalues $0 < \mu_1(x) \leq \mu_2(x)$ and eigenvectors $\mathbf{q}_j(x)$, $j = 1,2$ (written as vector columns) of the coefficient (diffusion) matrix $\mathbf{a}(x)$. Then

$$\mathbf{a}(x) = \mu_1(x)\mathbf{q}_1(x)\mathbf{q}_1(x)^T + \mu_2(x)\mathbf{q}_2(x)\mathbf{q}_2(x)^T.$$

In this notation, obviously,

$$\mu_1(x)\mathbf{q}^T\mathbf{q} \leq \mathbf{q}^T\mathbf{a}(x)\mathbf{q} \leq \mu_2(x)\mathbf{q}^T\mathbf{q}, \qquad \forall \mathbf{q} \in \mathbb{R}^2.$$

Depending on the variation of the eigenvalues $\mu_j(x)$, we may have various scenarios of highly anisotropic materials. The aspect ratio of coefficient anisotropy is introduced as

$$\kappa(\mathbf{a}) = \max_x \frac{\mu_2(x)}{\mu_1(x)} = \max_e \frac{\mu_2(e)}{\mu_1(e)}.$$

The direction of dominating anisotropy is determined by the eigenvector $\mathbf{q}_2(x)$.

The following simple examples illustrate the general mathematical concept of high anisotropy, where $\eta \gg 1$: (a) In the case of orthotropic problem we have, e.g., $\mathbf{a} = [1,0;0,\eta]$. Then $\mu_1 = 1$, $\mu_2 = \eta$, consequently $\kappa = \eta$, and $\mathbf{q}_2 = [0,1]^T$, i.e., the direction of dominating anisotropy is along the coordinate y-axis. (b) Let $\mathbf{a} = [1+\eta,\eta-1;\eta-1,1+\eta]$. Then $\mu_1 = 2$, $\mu_2 = 2\eta$, and $\kappa = \eta$. The direction of dominating anisotropy is determined by $\mathbf{q}_2 = [1+\eta,1-\eta]^T/\sqrt{2(1+\eta^2)}$ where $\mathbf{q}_2 \to [1,-1]^T/\sqrt{2}$ when $\eta \to \infty$.

In what follows later, two representative variants of the coefficient $\mathbf{a}(e)$ are considered:

(a) The *isotropic/orthotropic problem* associated with

$$\mathbf{a}(e) = \begin{bmatrix} 1 & 0 \\ 0 & \varepsilon \end{bmatrix}. \tag{6}$$

(b) The *rotated diffusion* problem associated with

$$\mathbf{a}(e) = \begin{bmatrix} \cos\theta & -\sin\theta \\ \sin\theta & \cos\theta \end{bmatrix} \begin{bmatrix} 1 & \\ & \varepsilon \end{bmatrix} \begin{bmatrix} \cos\theta & -\sin\theta \\ \sin\theta & \cos\theta \end{bmatrix}^T, \tag{7}$$

where $\varepsilon > 0$ and $\theta = \theta_e$ is a piecewise constant angle.

The setting of (7) allows to study problems with a given fixed or varying direction (angle) of anisotropy. Nongrid-aligned anisotropy in general is much more difficult to handle than orthotropy (or grid-aligned anisotropy) thus far.

The remainder of the paper is organized as follows. The theoretical background of the AMLI methods is presented next. Together with the classical formulations, Sect. 2 contains the main convergence results for linear and nonlinear AMLI methods. A complete set of robustness results for anisotropic linear FEM systems is presented in Sect. 3, where the HB AMLI method is considered. The estimates are robust with respect to coefficient and mesh anisotropy for both conforming and nonconforming elements. Section 4 is devoted to preconditioning of quadratic FEM systems. It starts with a few comments on HB splittings, which are not robust in this case. Then some very recent results are presented on an AMLI method based on ASCA. In the latter method two additional stabilizing components are incorporated, namely, augmented coarse grids and a global (block) smoothing. The numerical results in Sect. 5 demonstrate the potential of this approach for complicated and more realistic problems which are still beyond the scope of rigorous theory. The survey concludes with final remarks given in Sect. 6.

2 Algebraic Multilevel Methods

The AMLI methods have originally been introduced and studied in a multiplicative form; see [6, 7]. The presentation in this section follows [27]. Consider the linear system (4) where $A_h =: A^{(0)}$ is the fine-grid stiffness matrix. We assume that the standard components of a multigrid (MG) method, that is, the kth-level matrices $A^{(k)}$, smoothers $M^{(k)}$, and coarse-to-fine interpolation matrices $P^{(k)}$, have been defined and that the Galerkin relation $A^{(k+1)} = {P^{(k)}}^T A^{(k)} P^{(k)}$ holds for $k = 0, 1, \ldots, \ell - 1$.

The AMLI preconditioner $B^{(k)}$ is defined recursively via its inverse. On the coarsest level ℓ we set

$$ {B^{(\ell)}}^{-1} = {A^{(\ell)}}^{-1}. \tag{8} $$

Then, assuming that ${B^{(k+1)}}^{-1}$ has already been defined for $k+1 \leq \ell$, one constructs ${B^{(k)}}^{-1}$ in two steps. First, an approximation $Z^{(k+1)}$ of $A^{(k+1)}$ is defined by

$$ Z^{(k+1)} := A^{(k+1)} \left(I - p^{(k)} ({B^{(k+1)}}^{-1} A^{(k+1)}) \right)^{-1}, \tag{9} $$

where $p^{(k)}$ denotes a polynomial of degree $\nu = \nu_k$, satisfying

$$ p^{(k)}(0) = 1. \tag{10} $$

It is important to note that in view of (10) Eq. (9) is equivalent to

$$ {B_\nu^{(k+1)}}^{-1} := {Z^{(k+1)}}^{-1} = {B^{(k+1)}}^{-1} q^{(k)} (A^{(k+1)} {B^{(k+1)}}^{-1}) \tag{11} $$

where the polynomial $q^{(k)}$ is given by

$$ q^{(k)}(x) = \frac{1 - p^{(k)}(x)}{x} \tag{12} $$

showing that the application of ${B_\nu^{(k+1)}}^{-1} = {Z^{(k+1)}}^{-1}$ requires only applications of $A^{(k+1)}$ and ${B^{(k+1)}}^{-1}$ but not of the inverse of the coarse-level matrix $A^{(k+1)}$ (as this is the case in the exact two-level method). Second, the AMLI preconditioner $B^{(k)}$ at level k is defined by

$$ \begin{aligned} {B^{(k)}}^{-1} &:= {\bar{M}^{(k)}}^{-1} \\ &+ \left(I - {M^{(k)}}^{-T} A^{(k)} \right) P^{(k)} {B_\nu^{(k+1)}}^{-1} {P^{(k)}}^T \left(I - A^{(k)} {M^{(k)}}^{-1} \right) \end{aligned} \tag{13} $$

where $B_\nu^{(k+1)^{-1}}$ is given by (11) and $\bar{M}^{(k)}$ denotes the *symmetrized smoother* at level k, that is,

$$\bar{M}^{(k)^{-1}} = M^{(k)^{-1}} + M^{(k)^{-T}} - M^{(k)^{-T}} A^{(k)} M^{(k)^{-1}}. \tag{14}$$

We observe that the multilevel preconditioner defined via (11) and (13) is getting close to an exact two-level method when the polynomial (12) approximates well $1/x$ in which case $B_\nu^{(k+1)^{-1}} \approx A^{(k+1)^{-1}}$. In order to obtain an efficient multilevel method, the action of $B_\nu^{(k+1)^{-1}}$ on an arbitrary vector should be much cheaper to compute (in terms of the number of arithmetic operations) than the action of $A^{(k+1)^{-1}}$. Optimal order solution algorithms typically require the arithmetic work for one application of $B_\nu^{(k+1)^{-1}}$ to be of the order $\mathcal{O}(N_{k+1})$ where N_{k+1} denotes the number of unknowns at level $k+1$.

In the classical AMLI method, as it has been introduced in [6,7], the coarse-grid matrix $A^{(k+1)}$ is retrieved from a (two-level) hierarchical basis transformation of $A^{(k)}$. The preconditioner $\tilde{B}^{(k)}$ (in its multiplicative variant) then is defined by

$$\begin{aligned}
(\tilde{B}^{(k)})^{-1} &= \begin{bmatrix} B_{11}^{(k)^{-1}} & 0 \\ 0 & 0 \end{bmatrix} + \begin{bmatrix} -B_{11}^{(k)^{-1}} \hat{A}_{12}^{(k)} \\ I \end{bmatrix} B_\nu^{(k+1)^{-1}} \left[-\hat{A}_{21}^{(k)} B_{11}^{(k)^{-1}}, I \right] \\
&= \begin{bmatrix} B_{11}^{(k)^{-1}} & 0 \\ 0 & 0 \end{bmatrix} + (\tilde{L}^{(k)})^T \begin{bmatrix} 0 \\ I \end{bmatrix} B_\nu^{(k+1)^{-1}} [0, I]\, \tilde{L}^{(k)}
\end{aligned}$$

where

$$\tilde{L}^{(k)} = \begin{bmatrix} I - A_{11}^{(k)} B_{11}^{(k)^{-1}} & 0 \\ -\hat{A}_{21}^{(k)} B_{11}^{(k)^{-1}} & I \end{bmatrix}.$$

Writing the equation above in the form (13), one finds that

$$M^{(k)^{-1}} = M^{(k)^{-T}} = \begin{bmatrix} B_{11}^{(k)^{-1}} & 0 \\ 0 & 0 \end{bmatrix} \tag{15}$$

is a smoother that acts only on the hierarchical complement of the coarse space, where $B_{11}^{(k)}$ is a proper approximation of $A_{11}^{(k)}$. The corresponding symmetrized smoother then is given by

$$\bar{M}^{(k)^{-1}} = \begin{bmatrix} 2B_{11}^{(k)^{-1}} - B_{11}^{(k)^{-1}} A_{11}^{(k)} B_{11}^{(k)^{-1}} & 0 \\ 0 & 0 \end{bmatrix}, \tag{16}$$

and $P^{(k)}$ takes the simple form

$$P^{(k)} = \begin{bmatrix} 0 \\ I \end{bmatrix}. \tag{17}$$

The latter is due to the fact that the (classical) AMLI preconditioner is defined for the hierarchical two-level matrix $\hat{A}^{(k)}$, which contains the coarse-level matrix as a sub-matrix in its lower right block, i.e.,

$$A^{(k+1)} = [0,I]\hat{A}^{(k)} \begin{bmatrix} 0 \\ I \end{bmatrix}.$$

This, however, is in agreement with the Galerkin relation $A^{(k+1)} = {P^{(k)}}^T A^{(k)} P^{(k)}$ as is used in (algebraic) multigrid methods.

The convergence theory of the classical AMLI methods (in the multiplicative variant) is based on the spectral equivalence of the k-th level hierarchical matrix $\hat{A}^{(k)}$ and its (multiplicative) two-level preconditioner

$$\hat{B}^{(k)} = \begin{bmatrix} B_{11}^{(k)} & 0 \\ \hat{A}_{21}^{(k)} & \hat{A}_{22}^{(k)} \end{bmatrix} \begin{bmatrix} I & {B_{11}^{(k)}}^{-1} \hat{A}_{12}^{(k)} \\ 0 & I \end{bmatrix}, \tag{18}$$

that is,

$$\hat{\vartheta}_k \hat{B}^{(k)} \leq \hat{A}^{(k)} \leq \hat{B}^{(k)}, \quad k = \ell-1,\ldots,0. \tag{19}$$

Note that if $B_{11}^{(k)} = A_{11}^{(k)}$ then $\hat{\vartheta}_k = 1 - \gamma_k^2$ where γ_k is the constant in the strengthened Cauchy–Bunyakovsky–Schwarz (CBS) inequality associated with the hierarchical matrix $\hat{A}^{(k)}$. The subscript of γ is usually skipped when uniform estimate of the CBS constant with respect to the refinement level k is assumed (see, e.g., (35)). We conclude that the polynomial acceleration techniques described in this paper can be exploited in various implementations of AMLI preconditioners, which can be viewed as inexact two-level methods. The performance of these methods crucially depends on the particular choice of the polynomial $q^{(k)}$ in Eq. (11) and on two-level estimates like (19) or (31).

2.1 *Condition Number Estimates for AMLI Preconditioners*

Let us first summarize the main result of the analysis of the AMLI-cycle multigrid preconditioner as presented in [27].

The AMLI-cycle is a ν-fold multigrid (MG) cycle with variable $\nu = \nu_k$. In the following, let $\nu \geq 1$ and $k_0 \geq 1$ be two fixed integers. We set $\nu_{sk_0} = \nu > 1$ for $s = 1,2,3,\ldots$ and $\nu_k = 1$ otherwise. That is, we let

$${B_\nu^{((s+1)k_0)}}^{-1} = {B^{((s+1)k_0)}}^{-1} q_{\nu-1}(A^{((s+1)k_0)} {B^{((s+1)k_0)}}^{-1}) \tag{20}$$

if $k+1 = (s+1)k_0$, and

$${B_\nu^{(k+1)}}^{-1} = {B^{(k+1)}}^{-1} \tag{21}$$

otherwise. Then for the AMLI-cycle MG preconditioner $B^{(k)}$ defined in (13), the following result can be proven (cf. Theorem 5.29 in [27]).

Theorem 1 ([27]). *With a proper choice of the parameters k_0 and ν, and for a proper choice of the polynomial $p^{(k)}(x) = p_\nu(x)$ satisfying (10), the condition number of ${B^{(k)}}^{-1}A^{(k)}$ can be uniformly bounded provided the V-cycle preconditioners with bounded level difference $\ell - k \leq k_0$ have uniformly bounded condition numbers $K_{MG}^{\ell \mapsto k}$.*

More specifically, for a fixed k_0, and $\nu > \sqrt{K_{MG}^{\ell \mapsto k}}$, we can choose $\alpha > 0$ such that

$$\alpha K_{MG}^{\ell \mapsto k} + K_{MG}^{\ell \mapsto k} \frac{(1-\alpha)^\nu}{\left[\sum_{j=1}^{\nu}(1+\sqrt{\alpha})^{\nu-j}(1-\sqrt{\alpha})^{j-1}\right]^2} \leq 1$$

and employ the polynomial

$$p_\nu(x) = \frac{1 + T_\nu\left(\frac{1+\alpha-2x}{1-\alpha}\right)}{1 + T_\nu\left(\frac{1+\alpha}{1-\alpha}\right)}$$

where T_ν is the Chebyshev polynomial of the first kind of degree ν.

Alternatively, we can choose $\alpha \in (0,1)$ such that

$$\alpha K_{MG}^{\ell \mapsto k} + K_{MG}^{\ell \mapsto k} \frac{(1-\alpha)^\nu}{\sum_{j=1}^{\nu}(1-\alpha)^{j-1}} \leq 1$$

and use the polynomial $p_\nu(x) = (1-x)^\nu$ to define $q_{\nu-1}(x) := (1 - p_\nu(x))/x$ in (20).

Then for both choices of the polynomial p_ν (respectively $q_{\nu-1}$), the resulting AMLI-cycle preconditioner $B = B^{(0)}$, as defined via (8)–(13), is spectrally equivalent to the matrix $A = A^{(0)}$, and the following estimate holds

$$\mathbf{v}^T A \mathbf{v} \leq \mathbf{v}^T B \mathbf{v} \leq \frac{1}{\alpha} \mathbf{v}^T A \mathbf{v} \quad \forall \mathbf{v}, \tag{22}$$

with the respective $\alpha \in (0,1]$ depending on the choice of the polynomial.

2.2 Nonlinear AMLI-Cycle Method

Consider a sequence of two-by-two block matrices

$$A^{(k)} = \begin{bmatrix} A_{11}^{(k)} & 0 \\ A_{21}^{(k)} & S^{(k)} \end{bmatrix} \begin{bmatrix} I & {A_{11}^{(k)}}^{-1} A_{12}^{(k)} \\ 0 & I \end{bmatrix} = \begin{bmatrix} A_{11}^{(k)} & A_{12}^{(k)} \\ A_{21}^{(k)} & S^{(k)} + A_{21}^{(k)} {A_{11}^{(k)}}^{-1} A_{12}^{(k)} \end{bmatrix} \tag{23}$$

associated with a (nested) sequence of meshes $\mathcal{T}_k$, $k = 0,1,2,\ldots,\ell$, where $\mathcal{T}_\ell$ denotes the coarsest mesh (and $A^{(k)}$ could also be in hierarchical basis). Let $S^{(k)}$ be the Schur complement in the exact block factorization (23) of $A^{(k)}$. Moreover, the following abstract (linear) multiplicative two-level preconditioner

$$\bar{B}^{(k)} = \begin{bmatrix} B_{11}^{(k)} & 0 \\ A_{21}^{(k)} & Q^{(k)} \end{bmatrix} \begin{bmatrix} I & B_{11}^{(k)^{-1}} A_{12}^{(k)} \\ 0 & I \end{bmatrix} = \begin{bmatrix} B_{11}^{(k)} & A_{12}^{(k)} \\ A_{21}^{(k)} & Q^{(k)} + A_{21}^{(k)} B_{11}^{(k)^{-1}} A_{12}^{(k)} \end{bmatrix} \tag{24}$$

to $A^{(k)}$ is defined at levels $k = 0,1,2,\ldots,\ell-1$. Here $B_{11}^{(k)}$ is a preconditioner to $A_{11}^{(k)}$ and $Q^{(k)}$ is a sparse approximation of $S^{(k)}$. In order to relate the two sequences $(A^{(k)})_{k=0,1,2,\ldots,\ell-1}$ and $(\bar{B}^{(k)})_{k=0,1,2,\ldots,\ell-1}$ to each other, one sets

$$A^{(0)} := A_h = A, \tag{25}$$

where A_h is the stiffness matrix in (4), and defines

$$A^{(k+1)} := Q^{(k)}, \quad k = 0,1,2,\ldots,\ell-1. \tag{26}$$

Next the nonlinear AMLI-cycle preconditioner $B^{(k)}[\cdot] : \mathbb{R}^{N_k} \mapsto \mathbb{R}^{N_k}$ for $k = \ell - 1,\ldots,0$ is defined recursively by

$$B^{(k)^{-1}}[\mathbf{y}] := U^{(k)} D^{(k)} [L^{(k)} \mathbf{y}], \tag{27}$$

where

$$L^{(k)} := \begin{bmatrix} I & 0 \\ -A_{21}^{(k)} B_{11}^{(k)^{-1}} & I \end{bmatrix}, \tag{28}$$

$U^{(k)} = L^{(k)^T}$, and

$$D^{(k)}[\mathbf{z}] = \begin{bmatrix} B_{11}^{(k)^{-1}} \mathbf{z}_1 \\ Z^{(k+1)^{-1}}[\mathbf{z}_2] \end{bmatrix}. \tag{29}$$

The (nonlinear) mapping $Z^{(k+1)^{-1}}[\cdot]$ is defined by

$$\begin{aligned} Z^{(\ell)^{-1}}[\cdot] &= A^{(\ell)^{-1}}, \\ Z^{(k)^{-1}}[\cdot] &:= B^{(k)^{-1}}[\cdot] \quad \text{if } \nu = 1 \text{ and } k < \ell, \\ Z^{(k)^{-1}}[\cdot] &:= B_\nu^{(k)^{-1}}[\cdot] \quad \text{if } \nu > 1 \text{ and } k < \ell, \end{aligned} \tag{30}$$

with

$$B_\nu^{(k)^{-1}}[\mathbf{d}] := \mathbf{x}_{(\nu)}$$

where $\mathbf{x}_{(\nu)}$ is the ν-th iterate obtained when applying the generalized conjugate gradient (GCG) algorithm (see [8]) to the linear system $A^{(k)}\mathbf{x} = \mathbf{d}$ using $B^{(k)}[\cdot]$

as a preconditioner and starting with the initial guess $\mathbf{x}_{(0)} = \mathbf{0}$. The vector $\boldsymbol{\nu} = (\nu_1, \nu_2, \ldots, \nu_{\ell-1})^T$ specifies how many inner GCG iterations are performed at each of the levels $k = \ell - 1, \ldots, 1$, and $\nu_0 = m_{\max}$ denotes the maximum number of orthogonal search directions at level 0. Typically the algorithm is restarted after every $m_{\max}$ iterations. If a fixed number ν of inner GCG-type iterations is performed at every intermediate level, i.e., $\nu_k = \nu$ for $k = \ell - 1, \ldots, 1$, the method is referred to as (nonlinear) ν-fold W-cycle AMLI method.

Convergence: Next the main convergence result from [16] is presented.

Denoting by $\mathbf{x}_{(i)}$ the i-th iterate generated by the nonlinear AMLI method, the goal is to derive a bound for the error reduction factor in A norm. This can be done by assuming, for example, that the two-level preconditioners (24) and the matrices (23) are spectrally equivalent, i.e.,

$$\underline{\vartheta}_k \bar{B}^{(k)} \le A^{(k)} \le \overline{\vartheta}_k \bar{B}^{(k)}, \quad k = \ell - 1, \ldots, 0. \tag{31}$$

A slightly different approach to analyze the nonlinear AMLI-cycle method is based on the assumption that all fixed-length V-cycle multilevel methods from any coarse-level $k + k_0$ to level k with exact solution at level $k + k_0$ are uniformly convergent in k with an error reduction factor $\delta_{k_0} \in [0, 1)$; see [26,27]. Both approaches, however, are based on the idea to estimate the deviation of the nonlinear preconditioner $B^{(k)}[\cdot]$ from an SPD matrix $\bar{B}^{(k)}$.

The following theorem (see [16, 18]) summarizes the main convergence result.

Theorem 2 ([16]). *Consider the linear system $A^{(0)}\mathbf{x} = \mathbf{d}^{(0)}$ where $A^{(0)}$ is an SPD stiffness matrix, and let $\mathbf{x}_{(i)}$ be the sequence of iterates generated by the nonlinear AMLI algorithm. Further, assume that the approximation property (31) holds and let $\vartheta := \max_{0 \le k < \ell} \overline{\vartheta}_k / \underline{\vartheta}_k$. If ν, the number of inner GCG iterations at every coarse level (except level ℓ where $Z^{(\ell)^{-1}}[\cdot] = A^{(\ell)^{-1}}$) is chosen such that*

$$\delta(\nu) := \left(1 - \frac{4\vartheta(1-\varepsilon)^2}{(1 + \vartheta - 2\varepsilon + \vartheta\varepsilon^2)^2}\right)^{\nu/2} \le \varepsilon \tag{32}$$

for some positive $\varepsilon < 1$ then

$$\frac{\|\mathbf{x} - \mathbf{x}_{(i+1)}\|_{A^{(0)}}}{\|\mathbf{x} - \mathbf{x}_{(i)}\|_{A^{(0)}}} \le \sqrt{1 - \frac{4\vartheta(1-\varepsilon)^2}{(1 + \vartheta - 2\varepsilon + \vartheta\varepsilon^2)^2}} = \delta(1) =: \delta < 1. \tag{33}$$

Remark 1. Note that the relative condition number $\kappa(Q^{(k)^{-1}} S^{(k)})$ affects the approximation property (31). In the simplest case in which the multiplicative two-level preconditioner (24) is considered under the assumption $B_{11}^{(k)} = A_{11}^{(k)}$, this results in $\vartheta = \kappa(Q^{(k)^{-1}} S^{(k)})$.

2.3 Optimality Conditions

As has been stated in Theorem 2, uniform convergence of the AMLI method can be proven under the assumption (31), which guarantees that the (multiplicative) two-level preconditioner satisfies a certain approximation property. Equivalently, uniform convergence of the multilevel V-cycle preconditioner ($\nu = 1$) with bounded level difference can be required as the basic assumption to prove uniform convergence of the AMLI method for unbounded level difference as this was done in Theorem 1 in case of the linear preconditioner. In many cases these assumptions can be verified by studying the angle between the coarse space and its hierarchical complement. In fact, and this was shown in the original convergence analysis of linear AMLI methods [7], a stabilization of the condition number of the (multiplicative) multilevel preconditioner can be achieved under the assumption

$$A_{11}^{(k)} \leq B_{11}^{(k)} \leq \omega A_{11}^{(k)} \tag{34}$$

on the approximation of the pivot block $A_{11}^{(k)}$ if

$$\frac{1}{\sqrt{1-\gamma^2}} < \nu. \tag{35}$$

Assuming now that we have a fully stabilized multilevel method, i.e., the solutions for a repeatedly refined mesh (in principle for any number of regular refinement steps) are obtained at a constant number of iterations. Then the second condition to be fulfilled for an optimal order solution process is that the computational cost of each single iteration is proportional to the total number of degrees of freedom (DOF).

The computational work (operation count) of the ν-fold W-cycle of either linear or nonlinear AMLI at level 0 (associated with the finest mesh) can be estimated by

$$\begin{aligned} w^{(0)} &\leq c\,(N_0 + \nu N_1 + \ldots + \nu^\ell N_\ell) \\ &= c N_0 \left(1 + \frac{\nu}{\rho} + \left(\frac{\nu}{\rho}\right)^2 + \ldots + \left(\frac{\nu}{\rho}\right)^\ell\right) = c N_0 \frac{1-\left(\frac{\nu}{\rho}\right)^{\ell+1}}{1-\frac{\nu}{\rho}}. \end{aligned}$$

Assuming that the number of DOF at level $k+1$ is (approximately) $1/\rho$ times the number of DOF at level k, each visit of level k must induce less than ρ visits of level $k+1$ (at least in average). This means that if the coarsening ratio is, for example, four, i.e., $\rho = 4$, then two but also three inner GCG iterations, or, alternatively, the employment of second- but also third-degree matrix polynomials at every intermediate level, result in a computational complexity $O(N) = O(N_0)$ of one (outer) iteration. The condition for optimal order single iterations is thus

$$\nu < \rho, \tag{36}$$

which combined with (35) results in the (combined) optimality conditions

$$\frac{1}{\sqrt{1-\gamma^2}} < \nu < \rho. \tag{37}$$

In what follows, we assume that the default meaning of AMLI is the multiplicative one.

Remark 2. The optimality conditions for the symmetric preconditioner of block-diagonal (additive) form are given by

$$\sqrt{\frac{1+\gamma}{1-\gamma}} < \nu < \rho. \tag{38}$$

Stabilization techniques for additive multilevel iteration methods and nearly optimal order parameter-free block-diagonal preconditioners of AMLI type are discussed in [4, 5].

3 Linear Elements

The material selected in this section follows the spirit of the robust AMLI methods as originally presented in [3, 4, 9, 10, 22, 24] as well as the earlier survey paper [21]. The hierarchical basis approach is followed for both conforming and nonconforming elements. This allows systematically to use local constructions and analysis at the level of element and macroelement matrices.

3.1 Conforming Elements

Some Basic Relations: Let us remind that the analysis for an arbitrary triangle (e) can be done on the reference triangle $(\tilde{e})$. Transforming the finite element functions between these triangles, the element bilinear form $\mathcal{A}_e(.,.)$ takes the form

$$\mathcal{A}_{\tilde{e}}(\tilde{u},\tilde{v}) = \int_{\tilde{e}} \sum_{i,j} \tilde{a}_{ij} \frac{\partial \tilde{u}}{\partial \tilde{x}_i} \frac{\partial \tilde{v}}{\partial \tilde{x}_j}, \tag{39}$$

where the coefficients $\tilde{a}_{ij}$ depend on both the coordinates in e and the coefficients a_{ij} in the differential operator.

The important conclusion is that it suffices for the local analysis to consider the (macro)element stiffness matrices for the reference triangle and arbitrary anisotropic coefficients $[a_{ij}]$ or, alternatively, for the isotropic operator $-\Delta$ and an arbitrary triangle e. In this sense, the mesh and coefficient anisotropy are equivalent, which obviously holds true for any conforming or nonconforming triangular finite element.

Following the FEM assembling procedure, we write the global stiffness matrix A in the form

$$A = \sum_{e \in \mathcal{T}_k} R_e^T A_e R_e, \tag{40}$$

where A_e is the element stiffness matrix and R_e stands for the restriction mapping of the global vector of unknowns to the local one corresponding to element $e \in \mathcal{T}_k$.

Consider now the Laplace operator and an arbitrary shaped linear triangular finite element (mesh anisotropy). Then, the element stiffness matrix A_e can be written in the form

$$A_e = \frac{1}{2} \begin{bmatrix} b+c & -c & -b \\ -c & a+c & -a \\ -b & -a & a+b \end{bmatrix}, \tag{41}$$

where a, b, and c equal the cotangent of the angles in $e \in \mathcal{T}_h$. Without loss of generality, we assume in the local analysis that $|a| \leq b \leq c$, which follows from the next lemma; see, e.g., [3].

Lemma 1. *Let $\theta_1, \theta_2, \theta_3$ be the angles in an arbitrary triangle. Then with $a = \cot\theta_1$, $b = \cot\theta_2$, $c = \cot\theta_3$, it holds*

(i) $a = (1 - bc)/(b + c)$
(ii) If $\theta_1 \geq \theta_2 \geq \theta_3$ *then* $|a| \leq b \leq c$
(iii) $a + b > 0$.

Applying Lemma 1, we simply get the scaled representation of the element stiffness matrix:

$$A_e = \frac{c}{2} \begin{bmatrix} \beta+1 & -1 & -\beta \\ -1 & \alpha+1 & -\alpha \\ -\beta & -\alpha & \alpha+\beta \end{bmatrix}, \tag{42}$$

$\alpha = a/c$, $\beta = b/c$, and $(\alpha, \beta) \in D$, where

$$D = \{(\alpha, \beta) \in \mathbb{R}^2 : -\frac{1}{2} < \alpha \leq 1, \max\{-\frac{\alpha}{\alpha+1}, |\alpha|\} \leq \beta \leq 1\}. \tag{43}$$

The local analysis in terms of (α, β) belonging to the convex curvilinear triangle D plays a key role in the derivation of robust estimates for anisotropic problems; see [3, 10].

Uniform Estimates of the Constant in the Strengthened CBS Inequality: Consider two consecutive meshes $\mathcal{T}_{k+1} \subset \mathcal{T}_k$. A uniform refinement procedure is set as a default assumption where the current coarse triangle $e \in \mathcal{T}_{k+1}$ is subdivided in four congruent triangles by joining the mid-edge nodes to get the macroelement $E \in \mathcal{T}_k$. The related macroelement stiffness matrix consists of blocks which are 3×3

matrices and, the local eigenproblem to compute γ_E has a reduced dimension of 2×2.

In the so-arising six node-points of the macroelement, we can also use hierarchical basis functions, where we keep the linear basis functions in the vertex nodes and add piecewise quadratic basis functions in the mid-edge nodes with support on the whole triangle. Let us denote by $\hat{\gamma}_E$ the corresponding CBS constant. The following relation between γ_E and $\hat{\gamma}_E$ holds.

Theorem 3 ([23]). *Let us consider a piecewise Laplacian elliptic problem on an arbitrary finite element triangular mesh $\mathcal{T}_{k+1}$, and let each element from $\mathcal{T}_{k+1}$ be refined into four congruent elements to get $\mathcal{T}_k$. Then*

$$\hat{\gamma}_E^2 = \frac{4}{3}\gamma_E^2, \tag{44}$$

where $\hat{\gamma}_E$, γ_E are the local CBS constants for the hierarchical piecewise quadratic and the piecewise linear finite elements, respectively.

Taking into account that $\hat{\gamma}_E < 1$, we get the local estimate

$$\gamma_E^2 < \frac{3}{4} \tag{45}$$

which holds uniformly with respect to the mesh anisotropy. Then, the next fundamental result follows directly from the local estimate (45), the equivalence relation (39), and the inequality $\gamma \leq \max_E \gamma_E$.

Theorem 4. *Consider the problem (3) discretized by conforming linear finite elements, where the coarsest grid $\mathcal{T}_\ell$ is aligned with the discontinuities of the coefficient $\mathbf{a}(e)$, $e \in \mathcal{T}_\ell$. Let us assume also that $\mathcal{T}_{k+1} \subset \mathcal{T}_k$ are two consecutive meshes where each element from $\mathcal{T}_{k+1}$ is refined into four congruent elements to get $\mathcal{T}_k$. Then, the estimate*

$$\gamma^2 < \frac{3}{4} \tag{46}$$

of the CBS constant holds uniformly with respect to the coefficient jumps, mesh or/and coefficient anisotropy, and the refinement level k.

Preconditioning of the Pivot Block: When applicable, we will skip the superscripts of the pivot block and its approximation. Here, we will write A_{11}, B_{11}, instead of $A_{11}^{(k)}$, $B_{11}^{(k)}$. The construction and the analysis of the preconditioners B_{11} are based on a macroelement-by-macroelement assembling procedure. Following (40), we write A_{11} in the form

$$A_{11} = \sum_{E \in \mathcal{T}_{k+1}} R_E^T A_{E:11} R_E. \tag{47}$$

Following the scaled representation (42), we get

$$A_{E:11} = r_T c_T \begin{bmatrix} \alpha+\beta+1 & -1 & -\beta \\ -1 & \alpha+\beta+1 & -\alpha \\ -\beta & -\alpha & \alpha+\beta+1 \end{bmatrix}. \tag{48}$$

Then, the additive preconditioner of A_{11} is defined as follows:

$$B_{11}^{(A)} = \sum_{E \in \mathcal{T}_{k+1}} R_E^T B_{E:11}^{(A)} R_E, \tag{49}$$

where

$$B_{E:11}^{(A)} = 2 r_T c_T \begin{bmatrix} \alpha+\beta+1 & -1 & 0 \\ -1 & \alpha+\beta+1 & 0 \\ 0 & 0 & \alpha+\beta+1 \end{bmatrix}. \tag{50}$$

As one can see, the local matrix $B_{E:11}^{(A)}$ is obtained by preserving only the *strongest* off-diagonal entries. Alternatively, the multiplicative preconditioner $B_E^{(M)}$ is defined as a symmetric block Gauss–Seidel preconditioner of A_{11} subject to a proper node numbering (see, e.g., [3]).

Theorem 5 ([3, 4]). *The additive and multiplicative preconditioners of A_{11} are uniform, i.e.,*

$$\kappa\left(B_{11}^{(A)^{-1}} A_{11}\right) < \frac{1}{4}(11+\sqrt{105}) \approx 5.31, \tag{51}$$

$$\kappa\left(B_{11}^{(M)^{-1}} A_{11}\right) < \frac{15}{8} = 1.875. \tag{52}$$

These condition number bounds hold independently on shape and size of each element (mesh anisotropy) and on the coefficient matrix $\mathbf{a}(e)$ of the FEM problem (coefficient anisotropy).

3.2 Nonconforming Elements

For the nonconforming Crouzeix–Raviart finite element, where the nodal basis functions are defined at the midpoints along the edges of the triangle rather than at its vertices (cf. Fig. 1), the natural vector spaces $\mathcal{V}_H(E) := \text{span}\{\phi_I, \phi_{II}, \phi_{III}\}$ and $\mathcal{V}_h(E) := \text{span}\{\phi_i\}_{i=1}^9$ (cf. the macroelement in Fig.1) are no longer nested, i.e., $\mathcal{V}_H(E) \nsubseteq \mathcal{V}_h(E)$. A simple computation shows that the element stiffness matrix for the Crouzeix–Raviart (CR) element, A_e^{CR}, coincides with that of the corresponding conforming linear element up to a factor 4, i.e.,

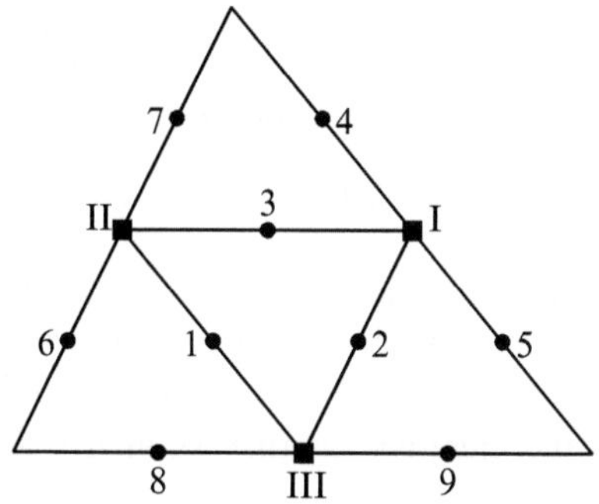

Fig. 1 Macroelement composed of four Crouzeix–Raviart elements

$$A_e^{\mathrm{CR}} = 2 \begin{bmatrix} b+c & -c & -b \\ -c & a+c & -a \\ -b & -a & a+b \end{bmatrix}, \tag{53}$$

(cf., (41)). The construction of the hierarchical stiffness matrix at macroelement level starts with the assembly of four such matrices according to the numbering of the nodal points, as shown in Fig. 1. It further utilizes a transformation, which is based on a proper decomposition of the vector space $\mathcal{V}(E) = \mathcal{V}_h(E)$, which is associated with the fine-grid basis functions related to this macroelement E. We consider hierarchical splittings, which make use of half-difference and half-sum basis functions. Let us denote by $\Phi_E := \{\phi^{(i)}\}_{i=1}^{9}$ the set of the "midpoint" basis functions of the four congruent elements in the macroelement E, as depicted in Fig. 1. The splitting of $\mathcal{V}(E)$ can be defined in the general form (see [22]):

$$\begin{aligned} \mathcal{V}_1(E) &:= \operatorname{span}\{\phi_1, \phi_2, \phi_3, \phi_1^D + \phi_4 - \phi_5, \phi_2^D + \phi_6 - \phi_7, \phi_3^D + \phi_8 - \phi_9\}, \\ \mathcal{V}_2(E) &:= \operatorname{span}\{\phi_1^C + \phi_4 + \phi_5, \phi_2^C + \phi_6 + \phi_7, \phi_3^C + \phi_8 + \phi_9\}, \end{aligned} \tag{54}$$

where $\phi_i^D := \sum_k d_{ik}\phi_k$ and $\phi_i^C := \sum_k c_{ik}\phi_k$ with $i,k \in \{1,2,3\}$. The transformation matrix is given by

$$J_E^T = J_E^T(C,D) = \begin{bmatrix} I_3 & D & C \\ 0 & J_- & J_+ \end{bmatrix} \quad (\in \mathbb{R}^{9\times 9}), \tag{55}$$

where I_3 denotes the 3×3 identity matrix and C and D are 3×3 matrices whose entries c_{ij}, respectively, d_{ij} are to be specified later. The 3×6 matrices

$$J_{-} := \frac{1}{2} \begin{bmatrix} 1 & -1 & & & & \\ & & 1 & -1 & & \\ & & & & 1 & -1 \end{bmatrix}^T \quad \text{and} \quad J_{+} := \frac{1}{2} \begin{bmatrix} 1 & 1 & & & & \\ & & 1 & 1 & & \\ & & & & 1 & 1 \end{bmatrix}^T \tag{56}$$

introduce the so-called half-difference and half-sum basis functions associated with the sides of the macroelement triangle. The matrix J_E transforms the vector of the macroelement basis functions $\phi_E := (\phi^{(i)})_{i=1}^9$ to the hierarchical basis vector $\tilde{\phi}_E := (\tilde{\phi}^{(i)})_{i=1}^9 = J_E^T \phi_E$, and the hierarchical stiffness matrix at macroelement level is obtained as

$$\tilde{A}_E = J_E^T A_E J_E = \begin{bmatrix} \tilde{A}_{E:11} & \tilde{A}_{E:12} \\ \tilde{A}_{E:12}^T & \tilde{A}_{E:22} \end{bmatrix} \begin{matrix} \} \in \mathcal{V}_1(E) \\ \} \in \mathcal{V}_2(E) \end{matrix} . \tag{57}$$

The related global stiffness matrix is obtained as $\tilde{A}_h := \sum_{E \in \mathcal{T}_H} R_E^T \tilde{A}_E R_E$.

The transformation matrix $J = J(C,D)$ such that $\tilde{\phi} = J^T \phi$ is then used for the transformation of the global matrix A_h to its hierarchical form $\tilde{A}_h = J^T A_h J$, and (by a proper permutation of rows and columns) the latter admits the 3×3-block representation:

$$\tilde{A}_h = \begin{bmatrix} \tilde{A}_{11} & \tilde{A}_{12} & \tilde{A}_{13} \\ \tilde{A}_{12}^T & \tilde{A}_{22} & \tilde{A}_{23} \\ \tilde{A}_{13}^T & \tilde{A}_{23}^T & \tilde{A}_{33} \end{bmatrix} \begin{matrix} \left.\vphantom{\begin{matrix} a \\ b \end{matrix}}\right\} \in \mathcal{V}_1 \\ \} \in \mathcal{V}_2 \end{matrix} \tag{58}$$

according to the interior, half-difference, and half-sum basis functions, which are associated with (54). The next two variants follow [9].

Definition 2 (Differences and Aggregates (DA)). The splitting based on differences and aggregates corresponds to $D = 0$ and $C = \frac{1}{2}\,\mathrm{diag}(1,1,1)$.

Definition 3 (First Reduce (FR) Splitting). The splitting based on differences and aggregates incorporating a "first reduce" (static condensation) step is characterized by setting $D = 0$ and $C = -A_{11}^{-1}\tilde{A}_{13}$ in (55).

Theorem 6 ([22]). *Consider the problem (3) discretized by nonconforming linear finite elements, where the multilevel meshes satisfy the conditions from Theorem 4. Then, the estimate*

$$\gamma_{FR}^2 \le \gamma_{DA}^2 \le \frac{3}{4} \tag{59}$$

of the CBS constants corresponding to FR and DA splittings holds uniformly with respect to the coefficient jumps, the mesh or/and coefficient anisotropy, and the refinement level k.

The preconditioning of the pivot blocks for FR and DA splittings is studied in [10]. The structure of the related systems (after a static condensation for the case of DA) coincides with those for conforming linear elements. Although the derivation of the related condition number estimates is rather different, it is based again on a macroelement analysis in terms of (α,β) belonging to the convex curvilinear triangle D; see (43). In both FR and DA cases, the construction of the additive and multiplicative preconditioners and the related robust upper bounds are the same as in the case of conforming elements; see Theorem 6.

Let us summarize the main results in this section. The results of Theorems 4–5 for the case of conforming elements, Theorem 6 and the analogue of Theorem 5 for nonconforming elements, in combination with the optimal solvers for systems with the additive and multiplicative preconditioners for the corresponding pivot blocks (see for more details [3]), ensure the optimal complexity of the related W-cycle AMLI algorithms with polynomial degree $\beta \in \{2,3\}$. All presented results are robust with respect to both mesh and/or coefficient anisotropy.

4 Quadratic Elements

In [23] and [1], it has been demonstrated that the standard (P_2 to P_1) hierarchical two-level splitting of piecewise quadratic basis functions does not result in robust two- and multilevel methods for highly anisotropic elliptic problems in general. A more recent paper, [19], proves that for orthotropic problems it is possible to construct a robust two-level preconditioner for FEM discretizations using conforming quadratic elements via the HB approach. In the general setting of an arbitrary elliptic operator, however, the standard techniques, based on HB two-level splittings (cf. [13]) and on the direct assembly of local Schur complements (cf. [14]), do not result in splittings in which the angle between the coarse space and its (hierarchical) complement is uniformly bounded with respect to the mesh and/or coefficient anisotropy.

One way to overcome this problem has been suggested in [20]. The idea is to construct a multilevel approximate block factorization based on ASCA and to combine the standard (nonlinear) AMLI with a block smoother. The recursive application of ASCA on a sequence of augmented coarse grids will be described in some more detail in the remainder of this section.

4.1 Notation

Let $H_A = (V_A, E_A)$ denote the (undirected) graph of a matrix $A \in \mathbb{R}^{N\times N}$. The set of vertices (nodes) of A is denoted by $V_A := \{v_i : 1 \leq i \leq N\}$ and the set of edges by $E_A := \{e_{ij} : 1 \leq i < j \leq N \text{ and } a_{ij} \neq 0\}$.

Definition 4. Any subgraph F of H_A is referred to as a structure. The set of structures whose relevant (local) structure matrices A_F satisfy the assembling property

$$\sum_{F\in\mathcal{F}} R_F^T A_F R_F = A. \tag{60}$$

is denoted by $\mathcal{F}$.

Definition 5. Any union G of structures $F \in \mathcal{F}$ is referred to as a macrostructure. The set of macrostructures is denoted by $\mathcal{G}$. It is assumed that any set of corresponding macrostructure matrices $\mathcal{A}_G = \{A_G : G \in \mathcal{G}\}$ has the assembling property

$$\sum_{G\in\mathcal{G}} R_G^T A_G R_G = A. \tag{61}$$

Definition 6. If $F_i \cap F_j = \emptyset$ (or $G_i \cap G_j = \emptyset$) for all $i \neq j$, we refer to the set $\mathcal{F}$ (or $\mathcal{G}$) as a nonoverlapping covering; otherwise, we call $\mathcal{F}$ (or $\mathcal{G}$) an overlapping covering.

4.2 Additive Schur Complement Approximation

Let $S = S^{(k)}$ be the exact Schur complement of $A = A^{(k)}$ that we wish to approximate on a specific (augmented) coarse grid, and let us denote the corresponding graph by H. To give an example, in case of a uniform mesh as illustrated in Fig. 3c, we construct overlapping coverings of H by structures F and macrostructures G where each macrostructure $G \in \mathcal{G}$ is composed of nine 13-node structures $F \in \mathcal{F}$ which overlap with half of their width or height as shown on Fig. 2. Then the following algorithm for approximating Q can be applied (see [18]):

1. For all $G \in \mathcal{G}$ assemble the macrostructure matrix A_G.
2. To each A_G perform a permutation of the rows and columns according to the global two-level splitting of the DOF and compute the Schur complement:

$$S_G = A_{G:22} - A_{G:21} A_{G:11}^{-1} A_{G:12}.$$

3. Assemble a sparse approximation Q to the exact global Schur complement $S = A_{22} - A_{21}(A_{11})^{-1} A_{12}$ from the local macrostructure Schur complements:

$$Q := S_{\mathcal{G}} = \sum_{G\in\mathcal{G}} R_{G:2}^T S_G R_{G:2}.$$

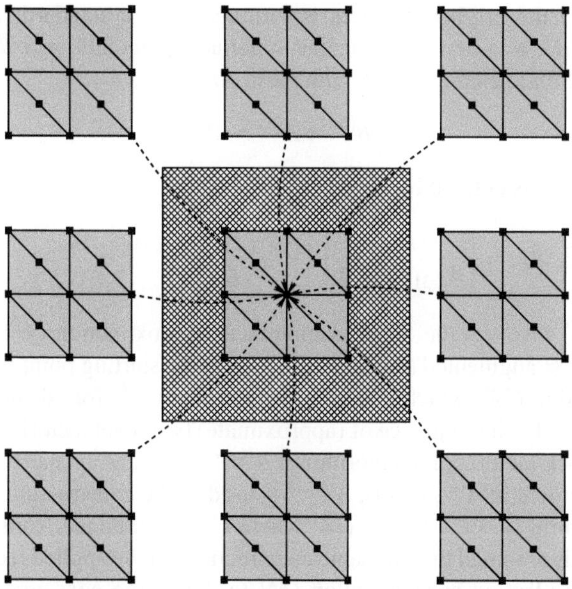

Fig. 2 One macrostructure G_i used in the computation of Q; G_i is composed of nine overlapping structures, $F_{i_1}, F_{i_2}, \ldots, F_{i_9}$

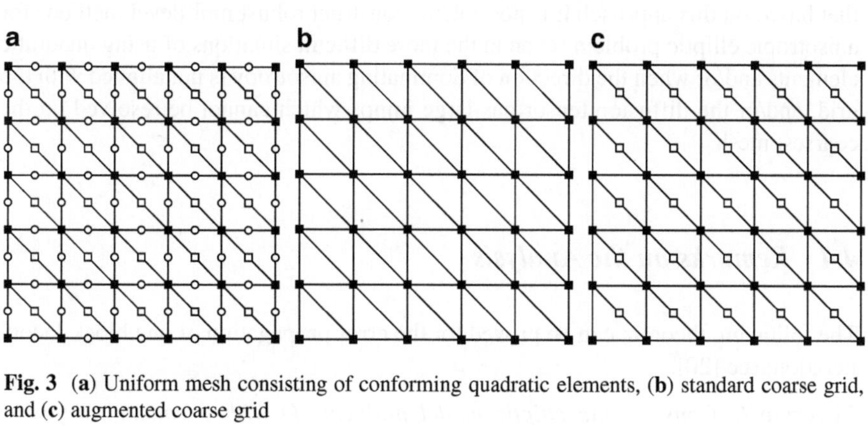

Fig. 3 (**a**) Uniform mesh consisting of conforming quadratic elements, (**b**) standard coarse grid, and (**c**) augmented coarse grid

4.3 Recursive Approximate Block Factorization on Augmented Grids

Consider a uniform fine mesh as depicted in Fig. 3a, the standard coarse grid as depicted in Fig. 3b, and the augmented coarse grid as illustrated in Fig. 3c.

Then, since the original problem is formulated on a standard (and not on an augmented) grid, as a first step, one has to define a preconditioner $\bar{B}^{(0)}$ at the level of the original finite element mesh with mesh size h, i.e.,

$$\bar{B}^{(0)} \approx A^{(0)} := A_h \tag{62}$$

where $\bar{B}^{(0)} := \bar{B}_h$ is defined by

$$\bar{B}_h := \begin{bmatrix} I & \\ A_{h:21}A_{h:11}^{-1} & I \end{bmatrix} \begin{bmatrix} A_{h:11} & \\ & Q_h \end{bmatrix} \begin{bmatrix} I & A_{h:11}^{-1}A_{h:12} \\ & I \end{bmatrix}. \tag{63}$$

Note that (63) involves the Schur complement approximation $Q^{(0)} := Q_h$, which refers to the first augmented (coarse) grid. This is the starting point for constructing $\bar{B}^{(k)}$ as defined in (24), which is used to approximate $A^{(k)}$ for all subsequent levels $k = 1, 2, \ldots, \ell - 1$. The sequence of (approximate) two-level factorizations defines a multilevel block factorization algorithm if $A^{(k+1)}$ serves as an approximation to the Schur complement of $A^{(k)}$, that is, $A^{(k+1)}$ is used in the construction of $\bar{B}^{(k)}$. Hence it is quite natural to set $A^{(k+1)} = Q^{(k)}$ where $Q^{(k)}$ is obtained from ASCA for all $k \geq 0$. Here it is assumed that the same construction can be applied recursively using the Schur complement approximation $Q^{(k)}$ to define the next *coarse(r) problem*; see (26). At levels $k \geq 1$ the use of the augmented coarse grids is advocated since it results in a very efficient combined AMLI algorithm with block (line) smoothing at every coarse level. The numerical experiments presented in Sect. 5 demonstrate that based on this approach it is possible to construct robust multilevel methods for anisotropic elliptic problems even in the more difficult situations of using quadratic elements and/or when the direction of dominating anisotropy is not aligned with the grid, and/or the diffusion tensor has large jumps which cannot be resolved on the coarsest mesh.

4.4 Remarks on the Analysis

The following theorem can be proved for the error propagation of one block Jacobi iteration; see [20].

Theorem 7. *Consider the elliptic model problem (1) with a constant diffusion coefficient* $\mathbf{a}(x) = (a_{ij})_{i,j=1}^2$ *scaled such that* $a_{11} = 1$*, and discretized on a uniform mesh with mesh size h, and Dirichlet boundary conditions. Further, let* $Q = D + L + L^T$ *denote the related ASCA where D and L are the block-diagonal and lower block-triangular parts of Q. Then the following bound holds for the iteration matrix of the block Jacobi method:*

$$\|I - D^{-1}Q\|_Q^2 \leq 1 - \frac{1}{1 + c_0} =: 1 - c_1, \tag{64}$$

where $c_0 := (a_{22} + |a_{12}|)/(ch^2)$ *and hence* c_1 *is in the interval* $(0, 1)$.

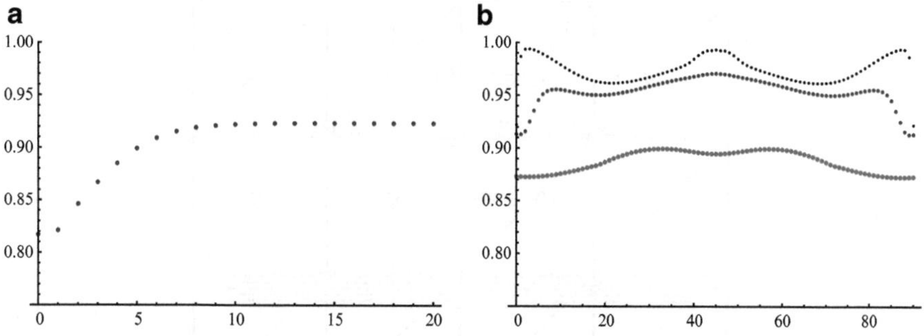

Fig. 4 (**a**) Grid-aligned anisotropy, $\varepsilon = 2^{-t}, t \in \{0,1,\ldots,20\}$: estimated convergence factor $1-\alpha$ plotted against t. (**b**) Rotated diffusion problem, $\varepsilon \in \{10^{-1}, 10^{-2}, 10^{-3}\}$, $\theta \in \{0°, 1°, \ldots, 90°\}$: estimated convergence factor $1-\alpha$ plotted against θ

The norm of the error propagation matrix of the two-level method corresponding to the preconditioner $\bar{B}$ as defined in (24) but with $B_{11} = A_{11}$ satisfies

$$\|I - \bar{B}^{-1}A\|_A^2 \le 1 - \alpha\lambda_{\min}(Q^{-1}S) \le 1 - \alpha, \tag{65}$$

where α can be estimated locally.

In Fig. 4 it is plotted a local estimate of the error reduction factor of the two-level method when considering grid-aligned and nongrid-aligned anisotropy. As it can be seen, in the first case, there is uniform convergence, i.e., the method is robust with respect to the parameter ε in (6) that has been varied in the range from 2^0 to 2^{-20}. However, the results are worse for the rotated diffusion problem associated with (7) where the convergence estimate in general deteriorates when ε tends to 0. Still, for a (moderate) fixed value of ε, the estimate is uniform with respect to the angle of the direction of strong anisotropy.

5 Numerical Tests

In this section 2D numerical results are presented for the studied FEM discretizations based on conforming linear (P_1) and quadratic (P_2) finite elements. On the level of the coarsest discretization, the considered domain $\Omega = [0,1] \times [0,1]$ is split into $2 \times 8 \times 8 = 2 \times 2^3 \times 2^3$ linear elements or, alternatively, into $2 \times 4 \times 4 = 2 \times 2^2 \times 2^2$ quadratic elements. Dirichlet boundary conditions are imposed upon the entire boundary $\Gamma = \partial\Omega$. The finest mesh in all experiments is obtained via $\ell = 2,\ldots,7$ steps of uniform mesh refinement resulting in $2 \times 2^{\ell+3} \times 2^{\ell+3}$ linear elements or $2 \times 2^{\ell+2} \times 2^{\ell+2}$ quadratic elements.

The numerical tests demonstrate the performance of the nonlinear AMLI W-cycle algorithm with 2 inner GCG iterations and an optional pre-smoothing step at every coarse level. The underlying multilevel block factorization is constructed

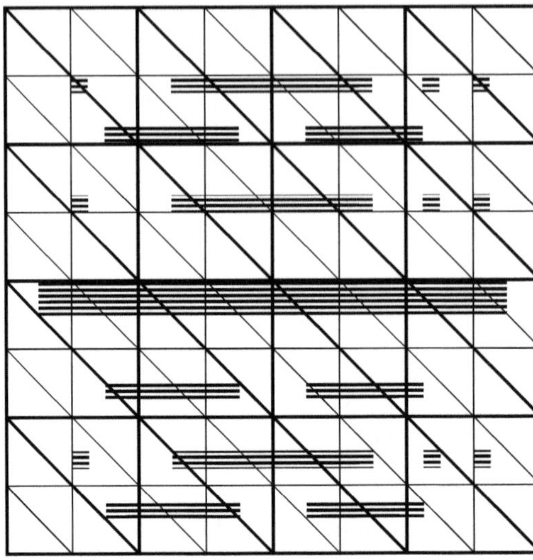

Fig. 5 Coarse mesh and coefficient, Example 1

based on the ASCA described in Sect. 4; see also [18,20]. The following variants of complementary subspace correction are tested:

(a) One-point Gauss–Seidel (PGS) iteration
(b) One-line Gauss–Seidel (LGS) iteration
(c) One-tree Gauss–Seidel (TGS) iteration

The blocks in variant (b) correspond to grid lines parallel to the x-axis. The blocks in variant (c) are constructed algebraically, by extracting *strong* paths from a previously computed nearly maximum spanning tree. The tree is constructed via a modified version of Kruskal's algorithm in which the global sorting of the edges according to their weights is replaced by a partial (local) sorting (cf. [16]). For a given edge $e_{ij} = (i, j)$, its weight w_{ij} is defined by $w_{ij} := |A_{ij}|/\sqrt{A_{ii}A_{jj}}$ (cf. [11]). If $w_{ij} > \rho$ for some threshold ρ, e.g., $\rho = 0.25$, then e_{ij} is called a strong edge.

Example 1. In the first set of experiments we consider a permeability field with inclusions and channels on a background of conductivity one, as shown in Fig. 5. The diffusion tensor $\mathbf{a}(x)$ equals the identity matrix outside the channels, whereas inside the channels it corresponds to highly anisotropic material and is determined by $\{a_{11}, a_{12}, a_{22}\} = \{10^5, 0, 1\}$.

The results presented in Table 1 show that no additional smoothing (complementary subspace correction) is required when the direction of dominating anisotropy is aligned with the grid. The method performs absolutely robustly and very similar for both P_1 and P_2 elements, although the channels are NOT resolved on the coarsest mesh!

Table 1 Number of iterations for residual reduction by a factor 10^8. Nonlinear AMLI W-cycle without additional smoothing, Example 1

Type of element	$\ell=3$	$\ell=4$	$\ell=5$	$\ell=6$	$\ell=7$
P_1	8	8	8	8	8
P_2	8	8	8	8	8

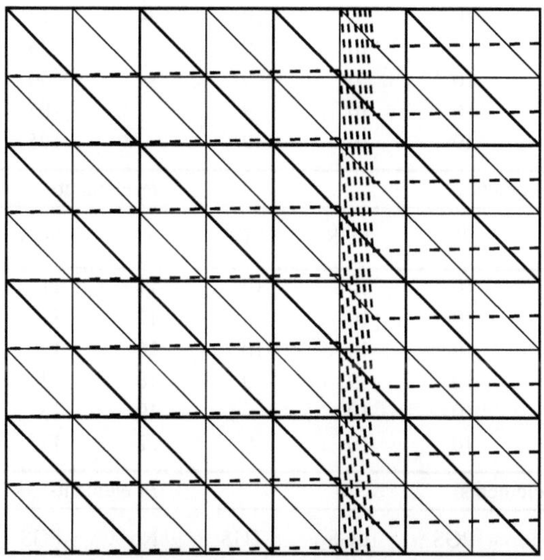

Fig. 6 Coarse mesh and direction of dominating anisotropy, Example 2

Example 2. In the second set of experiments the domain is split into three nonoverlapping parts $\Omega = \Omega_1 \bigcup \Omega_2 \bigcup \Omega_3$ where $\Omega_1 = [0, 5/8] \times [0, 1]$, $\Omega_2 = [5/8, 11/16] \times [0, 1]$, and $\Omega_3 = [11/16, 1] \times [0, 1]$ as shown on Fig. 6. We consider the rotated diffusion problem (7) where the angle $\theta_e = 1°$ over the left and right subdomains, while in the middle one $\theta_e = -85°$.

The results in Table 2 show that while the method performs robustly without additional smoothing in case of P_1 elements, the convergence deteriorates without a complementary subspace correction step in case of P_2 elements; however, it can be improved significantly by introducing a proper block Gauss–Seidel pre-smoothing step.

Example 3. The third set of experiments presents the performance of the nonlinear AMLI algorithm for the case of rotated diffusion problem with θ varied smoothly from the left to the right border of the domain $\Omega = [0, 1] \times [0, 1]$ according to the function $\theta = -\pi(1 - |2x - 1|)/6$ for $x \in (0, 1)$ (Fig. 7).

The results are very similar to those for the second test problem. Note that all numerical experiments were designed in such a way that the coarsest mesh does not resolve the arising jumps of the coefficient (Table 3).

Table 2 Number of iterations for residual reduction by a factor 10^8. Nonlinear AMLI W-cycle, Example 2

$\varepsilon = 10^{-6}$	P_1 elements				P_2 elements			
ℓ \ sm.	No	PGS	LGS	TGS	No	PGS	LGS	TGS
2	10	10	10	10	12	11	10	11
3	11	10	10	10	34	17	11	11
4	11	10	10	10	73	25	14	14
5	11	10	10	9	*	50	18	15
6	11	10	10	9	*	105	51	22
7	12	10	10	9	*	195	91	31
$\varepsilon = 10^{-4}$	P_1 elements				P_2 elements			
ℓ \ sm.	No	PGS	LGS	TGS	No	PGS	LGS	TGS
2	10	10	10	10	11	11	10	10
3	11	10	10	10	12	11	10	10
4	11	10	10	10	16	12	10	10
5	11	10	9	9	18	13	12	10
6	11	10	10	9	19	15	14	11
7	11	10	10	9	22	17	16	12
$\varepsilon = 10^{-2}$	P_1 elements				P_2 elements			
ℓ \ sm.	No	PGS	LGS	TGS	No	PGS	LGS	TGS
2	10	10	10	10	10	10	10	10
3	10	10	10	10	10	10	10	10
4	10	10	10	10	10	10	10	10
5	10	10	9	9	10	10	10	10
6	10	10	9	9	10	10	9	9
7	10	9	9	9	10	10	9	9

6 Concluding Remarks

The theory of robust AMLI methods based on HB techniques is well established for conforming and nonconforming linear finite element discretizations of anisotropic second-order elliptic problems under the fundamental assumption that variations of the coefficient tensor can be resolved on the coarsest mesh. However, in many practical applications, this is a too strong restriction. Hence, alternative methods, e.g., based on energy-minimizing coarse spaces or robust Schur complement approximations, have recently been moving into the center of interest.

Here we describe a class of nonlinear AMLI methods that are based on ASCA and, though not fully analyzed yet, have been shown to be very efficient for problems with highly heterogeneous and anisotropic media. In case of conforming FEM and P_1 elements, this method performs robustly (even without additional smoothing). Using P_2 elements the numerical results demonstrate that in certain

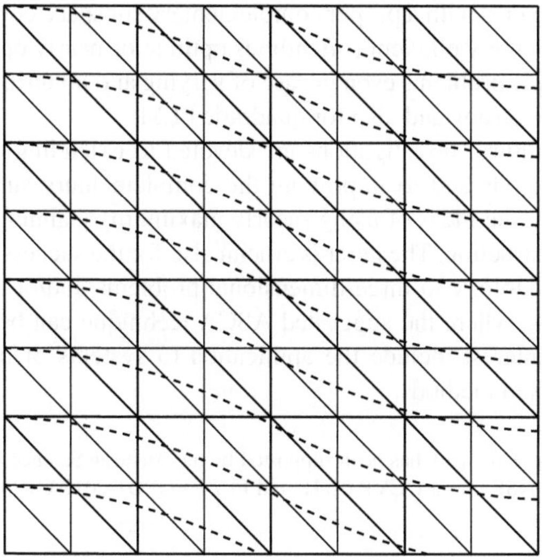

Fig. 7 Coarse mesh and direction of dominating anisotropy, Example 3

Table 3 Number of iterations for residual reduction by a factor 10^8. Nonlinear AMLI W-cycle, Example 3

$\varepsilon = 10^{-5}$	P_1 elements			P_2 elements		
ℓ \ sm.	No	PGS	TGS	No	PGS	TGS
2	11	11	11	13	12	12
3	11	11	11	13	12	12
4	11	11	11	18	13	13
5	11	11	11	28	16	14
6	12	11	11	46	23	19
7	12	11	11	59	30	25
$\varepsilon = 10^{-4}$	P_1 elements			P_2 elements		
ℓ \ sm.	No	PGS	TGS	No	PGS	TGS
2	11	11	11	13	12	12
3	11	11	11	13	12	12
4	11	11	11	16	13	13
5	11	11	11	22	15	14
6	11	11	11	25	18	16
7	12	11	11	28	20	18

situations (when the convergence deteriorates) an additional smoothing step can improve the performance of the AMLI algorithm considerably. The construction of block smoothers based on graph concepts such as spanning trees seems to be very promising in this context (cf. Examples 2 and 3). The combination of an augmented coarse grid with a proper complementary subspace correction step is the key to obtain extremely efficient (oftentimes optimal or nearly optimal) solvers for strongly anisotropic problems even in case of varying and nongrid-aligned direction of dominating anisotropy and also for quadratic FEM.

Current (and future) investigations are devoted to extending the theory of this new class of methods and to improving the complementary subspace correction step(s) by refining the ideas of using (nearly maximum) spanning trees and strong paths in their construction. The latter is crucial also for the successful generalization of the new methodology to three-dimensional problems and/or discretizations on unstructured grids, where the suggested ASCA technique can be applied directly. Other topics of interest include the application to systems of partial differential equations and mixed methods.

Acknowledgements This work has been supported by the Austrian Science Fund (grant P22989-N18), the Bulgarian NSF (grant DCVP 02/1), and FP7-REGPOT-2012-CT2012-316087-AComIn Grant.

References

1. Axelsson, O., Blaheta, R.: Two simple derivations of universal bounds for the C.B.S. inequality constant. Appl. Math. **49**(1), 57–72 (2004)
2. Axelsson, O., Blaheta, R., Neytcheva, M.: Preconditioning of boundary value problems using elementwise Schur complements. SIAM J. Matrix Anal. Appl. **31**(2), 767–789 (2009)
3. Axelsson, O., Margenov, S.: On multilevel preconditioners which are optimal with respect to both problem and discretization parameters. Comput. Methods Appl. Math. **3**(1), 6–22 (2003)
4. Axelsson, O., Padiy, A.: On the additive version of the algebraic multilevel iteration method for anisotropic elliptic problems. SIAM J. Sci. Comput. **20**(5), 1807–1830 (1999)
5. Axelsson, O.: Stabilization of algebraic multilevel iteration methods; Additive methods. Numer. Algorithm **21**(1–4), 23–47 (1999)
6. Axelsson, O., Vassilevski, P.: Algebraic multilevel preconditioning methods I. Numer. Math. **56**(2–3), 157–177 (1989)
7. Axelsson, O., Vassilevski, P.: Algebraic multilevel preconditioning methods II. SIAM J. Numer. Anal. **27**(6), 1569–1590 (1990)
8. Axelsson, O., Vassilevski, P.: Variable-step multilevel preconditioning methods, I: Self-adjoint and positive definite elliptic problems. Numer. Linear Algebra Appl. **1**(1), 75–101 (1994)
9. Blaheta, R., Margenov, S., Neytcheva, M.: Uniform estimate of the constant in the strengthened CBS inequality for anisotropic non-conforming FEM systems. Numer. Linear Algebra Appl. **11**(4), 309–326 (2004)
10. Blaheta, R., Margenov, S., Neytcheva, M.: Robust optimal multilevel preconditioners for non-conforming finite element systems. Numer. Linear Algebra Appl. **12**(5–6), 495–514 (2005)
11. Chan, T., Vanek, P.: Detection of strong coupling in algebraic multigrid solvers. Multigrid Methods VI, pp. 11–23. Springer, New York (2000)
12. Efendiev, Y., Galvis, J., Lazarov, R.D., Margenov, S., Ren, J.: Robust two-level domain decomposition preconditioners for high-contrast anisotropic flows in multiscale media. Comput. Methods Appl. Math. **12**(4), 1–22 (2012)
13. Georgiev, I., Lymbery, M., Margenov, S.: Analysis of the CBS constant for quadratic finite elements. In: Dimov, I., Dimova, S., Kolkovska, N. (eds.) Lecture Notes in Computer Science, vol. 6046, NMA 2010, pp. 412–419. Springer, New York (2011)

14. Georgiev, I., Kraus, J., Lymbery, M., Margenov, S.: On two-level splittings for quadratic FEM anisotropic elliptic problems. 5th Annual meeting of BGSIAM'10, pp. 35–40 (2011)
15. Hestenes, M.R., Stiefel, E.: Methods of conjugate gradients for solving linear systems. J. Res. Nat. Bur. Stand. Sect. B **49**(6), 409–436 (1952)
16. Kraus, J.: An algebraic preconditioning method for M-matrices: Linear versus non-linear multilevel iteration. Numer. Linear Algebra Appl. **9**(8), 599–618 (2002)
17. Kraus, J.: Algebraic multilevel preconditioning of finite element matrices using local Schur complements. Numer. Linear Algebra Appl. **13**(1), 49–70 (2006)
18. Kraus, J.: Additive Schur complement approximation and application to multilevel preconditioning. SIAM J. Sci. Comput. **34**(6), A2872–A2895 (2012)
19. Kraus, J., Lymbery, M., Margenov, S.: On the robustness of two-level preconditioners for quadratic fe orthotropic elliptic problems. LNCS Proc. (LSSC'11 Conference) **7116**, 582–589 (2012)
20. Kraus, J., Lymbery, M., Margenov, S.: Robust multilevel methods for quadratic finite element anisotropic elliptic problems, to appear in Numer. Linear Algebra Appl.; Also available as RICAM-Report No. 2012–29 (2013)
21. Kraus, J., Margenov, S.: Multilevel methods for anisotropic elliptic problems. Lectures on Advanced Computational Methods in Mechanics, de Gruyter Radon Series. Comput. Appl. Math. **1**, 47–88 (2007)
22. Kraus, J., Margenov, S., Synka, J.: On the multilevel preconditioning of Crouzeix-Raviart elliptic problems. Numer. Linear Algebra. Appl. **15**(5), 395–416 (2009)
23. Maitre, J.F., Musy, S.: The contraction number of a class of two-level methods; An exact evaluation for some finite element subspaces and model problems. Lect. Note Math. **960**, 535–544 (1982)
24. Margenov, S., Vassilevski, P.S.: Algebraic multilevel preconditioning of anisotropic elliptic problems. SIAM J. Sci. Comput. **15**(5), 1026–1037 (1994)
25. Neytcheva, M.: On element-by-element Schur complement approximations. Linear Algebra Appl. **434**(11), 2308–2324 (2011)
26. Notay, Y., Vassilevski, P.: Recursive Krylov-based multigrid cycles. Numer. Linear Algebra Appl. **15**(5), 473–487 (2008)
27. Vassilevski, P.: Multilevel Block Factorization Preconditioners. Springer, New York (2008)

A Weak Galerkin Mixed Finite Element Method for Biharmonic Equations

Lin Mu, Junping Wang, Yanqiu Wang, and Xiu Ye

Abstract This article introduces and analyzes a weak Galerkin mixed finite element method for solving the biharmonic equation. The weak Galerkin method, first introduced by two of the authors (J. Wang and X. Ye) in (Wang et al., Comput. Appl. Math. 241:103–115, 2013) for second-order elliptic problems, is based on the concept of *discrete weak gradients*. The method uses completely discrete finite element functions, and, using certain discrete spaces and with stabilization, it works on partitions of arbitrary polygon or polyhedron. In this article, the weak Galerkin method is applied to discretize the Ciarlet–Raviart mixed formulation for the biharmonic equation. In particular, an a priori error estimation is given for the corresponding finite element approximations. The error analysis essentially

The research of Wang was supported by the NSF IR/D program, while working at the Foundation. However, any opinion, finding, and conclusions or recommendations expressed in this material are those of the author and do not necessarily reflect the views of the National Science Foundation.

This research was supported in part by National Science Foundation Grant DMS-1115097.

L. Mu
Department of Mathematics, Michigan State University, East Lancing, MI 48824, USA
e-mail: lxmu@ualr.edu

J. Wang (✉)
Division of Mathematical Sciences, National Science Foundation, Arlington, VA 22230, USA
e-mail: jwang@nsf.gov

Y. Wang
Department of Mathematics, Oklahoma State University, Stillwater, OK 74075, USA
e-mail: yqwang@math.okstate.edu

X. Ye
Department of Mathematics, University of Arkansas at Little Rock, Little Rock, AR 72204, USA
e-mail: xxye@ualr.edu

O.P. Iliev et al. (eds.), *Numerical Solution of Partial Differential Equations: Theory, Algorithms, and Their Applications*, Springer Proceedings in Mathematics & Statistics 45,
DOI 10.1007/978-1-4614-7172-1_13,

follows the framework of Babuška, Osborn, and Pitkäranta (Math. Comp. 35:1039–1062, 1980) and uses specially designed mesh-dependent norms. The proof is technically tedious due to the discontinuous nature of the weak Galerkin finite element functions. Some computational results are presented to demonstrate the efficiency of the method.

Keywords Weak Galerkin finite element methods • Discrete gradient • Biharmonic equations • Mixed finite element methods

AMS subject classifications. Primary, 65N15, 65N30

1 Introduction

In this paper, we are concerned with numerical methods for the following biharmonic equation with clamped boundary conditions:

$$\begin{aligned} \Delta^2 u &= f \qquad \text{in } \Omega, \\ u &= 0 \qquad \text{on } \partial\Omega, \\ \frac{\partial u}{\partial \mathbf{n}} &= 0 \qquad \text{on } \partial\Omega, \end{aligned} \tag{1}$$

where Ω is a bounded polygonal or polyhedral domain in $\mathbb{R}^d$ $(d = 2,3)$. To solve the problem (1) using a primal-based conforming finite element method, one would need C^1 continuous finite elements, which usually involve large degree of freedoms and hence can be computationally expensive. There are alternative numerical methods, for example, by using either nonconforming elements [2,25,28], the C^0 discontinuous Galerkin method [8, 14], or mixed finite element methods [6, 10, 11, 13, 20–22, 24–27]. One of the earliest mixed formulations proposed for (1) is the Ciarlet–Raviart mixed finite element formulation [11] which decomposes (1) into a system of second-order partial differential equations. In this mixed formulation, one introduces a dual variable $w = -\Delta u$ and rewrites the fourth-order biharmonic equation into two coupled second-order equations:

$$\begin{cases} w + \Delta u = 0, \\ -\Delta w = f. \end{cases} \tag{2}$$

In [11], the above system of second-order equations is discretized by using the standard H^1 conforming elements. However, only suboptimal order of error estimates is proved in [11] for quadratic or higher order of elements. Improved error estimates have been established in [5, 15, 19, 32] for quadratic or higher order of elements. In [5], Babuška, Osborn, and Pitkäranta pointed out that a suitable choice of norms are L^2 for w and H^2 (or H^2-equivalence) for u in order to use the standard LBB stability analysis. In this sense, one has "optimal" order of convergence in H^2 norm for u and in L^2 norm for w, for quadratic or higher order of elements. However, when equal-order approximation is used for both u and w, the "optimal" order of error estimate is restricted by the interpolation error in H^2 norm and thus may not be really optimal. Moreover, this standard technique does not apply to the piecewise linear discretization, since in this case the interpolation error cannot even be measured in H^2 norm. A solution to this has been proposed by Scholz [32] by using an L^∞ argument. Scholz was able to improve the convergence rate in L^2 norm for w by $h^{\frac{1}{2}}$, and this theoretical result is known to be sharp. Also, Scholz's proof works for all equal-order elements including piecewise linears.

The goal of this paper is to propose and analyze a weak Galerkin discretization method for the mixed formulation (2) with equal-order elements. The weak Galerkin method was recently introduced in [29, 35, 36] for second-order elliptic equations. It is an extension of the standard Galerkin finite element method where classical derivatives were substituted by weakly defined derivatives on functions with discontinuity. Error estimates of optimal order have been established for various weak Galerkin discretization schemes for second-order elliptic equations [29, 35, 36]. A numerical implementation of weak Galerkin was presented in [29, 30] for some model problems.

Some advantages of the weak Galerkin method have been identified in [29, 30, 36]. For example, the weak Galerkin method based on a stabilization works for finite element partitions of arbitrary polygon or polyhedron [29, 36]. Weak Galerkin methods use completely discrete finite element spaces and the resulting numerical scheme is symmetric, positive definite, and parameter-free if the original problem is. Weak Galerkin methods retain the mass conservation property as the original system. The unknowns in the interior of each element can be eliminated in parallel, yielding a discrete problem with much fewer number of unknowns that the original system and other competing algorithms. Nevertheless, the weak Galerkin method is still a very new method, and there remains a lot to explore for researchers. This paper shall demonstrate the portability of weak Galerkin to the biharmonic equation. Our future research will focus on a generalization of weak Galerkin to other numerically challenging equations.

Applying the weak Galerkin method to both second-order equations in (2) appears to be trivial and straightforward at first glance. However, the application turns out to be much more complicated than simply combining one weak Galerkin scheme with another one. The application is particularly non-trivial in the mathematical theory on error analysis. In deriving an a priori error estimate, we follow the framework as developed in [5] by using mesh-dependent norms. Many commonly used properties and inequalities for standard Galerkin finite element method need to be re-derived for weak Galerkin methods with respect to the mesh-dependent norms. Due to the discrete nature of the weak Galerkin functions, technical difficulties arise in the derivation of inequalities or estimates. The technical estimates and tools that we have developed in this paper should be essential to the analysis of weak

Galerkin methods for other type of modeling equations. They should also play an important role in future developments of preconditioning techniques for weak Galerkin methods. Therefore, we believe this paper provides useful technical tools for future research, in addition to introducing an efficient new method for solving biharmonic equations.

The paper is organized as follows. In Sect. 2, a weak Galerkin discretization scheme for the Ciarlet–Raviart mixed formulation of the biharmonic equation is introduced and proved to be well-posed. Section 3 is dedicated to defining and analyzing several technical tools, including projections, mesh-dependent norms, and some estimates. With the aid of these tools, an error analysis is presented in Sect. 4. Finally, in Sect. 5, we report some numerical results that show the efficiency of the method.

2 A Weak Galerkin Finite Element Scheme

For illustrative purpose, we consider only the two-dimensional case of (1), and the corresponding weak Galerkin method will be based on a shape-regular triangulation of the domain Ω. The analysis given in this paper can easily be generalized into two-dimensional rectangular meshes and with a few adaptations, also into three-dimensional tetrahedral and cubic meshes. Another issue we would like to clarify is that, although the weak Galerkin method using certain discrete spaces and with stabilization is known to work on partitions of arbitrary polygon or polyhedron [29, 36], here we choose to concentrate on a weak Galerkin discretization without stabilization. This discretization only works for triangular, rectangular, tetrahedral and cubic meshes, but the theoretical analysis would be considerably easier since there is no stabilization involved. We are confident that the technique introduced in this paper can be generalized to the stabilized weak Galerkin method on arbitrary meshes [29, 36]. But details need to be worked out in future research.

Let $D \subseteq \Omega$ be a polygon; we use the standard definition of Sobolev spaces $H^s(D)$ and $H_0^s(D)$ with $s \geq 0$ (e.g., see [1, 12] for details). The associated inner product, norm, and semi-norms in $H^s(D)$ are denoted by $(\cdot,\cdot)_{s,D}$, $\|\cdot\|_{s,D}$, and $|\cdot|_{r,D}, 0 \leq r \leq s$, respectively. When $s = 0$, $H^0(D)$ coincides with the space of square integrable functions $L^2(D)$. In this case, the subscript s is suppressed from the notation of norm, semi-norm, and inner products. Furthermore, the subscript D is also suppressed when $D = \Omega$. For $s < 0$, the space $H^s(D)$ is defined to be the dual of $H_0^{-s}(D)$.

Occasionally, we need to use the more general Sobolev space $W^{s,p}(\Omega)$, for $1 \leq p \leq \infty$, and its norm $\|\cdot\|_{W^{s,p}(\Omega)}$. The definition simply follows the standard one given in [1, 12]. When $s = 0$, the space $W^{s,p}(\Omega)$ coincides with $L^p(\Omega)$.

The above definition/notation can easily be extended to vector-valued and matrix-valued functions. The norm, semi-norms, and inner-product for such functions shall follow the same naming convention. In addition, all these definitions can be transferred from a polygonal domain D to an edge e, a domain with lower dimension. Similar notation system will be employed. For example, $\|\cdot\|_{s,e}$ and $\|\cdot\|_e$ would denote the norm in $H^s(e)$ and $L^2(e)$ etc. We also define the $H(div)$ space as follows:

$$H(div,\Omega)=\{\mathbf{q}:\ \mathbf{q}\in[L^2(\Omega)]^2,\ \nabla\cdot\mathbf{q}\in L^2(\Omega)\}.$$

Using notations defined above, the variational form of the Ciarlet–Raviart mixed formulation (2) seeks $u\in H_0^1(\Omega)$ and $w\in H^1(\Omega)$ satisfying

$$\begin{cases}(w,\phi)-(\nabla u,\nabla\phi)=0 & \text{for all } \phi\in H^1(\Omega),\\ (\nabla w,\nabla\psi)=(f,\psi) & \text{for all } \psi\in H_0^1(\Omega).\end{cases}\tag{3}$$

For any solution w and u of (3), it is not hard to see that $w=-\Delta u$. In addition, by choosing $\phi=1$ in the first equation of (3), we obtain

$$\int_\Omega w\,dx=0.$$

Define $\bar{H}^1(\Omega)\subset H^1(\Omega)$ by

$$\bar{H}^1(\Omega)=\{v:\ v\in H^1(\Omega),\ \int_\Omega v\,dx=0\},$$

which is a subspace of $H^1(\Omega)$ with mean-value-free functions. Clearly, the solution w of (3) is a function in $\bar{H}^1(\Omega)$.

One important issue in the analysis is the regularity of the solution u and w. For two-dimensional polygonal domains, this has been thoroughly discussed in [7]. According to their results, the biharmonic equation with clamped boundary condition (1) satisfies

$$\|u\|_{4-k}\le c\|f\|_{-k},\tag{4}$$

where c is a constant depending only on the domain Ω. Here, the parameter k is determined by

$k=1$ if all internal angles of Ω are less than 180°

$k=0$ if all internal angles of Ω are less than $126.283696\cdots^\circ$

The above regularity result indicates that the solution $u\in H^3(\Omega)$ when Ω is a convex polygon and $f\in H^{-1}(\Omega)$. It follows that the auxiliary variable $w\in H^1(\Omega)$. Moreover, if all internal angles of Ω are less than $126.283696\cdots^\circ$ and $f\in L^2(\Omega)$, then $u\in H^4(\Omega)$ and $w\in H^2(\Omega)$. The drawback of the mixed formulation (3) is that the auxiliary variable w may not possess the required regularity when the domain is non-convex. We shall explore other weak Galerkin methods to deal with such cases.

Next, we present the weak Galerkin discretization of the Ciarlet–Raviart mixed formulation. Let $\mathcal{T}_h$ be a shape-regular, quasi-uniform triangular mesh on a polygonal domain Ω, with characteristic mesh size h. For each triangle $K \in \mathcal{T}_h$, denote by K_0 and ∂K the interior and the boundary of K, respectively. Also denote by h_K the size of the element K. The boundary ∂K consists of three edges. Denote by $\mathcal{E}_h$ the collection of all edges in $\mathcal{T}_h$. For simplicity of notation, throughout the paper, we use "$\lesssim$" to denote "less than or equal to up to a general constant independent of the mesh size or functions appearing in the inequality."

Let j be a nonnegative integer. On each $K \in \mathcal{T}_h$, denote by $P_j(K_0)$ the set of polynomials with degree less than or equal to j. Likewise, on each $e \in \mathcal{E}_h$, $P_j(e)$ is the set of polynomials of degree no more than j. Following [35], we define a weak discrete space on mesh $\mathcal{T}_h$ by

$$V_h = \{v : v|_{K_0} \in P_j(K_0),\ K \in \mathcal{T}_h;\ v|_e \in P_j(e), e \in \mathcal{E}_h\}.$$

Observe that the definition of V_h does not require any continuity of $v \in V_h$ across the interior edges. A function in V_h is characterized by its value on the interior of each element plus its value on the edges/faces. Therefore, it is convenient to represent functions in V_h with two components, $v = \{v_0, v_b\}$, where v_0 denotes the value of v on all K_0 and v_b denotes the value of v on $\mathcal{E}_h$.

We further define an L^2 projection from $H^1(\Omega)$ onto V_h by setting $Q_h v \equiv \{Q_0 v, Q_b v\}$, where $Q_0 v|_{K_0}$ is the local L^2 projection of v in $P_j(K_0)$, for $K \in \mathcal{T}_h$, and $Q_b v|_e$ is the local L^2 projection in $P_j(e)$, for $e \in \mathcal{E}_h$. To take care of the homogeneous Dirichlet boundary condition, define

$$V_{0,h} = \{v \in V_h : v = 0 \text{ on } \mathcal{E}_h \cap \partial\Omega\}.$$

It is not hard to see that the L^2 projection Q_h maps $H_0^1(\Omega)$ onto $V_{0,h}$.

The weak Galerkin method seeks an approximate solution $[u_h; w_h] \in V_{0,h} \times V_h$ to the mixed form of the biharmonic problem (2). To this end, we first introduce a discrete L^2-equivalent inner-product and a discrete gradient operator on V_h. For any $v_h = \{v_0, v_b\}$ and $\phi_h = \{\phi_0, \phi_b\}$ in V_h, define an inner-product as follows:

$$(\!(v_h, \phi_h)\!) \triangleq \sum_{K\in\mathcal{T}_h} (v_0, \phi_0)_K + \sum_{K\in\mathcal{T}_h} h_K \langle v_0 - v_b, \phi_0 - \phi_b\rangle_{\partial K}.$$

It is not hard to see that $(\!(v_h, v_h)\!) = 0$ implies $v_h \equiv 0$. Hence, the inner-product is well defined. Notice that the inner-product $(\!(\cdot,\cdot)\!)$ is also well defined for any $v \in H^1(\Omega)$ for which $v_0 = v$ and $v_b|_e = v|_e$ is the trace of v on the edge e. In this case, the inner-product $(\!(\cdot,\cdot)\!)$ is identical to the standard L^2 inner-product.

The discrete gradient operator is defined element-wise on each $K \in \mathcal{T}_h$. To this end, let $RT_j(K)$ be a space of Raviart–Thomas element [31] of order j on triangle K. That is,

$$RT_j(K) = (P_j(K))^2 + \mathbf{x}P_j(K).$$

The degrees of freedom of $RT_j(K)$ consist of moments of normal components on each edge of K up to order j, plus all the moments in the triangle K up to order $(j-1)$. Define

$$\Sigma_h = \{\mathbf{q} \in (L^2(\Omega))^2 : \mathbf{q}|_K \in RT_j(K),\ K \in \mathcal{T}_h\}.$$

Note that Σ_h is not necessarily a subspace of $H(div,\Omega)$, since it does not require any continuity in the normal direction across any edge. A discrete weak gradient [35] of $v_h = \{v_0, v_b\} \in V_h$ is defined to be a function $\nabla_w v_h \in \Sigma_h$ such that on each $K \in \mathcal{T}_h$,

$$(\nabla_w v_h, \mathbf{q})_K = -(v_0, \nabla\cdot\mathbf{q})_K + \langle v_b, \mathbf{q}\cdot\mathbf{n}\rangle_{\partial K}, \quad \text{for all } \mathbf{q} \in RT_j(K), \tag{5}$$

where $\mathbf{n}$ is the unit outward normal on ∂K. Clearly, such a discrete weak gradient is always well defined. Also, the discrete weak gradient is a good approximation to the classical gradient, as demonstrated in [35]:

Lemma 2.1. *For any $v_h = \{v_0, v_b\} \in V_h$ and $K \in \mathcal{T}_h$, $\nabla_w v_h|_K = 0$ if and only if $v_0 = v_b =$ constant on K. Furthermore, for any $v \in H^{m+1}(\Omega)$, where $0 \le m \le j+1$, we have*

$$\|\nabla_w(Q_h v) - \nabla v\| \lesssim h^m \|v\|_{m+1}.$$

We are now in a position to present the weak Galerkin finite element formulation for the biharmonic problem (2) in the mixed form: *Find $u_h = \{u_0, u_b\} \in V_{0,h}$ and $w_h = \{w_0, w_b\} \in V_h$ such that*

$$\begin{cases} (\!(w_h, \phi_h)\!) - (\nabla_w u_h, \nabla_w \phi_h) = 0, & \text{for all } \phi_h = \{\phi_0, \phi_b\} \in V_h, \\ (\nabla_w w_h, \nabla_w \psi_h) = (f, \psi_0), & \text{for all } \psi_h = \{\psi_0, \psi_b\} \in V_{0,h}. \end{cases} \tag{6}$$

Theorem 2.2. *The weak Galerkin finite element formulation (6) has one and only one solution $[u_h; w_h]$ in the corresponding finite element spaces.*

Proof. For the discrete problem arising from (6), it suffices to show that the solution to (6) is trivial if $f = 0$; the existence of solution stems from its uniqueness.

Assume that $f = 0$ in (6). By taking $\phi_h = w_h$ and $\psi_h = u_h$ in (6) and adding the two resulting equations together, we immediately have $(\!(w_h, w_h)\!) = 0$, which implies $w_h \equiv 0$. Next, by setting $\phi_h = u_h$ in the first equation of (6), we arrive at $(\nabla_w u_h, \nabla_w u_h) = 0$. By using Lemma 2.1, we see that u_h must be a constant in Ω, which together with the fact that $u_h = 0$ on $\partial\Omega$ implies $u_h \equiv 0$ in Ω. This completes the proof of the theorem. □

One important observation of (6) is that the solution w_h has mean value zero over the domain Ω, which is a property that the exact solution $w = -\Delta u$ must possess. This can be seen by setting $\phi_h = 1$ in the first equation of (6), yielding

$$(w_h, 1) = (\!(w_h, 1)\!) = (\nabla_w u_h, \nabla_w 1) = 0,$$

where we have used the definition of $(\!(\cdot,\cdot)\!)$ and Lemma 2.1. For convenience, we introduce a space $\bar{V}_h \subset V_h$ defined as follows:

$$\bar{V}_h = \{v_h : \; v_h = \{v_0, v_b\} \in V_h, \; \int_\Omega v_0 \, dx = 0\}.$$

3 Technical Tools: Projections, Mesh-Dependent Norms, and Some Estimates

The goal of this section is to establish some technical results useful for deriving an error estimate for the weak Galerkin finite element method (6).

3.1 Some Projection Operators and Their Properties

Let $\mathbf{P}_h$ be the L^2 projection from $(L^2(\Omega))^2$ to Σ_h and $\mathbf{\Pi}_h$ be the classical interpolation [10] from $(H^\gamma(\Omega))^2, \gamma > \frac{1}{2}$, to Σ_h defined by using the degrees of freedom of Σ_h in the usual mixed finite element method. It follows from the definition of $\mathbf{\Pi}_h$ that $\mathbf{\Pi}_h \mathbf{q} \in H(div,\Omega) \cap \Sigma_h$ for all $\mathbf{q} \in (H^\gamma(\Omega))^2$. In other words, $\mathbf{\Pi}_h \mathbf{q}$ has continuous normal components across internal edges. It is also well known that $\mathbf{\Pi}_h$ preserves the boundary condition $\mathbf{q}\cdot\mathbf{n}|_{\partial\Omega} = 0$, if it were imposed on $\mathbf{q}$. The properties of $\mathbf{\Pi}_h$ have been well developed in the context of mixed finite element methods [10, 18]. For example, for all $\mathbf{q} \in (W^{m,p}(\Omega))^2$ where $\frac{1}{2} < m \le j+1$ and $2 \le p \le \infty$, we have

$$Q_0(\nabla\cdot\mathbf{q}) = \nabla\cdot\mathbf{\Pi}_h\mathbf{q}, \qquad \text{if in addition } \mathbf{q} \in H(div,\Omega), \tag{7}$$

$$\|\mathbf{q} - \mathbf{\Pi}_h\mathbf{q}\|_{L^p(\Omega)} \lesssim h^m \|\mathbf{q}\|_{W^{m,p}(\Omega)}. \tag{8}$$

It is also well known that for all $0 \le m \le j+1$,

$$\|\mathbf{q} - \mathbf{P}_h\mathbf{q}\| \lesssim h^m \|\mathbf{q}\|_m. \tag{9}$$

Using the above estimates and the triangle inequality, one can easily derive the following estimate:

$$\|\mathbf{\Pi}_h\nabla v - \mathbf{P}_h\nabla v\| \lesssim h^m \|v\|_{m+1} \tag{10}$$

for all $v \in H^{m+1}(\Omega)$ where $\frac{1}{2} < m \le j+1$.

Next, we shall present some useful relations for the discrete weak gradient ∇_w, the projection operator $\mathbf{P}_h$, and the interpolation $\mathbf{\Pi}_h$. The results can be summarized as follows.

Lemma 3.1. *Let $\gamma > \frac{1}{2}$ be any real number. The following results hold true.*

(i) For any $v \in H^1(\Omega)$, we have

$$\nabla_w(Q_h v) = \mathbf{P}_h(\nabla v). \tag{11}$$

(ii) For any $\mathbf{q} \in (H^\gamma(\Omega))^2 \cap H(div,\Omega)$ and $v_h = \{v_0, v_b\} \in V_h$, we have

$$(\nabla \cdot \mathbf{q},\, v_0) = -(\mathbf{\Pi}_h \mathbf{q},\, \nabla_w v_h) + \sum_{e \in \mathcal{E}_h \cap \partial\Omega} \langle (\mathbf{\Pi}_h \mathbf{q}) \cdot \mathbf{n}, v_b \rangle_e. \tag{12}$$

In particular, if either $v_h \in V_{0,h}$ or $\mathbf{q} \cdot \mathbf{n} = 0$ on $\partial\Omega$, then

$$(\nabla \cdot \mathbf{q},\, v_0) = -(\mathbf{\Pi}_h \mathbf{q},\, \nabla_w v_h). \tag{13}$$

Proof. To prove (11), we first recall the following well-known relation [10]:

$$\nabla \cdot RT_j(K) = P_j(K_0), \qquad RT_j(K) \cdot \mathbf{n}|_e = P_j(e).$$

Thus, for any $\mathbf{w} \in \Sigma_h$ and $K \in \mathcal{T}_h$, by the definition of ∇_w and properties of the L^2 projection, we have

$$\begin{aligned}
(\nabla_w Q_h v, \mathbf{w})_K &= -(Q_0 v, \nabla \cdot \mathbf{w})_K + \langle Q_b v, \mathbf{w} \cdot \mathbf{n} \rangle_{\partial K} \\
&= -(v, \nabla \cdot \mathbf{w})_K + \langle v, \mathbf{w} \cdot \mathbf{n} \rangle_{\partial K} \\
&= (\nabla v, \mathbf{w})_K \\
&= (\mathbf{P}_h \nabla v, \mathbf{w})_K,
\end{aligned}$$

which implies (11). As to (12), using the fact that $\nabla \cdot RT_j(K) = P_j(K_0)$, the property (7), and the definition of ∇_w, we obtain

$$\begin{aligned}
(\nabla \cdot \mathbf{q},\, v_0) &= (Q_0(\nabla \cdot \mathbf{q}),\, v_0) = (\nabla \cdot \mathbf{\Pi}_h \mathbf{q},\, v_0) \\
&= -\sum_{K \in \mathcal{T}_h} (\mathbf{\Pi}_h \mathbf{q}, \nabla_w v_h)_K + \sum_{K \in \mathcal{T}_h} \langle v_b, \mathbf{\Pi}_h \mathbf{q} \cdot \mathbf{n} \rangle_{\partial K} \\
&= -\sum_{K \in \mathcal{T}_h} (\mathbf{\Pi}_h \mathbf{q}, \nabla_w v_h)_K + \sum_{e \in \mathcal{T}_h \cap \partial\Omega} \langle (\mathbf{\Pi}_h \mathbf{q}) \cdot \mathbf{n}, v_b \rangle_e.
\end{aligned}$$

This completes the proof of (12). The equality (13) is a direct consequence of (12) since the boundary integrals vanish under the given condition. □

3.2 Discrete Norms and Inequalities

Let $v_h = \{v_0, v_b\} \in V_h$. Define on each $K \in \mathcal{T}_h$

$$\|v_h\|_{0,h,K}^2 = \|v_0\|_{0,K}^2 + h\|v_0 - v_b\|_{\partial K}^2,$$
$$\|v_h\|_{1,h,K}^2 = \|v_0\|_{1,K}^2 + h^{-1}\|v_0 - v_b\|_{\partial K}^2,$$
$$|v_h|_{1,h,K}^2 = |v_0|_{1,K}^2 + h^{-1}\|v_0 - v_b\|_{\partial K}^2.$$

Using the above quantities, we define the following discrete norms and semi-norms for the finite element space V_h:

$$\|v_h\|_{0,h} := \left(\sum_{K\in\mathcal{T}_h} \|v_h\|_{0,h,K}^2\right)^{1/2},$$
$$\|v_h\|_{1,h} := \left(\sum_{K\in\mathcal{T}_h} \|v_h\|_{1,h,K}^2\right)^{1/2},$$
$$|v_h|_{1,h} := \left(\sum_{K\in\mathcal{T}_h} |v_h|_{1,h,K}^2\right)^{1/2}.$$

It is clear that $\|v_h\|_{0,h}^2 = (\!(v_h, v_h)\!)$. Hence, $\|\cdot\|_{0,h}$ provides a discrete L^2 norm for V_h. It is not hard to see that $|\cdot|_{1,h}$ and $\|\cdot\|_{1,h}$ define a discrete H^1 semi-norm and a norm for V_h, respectively. Observe that $|v_h|_{1,h} = 0$ if and only if $v_h \equiv constant$. Thus, $|\cdot|_{1,h}$ is a norm in $V_{0,h}$ and $\bar{V}_h$.

For any $K \in \mathcal{T}_h$ and e being an edge of K, the following trace inequality is well known:

$$\|g\|_e^2 \lesssim h^{-1}\|g\|_K^2 + h^{2s-1}|g|_{s,K}^2, \quad \frac{1}{2} < s \le 1, \tag{14}$$

for all $g \in H^1(K)$. Here, $|g|_{s,K}$ is the semi-norm in the Sobolev space $H^s(K)$. The inequality (14) can be verified through a scaling argument for the standard Sobolev trace inequality in H^s with $s \in (\frac{1}{2}, 1]$. If g is a polynomial in K, then we have from (14) and the standard inverse inequality that

$$\|g\|_e^2 \lesssim h^{-1}\|g\|_K^2. \tag{15}$$

From (15) and the triangle inequality, it is not hard to see that for any $v_h \in V_h$ one has

$$\left(\sum_{K\in\mathcal{T}_h} (\|v_0\|_{0,K}^2 + h\|v_b\|_{\partial K}^2)\right)^{1/2} \lesssim \|v_h\|_{0,h} \lesssim \left(\sum_{K\in\mathcal{T}_h} (\|v_0\|_{0,K}^2 + h\|v_b\|_{\partial K}^2)\right)^{1/2}.$$

In the rest of this paper, we shall use the above equivalence without particular mentioning or referencing.

The following Lemma establishes an equivalence between the two semi-norms $|\cdot|_{1,h}$ and $\|\nabla_w \cdot\|$.

Lemma 3.2. *For any* $v_h = \{v_0, v_b\} \in V_h$, *we have*

$$|v_h|_{1,h} \lesssim \|\nabla_w v_h\| \lesssim |v_h|_{1,h}. \tag{16}$$

Proof. Using the definition of ∇_w, integration by parts, the Schwarz inequality, the inequality (15), and the Young's inequality, we have

$$\begin{aligned}
\|\nabla_w v_h\|_K^2 &= -(v_0, \nabla\cdot\nabla_w v_h)_K + \langle v_b, \nabla_w v_h\cdot\mathbf{n}\rangle_{\partial K}\\
&= \langle v_b - v_0, \nabla_w v_h\cdot\mathbf{n}\rangle_{\partial K} + (\nabla v_0, \nabla_w v_h)_K\\
&\le \|v_0 - v_b\|_{\partial K}\|\nabla_w v_h\cdot\mathbf{n}\|_{\partial K} + \|\nabla v_0\|_K\|\nabla_w v_h\|_K\\
&\lesssim \|v_0 - v_b\|_{\partial K} h^{-\frac{1}{2}}\|\nabla_w v_h\|_K + \|\nabla v_0\|_K\|\nabla_w v_h\|_K\\
&\lesssim \|\nabla_w v_h\|_K\left(\|\nabla v_0\|_K + h^{-\frac{1}{2}}\|v_0 - v_b\|_{\partial K}\right).
\end{aligned}$$

This completes the proof of $\|\nabla_w v_h\| \lesssim |v_h|_{1,h}$.

To prove $|v_h|_{1,h} \lesssim \|\nabla_w v_h\|$, let $K \in \mathcal{T}_h$ be any element and consider the following subspace of $RT_j(K)$:

$$D(j,K) := \{\mathbf{q}\in RT_j(K) :\ \mathbf{q}\cdot\mathbf{n} = 0 \text{ on } \partial K\}.$$

Note that $D(j,K)$ forms a dual of $(P_{j-1}(K))^2$. Thus, for any $\nabla v_0 \in (P_{j-1}(K))^2$, one has

$$\|\nabla v_0\|_K = \sup_{\mathbf{q}\in D(j,K)} \frac{(\nabla v_0, \mathbf{q})_K}{\|\mathbf{q}\|_K}. \tag{17}$$

It follows from the integration by parts and the definition of ∇_w that

$$(\nabla v_0, \mathbf{q})_K = -(v_0, \nabla\cdot\mathbf{q})_K = (\nabla_w v_h, \mathbf{q})_K,$$

which, together with (17) and the Cauchy–Schwarz inequality, gives

$$\|\nabla v_0\|_K \le \|\nabla_w v_h\|_K. \tag{18}$$

Note that for $j = 0$, we have $\nabla v_0 = 0$ and the above inequality is satisfied trivially.

Analogously, let e be an edge of K and denote by $D_e(j,K)$ the collection of all $\mathbf{q}\in RT_j(K)$ such that all degrees of freedom, except those for $\mathbf{q}\cdot\mathbf{n}|_e$, vanish. It is well known that $D_e(j,K)$ forms a dual of $P_j(e)$. Thus, we have

$$\|v_0 - v_b\|_e = \sup_{\mathbf{q}\in D_e(j,K)} \frac{\langle v_0 - v_b, \mathbf{q}\cdot\mathbf{n}\rangle_e}{\|\mathbf{q}\cdot\mathbf{n}\|_e}. \tag{19}$$

It follows from (5) and the integration by parts on $(v_0, \nabla\cdot\mathbf{q})_K$ that

$$(\nabla_w v_h, \mathbf{q})_K = (\nabla v_0, \mathbf{q})_K + \langle v_b - v_0, \mathbf{q}\cdot\mathbf{n}\rangle_{\partial K}, \qquad \forall\, \mathbf{q}\in RT_j(K). \tag{20}$$

In particular, for $\mathbf{q} \in D_e(j,K)$, we have

$$(\nabla v_0, \mathbf{q})_K = 0, \qquad \langle v_b - v_0, \mathbf{q}\cdot\mathbf{n}\rangle_{\partial K} = \langle v_b - v_0, \mathbf{q}\cdot\mathbf{n}\rangle_e.$$

Substituting the above into (20) yields

$$(\nabla_w v_h, \mathbf{q})_K = \langle v_b - v_0, \mathbf{q}\cdot\mathbf{n}\rangle_e, \qquad \forall\, \mathbf{q} \in D_e(j,K). \tag{21}$$

Using the Cauchy–Schwarz inequality we arrive at

$$|\langle v_b - v_0, \mathbf{q}\cdot\mathbf{n}\rangle_e| \le \|\nabla_w v_h\|_K \, \|\mathbf{q}\|_K,$$

for all $\mathbf{q} \in D_e(j,K)$. By the scaling argument, for such $\mathbf{q} \in D_e(j,K)$, we have $\|\mathbf{q}\|_K \lesssim h^{\frac{1}{2}} \|\mathbf{q}\cdot\mathbf{n}\|_e$. Thus, we obtain

$$|\langle v_b - v_0, \mathbf{q}\cdot\mathbf{n}\rangle_e| \lesssim h^{\frac{1}{2}} \|\nabla_w v_h\|_K \, \|\mathbf{q}\cdot\mathbf{n}\|_e, \qquad \forall \mathbf{q} \in D_e(j,K),$$

which, together with (19), implies the following estimate:

$$\|v_0 - v_b\|_e \lesssim h^{\frac{1}{2}} \|\nabla_w v_h\|_K.$$

Combining the above estimate with (18) gives a proof of $|v_h|_{1,h} \lesssim \|\nabla_w v_h\|$. This completes the proof of (16). □

The discrete semi-norms satisfy the usual inverse inequality, as stated in the following lemma.

Lemma 3.3. *For any $v_h = \{v_0, v_b\} \in V_h$, we have*

$$|v_h|_{1,h} \lesssim h^{-1} \|v_h\|_{0,h}. \tag{22}$$

Consequently, by combining (16) and (22), we have

$$\|\nabla_w v_h\| \lesssim h^{-1} \|v_h\|_{0,h}. \tag{23}$$

Proof. The proof follows from the standard inverse inequality and the definition of $\|\cdot\|_{0,h}$ and $|\cdot|_{1,h}$; details are thus omitted. □

Next, let us show that the discrete semi-norm $\|\nabla_w(\cdot)\|$, which is equivalent to $|\cdot|_{1,h}$ as proved in Lemma 3.2, satisfies a Poincaré-type inequality.

Lemma 3.4. *The Poincaré-type inequality holds true for functions in $V_{0,h}$ and $\bar{V}_h$. In other words, we have the following estimates:*

$$\|v_h\|_{0,h} \lesssim \|\nabla_w v_h\| \qquad \forall\, v_h \in V_{0,h}, \tag{24}$$

$$\|v_h\|_{0,h} \lesssim \|\nabla_w v_h\| \qquad \forall\, v_h \in \bar{V}_h. \tag{25}$$

Proof. For any $v_h \in V_{0,h}$, let $\mathbf{q} \in (H^1(\Omega))^2$ be such that $\nabla \cdot \mathbf{q} = v_0$ and $\|\mathbf{q}\|_1 \lesssim \|v_0\|$. Such a vector-valued function $\mathbf{q}$ exists on any polygonal domain [3]. One way to prove the existence of $\mathbf{q}$ is as follows. First, one extends v_h by zero to a convex domain which contains Ω. Secondly, one considers the Poisson equation on the enlarged domain and set $\mathbf{q}$ to be the flux. The required properties of $\mathbf{q}$ follow immediately from the full regularity of the Poisson equation on convex domains. By (7), we have

$$\|\mathbf{\Pi}_h \mathbf{q}\| \lesssim \|\mathbf{q}\|_1 \lesssim \|v_0\|.$$

Consequently, by (13) and the Schwarz inequality,

$$\|v_0\|^2 = (v_0, \nabla \cdot \mathbf{q}) = -(\mathbf{\Pi}_h \mathbf{q}, \nabla_w v_h) \lesssim \|v_0\| \|\nabla_w v_h\|.$$

It follows from Lemma 3.2 that

$$\sum_{K \in \mathcal{T}_h} h \|v_0 - v_b\|_{\partial K}^2 \lesssim \sum_{K \in \mathcal{T}_h} h^{-1} \|v_0 - v_b\|_{\partial K}^2 \leq |v_h|_{1,h}^2 \lesssim \|\nabla_w v_h\|^2.$$

Combining the above two estimates gives a proof of the inequality (24).

As to (25), since $v_h \in \bar{V}_h$ has mean value zero, one may find a vector-valued function $\mathbf{q}$ satisfying $\nabla \cdot \mathbf{q} = v_0$ and $\mathbf{q} \cdot \mathbf{n} = 0$ on $\partial\Omega$ (see [3] for details). In addition, we have $\|\mathbf{q}\|_1 \lesssim \|v_0\|$. The rest of the proof follows the same avenue as the proof of (24). □

Next, we shall introduce a discrete norm in the finite element space $V_{0,h}$ that plays the role of the standard H^2 norm. To this end, for any internal edge $e \in \mathcal{E}_h$, denote by K_1 and K_2 the two triangles sharing e, and by $\mathbf{n}_1$, $\mathbf{n}_2$ the outward normals with respect to K_1 and K_2. Define the jump on e by

$$[\![\nabla_w \psi_h \cdot \mathbf{n}]\!] = (\nabla_w \psi_h)|_{K_1} \cdot \mathbf{n}_1 + (\nabla_w \psi_h)|_{K_2} \cdot \mathbf{n}_2.$$

If the edge e is on the boundary $\partial\Omega$, then there is only one triangle K which admits e as an edge. The jump is then modified as

$$[\![\nabla_w \psi_h \cdot \mathbf{n}]\!] = (\nabla_w \psi_h)|_K \cdot \mathbf{n}.$$

For $\psi_h \in V_{0,h}$, define

$$|\!|\!|\psi_h|\!|\!| = \left(\sum_{K \in \mathcal{T}_h} \|\nabla \cdot \nabla_w \psi_h\|_K^2 + \sum_{e \in \mathcal{E}_h} h^{-1} \|[\![\nabla_w \psi_h \cdot \mathbf{n}]\!]\|_e^2) \right)^{1/2}. \tag{26}$$

Lemma 3.5. *The map $|\!|\!|\cdot|\!|\!| : V_{0,h} \to \mathbb{R}$, as given in (26), defines a norm in the finite element space $V_{0,h}$. Moreover, one has*

$$(\nabla_w v_h, \nabla_w \psi_h) \lesssim \|v_h\|_{0,h} |||\psi_h||| \qquad \forall\, v_h \in V_h,\ \psi_h \in V_{0,h}, \tag{27}$$

$$\sup_{v_h \in V_h} \frac{(\nabla_w v_h, \nabla_w \psi_h)}{\|v_h\|_{0,h}} \gtrsim |||\psi_h||| \qquad \forall\, \psi_h \in V_{0,h}. \tag{28}$$

Proof. To verify that $|||\cdot|||$ defines a norm, it is sufficient to show that $|||\psi_h||| = 0$ implies $\psi_h \equiv 0$. To this end, let $|||\psi_h||| = 0$. It follows that $\nabla \cdot \nabla_w \psi_h = 0$ on each element and $[[\nabla_w \psi_h \cdot \mathbf{n}]] = 0$ on each edge. The definition of the discrete weak gradient ∇_w then implies the following:

$$(\nabla_w \psi_h, \nabla_w \psi_h) = \sum_{K \in \mathcal{T}_h} \left(-(\psi_0, \nabla \cdot \nabla_w \psi_h)_K + \langle \psi_b, \nabla_w \psi_h \cdot \mathbf{n} \rangle_{\partial K}\right) = 0.$$

Thus, we have $\nabla_w \psi_h = 0$. Since $\psi_h \in V_{0,h}$, then $\nabla_w \psi_h = 0$ implies $\psi_h \equiv 0$. This shows that $|||\cdot|||$ defines a norm in $V_{0,h}$. The inequality (27) follows immediately from the following identity:

$$(\nabla_w v_h, \nabla_w \psi_h) = \sum_{K \in \mathcal{T}_h} \left(-(v_0, \nabla \cdot \nabla_w \psi_h)_K + \langle v_b, \nabla_w \psi_h \cdot \mathbf{n} \rangle_{\partial K}\right)$$

and the Schwarz inequality.

To verify (28), we chose a particular $v_h^* \in V_h$ such that

$$v_0^* = -\nabla \cdot \nabla_w \psi_h \qquad \text{in } K_0,$$

$$v_b^* = h^{-1} [[\nabla_w \psi_h \cdot \mathbf{n}]] \qquad \text{on edge } e.$$

It is not hard to see that $\|v_h^*\|_{0,h} \lesssim |||\psi_h|||$. Thus, we have

$$\begin{aligned}
\sup_{v_h \in V_h} \frac{(\nabla_w v_h, \nabla_w \psi_h)}{\|v_h\|_{0,h}} &\geq \frac{(\nabla_w v_h^*, \nabla_w \psi_h)}{\|v_h^*\|_{0,h}} \\
&= \frac{\sum_{K \in \mathcal{T}_h} \left(-(v_0^*, \nabla \cdot \nabla_w \psi_h)_K + \langle v_b^*, \nabla_w \psi_h \cdot \mathbf{n} \rangle_{\partial K}\right)}{\|v_h^*\|_{0,h}} \\
&= \frac{|||\psi_h|||^2}{\|v_h^*\|_{0,h}} \gtrsim |||\psi_h|||.
\end{aligned}$$

This completes the proof of the lemma. □

Remark 3.1. Using the boundedness (27) and the discrete Poincare inequality (24), we have the following estimate for all $\psi_h \in V_{0,h}$:

$$\|\nabla_w \psi_h\|^2 = (\nabla_w \psi_h, \nabla_w \psi_h) \lesssim \|\psi_h\|_{0,h} |||\psi_h||| \lesssim \|\nabla_w \psi_h\| \, |||\psi_h|||.$$

This implies that $\|\nabla_w \psi_h\| \lesssim |||\psi_h|||$. In other words, $|||\cdot|||$ is a norm that is stronger than $\|\cdot\|_{1,h}$. In fact, the norm $|||\cdot|||$ can be viewed as a discrete equivalence of the standard H^2 norm for smooth functions with proper boundary conditions.

Next, we shall establish an estimate for the L^2 projection operator Q_h in the discrete norm $\|\cdot\|_{0,h}$.

Lemma 3.6. *Let Q_h be the L^2 projection operator into the finite element space V_h. Then, for any $v \in H^m(\Omega)$ with $\frac{1}{2} < m \leq j+1$, we have*

$$\|v - Q_h v\|_{0,h} \lesssim h^m \|v\|_m. \tag{29}$$

Proof. For the L^2 projection on each element K, it is known that the following estimate holds true:

$$\|v - Q_0 v\|_K \lesssim h^m \|v\|_{m,K}. \tag{30}$$

Thus, it suffices to deal with the terms associated with the edges/faces given by

$$\sum_K h \|(v - Q_0 v) - (v - Q_b v)\|_{\partial K}^2 = \sum_K h \|Q_0 v - Q_b v\|_{\partial K}^2. \tag{31}$$

Since Q_b is the L^2 projection on edges, then we have

$$\|Q_0 v - Q_b v\|_{\partial K}^2 \leq \|v - Q_0 v\|_{\partial K}^2.$$

Let $s \in (\frac{1}{2}, 1]$ be any real number satisfying $s \leq m$. It follows from the above inequality and the trace inequality (14) that

$$\|Q_0 v - Q_b v\|_{\partial K}^2 \lesssim h^{-1} \|v - Q_0 v\|_K^2 + h^{2s-1} |v - Q_0 v|_{s,K}^2.$$

Substituting the above into (31) yields

$$\begin{aligned}\sum_K h \|(v - Q_0 v) - (v - Q_b v)\|_{\partial K}^2 &\lesssim \sum_K \left(\|v - Q_0 v\|_K^2 + h^{2s} |v - Q_0 v|_{s,K}^2\right) \\ &\lesssim h^{2m} \|v\|_m^2,\end{aligned}$$

which, together with (30), completes the proof of the lemma. □

3.3 Ritz and Neumann Projections

To establish an error analysis in the forthcoming section, we shall introduce and analyze two additional projection operators, the Ritz projection R_h and the Neumann projection N_h, by applying the weak Galerkin method to the Poisson equation with various boundary conditions.

For any $v \in H_0^1(\Omega) \cap H^{1+\gamma}(\Omega)$ with $\gamma > \frac{1}{2}$, the Ritz projection $R_h v \in V_{0,h}$ is defined as the unique solution of the following problem:

$$(\nabla_w (R_h v), \nabla_w \psi_h) = (\mathbf{\Pi}_h \nabla v, \nabla_w \psi_h), \qquad \forall\, \psi_h \in V_{0,h}. \tag{32}$$

Here, $\gamma > \frac{1}{2}$ in the definition of R_h is imposed to ensure that $\mathbf{\Pi}_h \nabla v$ is well defined. From the identity (13), clearly if $\Delta v \in L^2(\Omega)$, then $R_h v$ is identical to the weak Galerkin finite element solution [35] to the Poisson equation with homogeneous Dirichlet boundary condition for which v is the exact solution. Analogously, for any $v \in \bar{H}^1(\Omega) \cap H^{1+\gamma}(\Omega)$ with $\gamma > \frac{1}{2}$, we define the Neumann projection $N_h v \in \bar{V}_h$ as the solution to the following problem:

$$(\nabla_w (N_h v), \nabla_w \psi_h) = (\mathbf{\Pi}_h \nabla v, \nabla_w \psi_h), \qquad \forall\, \psi_h \in \bar{V}_h. \tag{33}$$

It is useful to note that the above equation holds true for all $\psi_h \in V_h$ as $\nabla_w 1 = 0$. Similarly, if $\Delta v \in L^2(\Omega)$ and in addition $\partial v / \partial \mathbf{n} = 0$ on $\partial\Omega$, then $N_h v$ is identical to the weak Galerkin finite element solution to the Poisson equation with homogeneous Neumann boundary condition, for which v is the exact solution. The well-posedness of R_h and N_h follows immediately from the Poincaré-type inequalities (24) and (25).

Using (11), it is easy to see that for all $\psi_h \in V_{0,h}$ we have

$$(\nabla_w (Q_h v - R_h v), \nabla_w \psi_h) = ((\mathbf{P}_h - \mathbf{\Pi}_h) \nabla v, \nabla_w \psi_h). \tag{34}$$

And similarly, for all $\psi_h \in \bar{V}_h$,

$$(\nabla_w (Q_h v - N_h v), \nabla_w \psi_h) = ((\mathbf{P}_h - \mathbf{\Pi}_h) \nabla v, \nabla_w \psi_h). \tag{35}$$

From the definitions of $\bar{V}_h$ and Q_h, clearly Q_h maps $\bar{H}^1(\Omega)$ into $\bar{V}_h$.

For convenience, let us adopt the following notation:

$$\{R_0 v, R_b v\} := R_h v, \qquad \{N_0 v, N_b v\} := N_h v,$$

where again the subscript "0" denotes the function value in the interior of triangles, while "b" denotes the trace on $\mathcal{E}_h$. For Ritz and Neumann projections, the following approximation error estimates hold true.

Lemma 3.7. *For $v \in H_0^1(\Omega) \cap H^{m+1}(\Omega)$ or $\bar{H}^1(\Omega) \cap H^{m+1}(\Omega)$, where $\frac{1}{2} < m \le j+1$, we have*

$$\|\nabla_w (Q_h v - R_h v)\| \lesssim h^m \|v\|_{m+1}, \tag{36}$$

$$\|\nabla_w (Q_h v - N_h v)\| \lesssim h^m \|v\|_{m+1}. \tag{37}$$

Moreover, assume $\Delta v \in L^2(\Omega)$ and that the Poisson problem in Ω with either the homogeneous Dirichlet boundary condition or the homogeneous Neumann boundary condition has H^{1+s} regularity, where $\frac{1}{2} < s \le 1$, then

$$\|Q_0 v - R_0 v\| \lesssim h^{m+s}\|v\|_{m+1} + h^{1+s}\|(I-Q_0)\Delta v\|, \tag{38}$$

$$\|Q_0 v - N_0 v\| \lesssim h^{m+\min(s,j+\frac{1}{2})}\|v\|_{m+1} + h^{1+s}\|(I-Q_0)\Delta v\|. \tag{39}$$

Proof. The estimates (36)–(37) follow immediately from (34)–(35), (10), and the Schwarz inequality. Next, we prove (39) by using the standard duality argument. Let $\phi \in \bar{H}^1(\Omega)$ be the solution of $-\Delta\phi = Q_0 v - N_0 v$ with boundary condition $\frac{\partial\phi}{\partial\mathbf{n}}\Big|_{\partial\Omega} = 0$. Note that ϕ is well defined since $Q_h v - N_h v \in \bar{V}_h$. According to the regularity assumption, we have $\phi \in H^{1+s}(\Omega)$ and $\|\phi\|_{1+s} \lesssim \|Q_0 v - N_0 v\|$. Then, by (13), (35), the Schwarz inequality and (10), we arrive at

$$\begin{aligned}
\|Q_0 v - N_0 v\|^2 &= (Q_0 v - N_0 v, -\Delta\phi) = (\boldsymbol{\Pi}_h\nabla\phi, \nabla_w(Q_h v - N_h v)) \\
&= (\boldsymbol{\Pi}_h\nabla\phi - \nabla_w(N_h\phi), \nabla_w(Q_h v - N_h v)) + ((\mathbf{P}_h - \boldsymbol{\Pi}_h)\nabla v, \nabla_w(N_h\phi)) \\
&\le \Big(\|\boldsymbol{\Pi}_h\nabla\phi - \mathbf{P}_h\nabla\phi\| + \|\nabla_w(Q_h\phi - N_h\phi)\|\Big)\|\nabla_w(Q_h v - N_h v)\| \\
&\quad + ((\mathbf{P}_h - \boldsymbol{\Pi}_h)\nabla v, \nabla_w(N_h\phi - Q_h\phi)) + ((\mathbf{P}_h - \boldsymbol{\Pi}_h)\nabla v, \mathbf{P}_h\nabla\phi) \\
&\lesssim h^{m+s}\|\phi\|_{1+s}\|v\|_{m+1} + ((I - \boldsymbol{\Pi}_h)\nabla v, \mathbf{P}_h\nabla\phi).
\end{aligned}$$

Using integration by parts, the triangular inequality and the definition of $\boldsymbol{\Pi}_h$, we have

$$\begin{aligned}
&((I-\boldsymbol{\Pi}_h)\nabla v, \mathbf{P}_h\nabla\phi) \\
=&((I-\boldsymbol{\Pi}_h)\nabla v, (\mathbf{P}_h - I)\nabla\phi) + ((I-\boldsymbol{\Pi}_h)\nabla v, \nabla\phi) \\
\lesssim& h^{m+s}\|\phi\|_{1+s}\|v\|_{m+1} + ((I-\boldsymbol{\Pi}_h)\nabla v\cdot\mathbf{n}, \phi)_{\partial\Omega} - (\nabla\cdot(I-\boldsymbol{\Pi}_h)\nabla v, \phi) \\
=& h^{m+s}\|\phi\|_{1+s}\|v\|_{m+1} + ((I-\boldsymbol{\Pi}_h)\nabla v\cdot\mathbf{n}, \phi - Q_b\phi)_{\partial\Omega} - ((I-Q_0)\Delta v, \phi) \\
\lesssim& h^{m+s}\|\phi\|_{1+s}\|v\|_{m+1} + (h^{m-\frac{1}{2}}\|v\|_{m+\frac{1}{2},\partial\Omega})(h^{\min(s+\frac{1}{2},j+1)}\|\phi\|_{s+\frac{1}{2},\partial\Omega}) \\
&- ((I-Q_0)\Delta v, (I-Q_0)\phi) \\
\lesssim& h^{m+\min(s,j+\frac{1}{2})}\|\phi\|_{1+s}\|v\|_{m+1} + h^{1+s}\|\phi\|_{1+s}\|(I-Q_0)\Delta v\|.
\end{aligned} \tag{40}$$

In the proof of (40), we have used the fact that $\Pi_h(\nabla v\cdot\mathbf{n})$ is exactly the L^2 projection of $\nabla v\cdot\mathbf{n}$ on $\partial\Omega$. Combining the above gives

$$\begin{aligned}
\|Q_0 v - N_0 v\|^2 &\lesssim \Big(h^{m+\min(s,j+\frac{1}{2})}\|v\|_{m+1} + h^{1+s}\|(I-Q_0)\Delta v\|\Big)\|\phi\|_{1+s} \\
&\lesssim \Big(h^{m+\min(s,j+\frac{1}{2})}\|v\|_{m+1} + h^{1+s}\|(I-Q_0)\Delta v\|\Big)\|Q_0 v - N_0 v\|.
\end{aligned}$$

This completes the proof of the estimate (39). The inequality (38) can be verified in a similar way by considering a function $\phi \in H_0^1(\Omega)$ satisfying a Poisson equation with homogeneous Dirichlet boundary condition. Observe that in this case, the boundary integral $((I - \mathbf{\Pi}_h)\nabla v \cdot \mathbf{n}, \phi)_{\partial\Omega}$ in inequality (40) shall vanish due to the vanishing value of ϕ. □

Remark 3.2. It is not hard to see from (40) that for the Neumann projection, if in addition we have $\frac{\partial v}{\partial n} = 0$ on $\partial\Omega$, then the term $((I - \mathbf{\Pi}_h)\nabla v \cdot \mathbf{n}, \phi)_{\partial\Omega}$ vanishes and one obtains the optimal order estimate of h^{m+s} instead of $h^{m+\min(s,j+\frac{1}{2})}$ for the Neumann projection operator.

Remark 3.3. If the Poisson equation has the full H^2 regularity in Ω, then for v satisfying the assumptions of Lemma 3.7, we have

$$\|Q_0 v - R_0 v\| \lesssim h^{m+1}\|v\|_{m+1} + h^2\|(I - Q_0)\Delta v\| \qquad \text{for } \frac{1}{2} < m \le j+1,$$

$$\|Q_0 v - N_0 v\| \lesssim \begin{cases} h^{m+\frac{1}{2}}\|v\|_{m+1} + h^2\|(I - Q_0)\Delta v\| & \text{for } j = 0,\ \frac{1}{2} < m \le 1, \\ h^{m+1}\|v\|_{m+1} + h^2\|(I - Q_0)\Delta v\| & \text{for } j \ge 1,\ \frac{1}{2} < m \le j+1. \end{cases}$$

Again, if in addition, $\frac{\partial v}{\partial n} = 0$ on $\partial\Omega$, then the Neumann projection has optimal order of error estimates, even for $j = 0$.

Remark 3.4. The duality argument used in Lemma 3.7 works only for $\|Q_0 v - R_0 v\|$ and $\|Q_0 v - N_0 v\|$. For $\|Q_h v - R_h v\|_{0,h}$ and $\|Q_h v - N_h v\|_{0,h}$ involving element boundary information, we currently have only suboptimal estimates. More precisely, for v satisfying the assumptions in Lemma 3.7, the following estimates hold true:

$$\|Q_h v - R_h v\|_{0,h} \lesssim \|\nabla_w (Q_h v - R_h v)\| \lesssim h^m \|v\|_{m+1} \qquad \text{for } \frac{1}{2} < m \le j+1,$$

$$\|Q_h v - N_h v\|_{0,h} \lesssim \|\nabla_w (Q_h v - N_h v)\| \lesssim h^m \|v\|_{m+1} \qquad \text{for } \frac{1}{2} < m \le j+1. \quad (41)$$

Although numerical experiments in [30] suggest an optimal order of convergence in the $\|\cdot\|_{0,h}$ norm, it remains to see if optimal order error estimates hold true or not theoretically.

Another important observation is that, for sufficiently smooth v, $\nabla_w R_h v$ is identical to the mixed finite element approximation of ∇v, discretized by using RT_j and discrete P_j elements. Indeed, we have the following lemma:

Lemma 3.8. *For any $v \in H_0^1 \cap H^{1+\gamma}(\Omega)$ with $\gamma > \frac{1}{2}$ and $\Delta v \in L^2(\Omega)$, let $\mathbf{q}_h \in \Sigma_h \cap H(div,\Omega)$ and $v_0 \in L^2(\Omega)$ be piecewise P_j polynomials solving*

$$\begin{cases} (\mathbf{q}_h, \boldsymbol{\chi}_h) - (\nabla \cdot \boldsymbol{\chi}_h, v_0) = 0 & \forall \boldsymbol{\chi}_h \in \Sigma_h \cap H(div, \Omega), \\ (\nabla \cdot \mathbf{q}_h, \psi_0) = (\Delta v, \psi_0) & \forall \psi_0 \in L^2(\Omega) \text{ piecewise } P_j \text{ polynomials.} \end{cases} \tag{42}$$

In other words, $\mathbf{q}_h$ and v_0 are the mixed finite element solution, discretized using the RT_j element, to the Poisson equation with homogeneous Dirichlet boundary condition for which v is the exact solution. Then, one has $\nabla_w R_h v = \mathbf{q}_h$.

Proof. We first show that $\nabla_w R_h v \in \Sigma_h \cap H(div, \Omega)$ by verifying that $(\nabla_w R_h v) \cdot \mathbf{n}$ is continuous across internal edges. Let $e \in \mathcal{E}_h \backslash \partial\Omega$ be an internal edge and K_1, K_2 be two triangles sharing e. Denote $\mathbf{n}_1$ and $\mathbf{n}_2$ the outward normal vectors on e, with respect to K_1 and K_2, respectively. Let $\psi_h \in V_{0,h}$ satisfy $\psi_b|_e \neq 0$ and ψ_0, ψ_b vanish elsewhere. By the definition of R_h, ∇_w and the fact that $\boldsymbol{\Pi}_h \nabla v \in H(div, \Omega)$, we have

$$\begin{aligned} 0 &= (\boldsymbol{\Pi}_h \nabla v - \nabla_w R_h v, \nabla_w \psi_h) \\ &= (\boldsymbol{\Pi}_h \nabla v - \nabla_w R_h v, \nabla_w \psi_h)_{K_1} + (\boldsymbol{\Pi}_h \nabla v - \nabla_w R_h v, \nabla_w \psi_h)_{K_2} \\ &= ((\boldsymbol{\Pi}_h \nabla v - \nabla_w R_h v)|_{K_1} \cdot \mathbf{n}_1 + (\boldsymbol{\Pi}_h \nabla v - \nabla_w R_h v)|_{K_2} \cdot \mathbf{n}_2, \psi_b)_e \\ &= -(\nabla_w R_h v|_{K_1} \cdot \mathbf{n}_1 + \nabla_w R_h v|_{K_2} \cdot \mathbf{n}_2, \psi_b)_e. \end{aligned}$$

The above equation holds true for all $\psi_b|_e \in P_j(e)$. Since $\nabla_w R_h v|_{K_1} \cdot \mathbf{n}_1 + \nabla_w R_h v|_{K_2} \cdot \mathbf{n}_2$ is also in $P_j(e)$, therefore it must be 0. This completes the proof of $\nabla_w R_h v \in H(div, \Omega)$.

Next, we prove that $\nabla_w R_h v$ is identical to the solution $\mathbf{q}_h$ of (42). Since the solution to (42) is unique, we only need to show that $\nabla_w R_h v$, together with a certain v_0, satisfies both equations in (42). Consider the test function $\psi_h \in V_{0,h}$ with the form $\psi_h = \{\psi_0, 0\}$. By the definition of ∇_w, Eqs. (32) and (13), we have

$$(\nabla \cdot \nabla_w R_h v, \psi_0) = -(\nabla_w R_h v, \nabla_w \psi_h) = -(\boldsymbol{\Pi}_h \nabla v, \nabla_w \psi_h) = (\Delta v, \psi_0).$$

Hence $\nabla_w R_h v$ satisfies the second equation of (42). Now, note that $\nabla \cdot$ is an onto operator from $\Sigma_h \cap H(div, \Omega)$ to the space of piecewise P_j polynomials, which allows us to define a v_0 that satisfies the first equation in (42) with $\mathbf{q}_h$ set to be $\nabla_w R_h v$. This completes the proof the lemma. □

Remark 3.5. Using the same argument and noticing that (33) holds for all $\psi_h \in V_h$, one can analogously prove that for $v \in \bar{H}^1(\Omega) \cap H^{1+\gamma}(\Omega)$ with $\gamma > \frac{1}{2}$ and $\Delta v \in L^2(\Omega)$,

$$\nabla_w N_h v \in \Sigma_h \cap H(div, \Omega),$$

and

$$\nabla \cdot \nabla_w N_h v = Q_0 \Delta v.$$

Because $\nabla_w R_h v$ is identical to the mixed finite element solution to the Poisson equation, by [18, 34], we have the following quasi-optimal order L^∞ estimate:

$$\|\nabla v - \nabla_w R_h v\|_{L^\infty(\Omega)} \lesssim h^{n+1} |\ln h| \|\Delta v\|_{W^{n,\infty}(\Omega)}, \tag{43}$$

for $0 \le n \le j$. Furthermore, for $j \ge 1$ and $v \in W^{j+2,\infty}(\Omega)$, we have the following optimal order error estimate:

$$\|\nabla v - \nabla_w R_h v\|_{L^\infty(\Omega)} \lesssim h^{n+1} \|v\|_{W^{n+2,\infty}(\Omega)}, \tag{44}$$

for $1 \le n \le j$.

Inspired by [32], using the above L^∞ estimates, we obtain the following lemma, which will play an essential role in the error analysis to be given in the next section.

Lemma 3.9. *The following quasi-optimal and optimal order error estimates hold true:*

(i) Let $0 \le n \le j$ and $v \in H_0^1(\Omega) \cap W^{n+2,\infty}(\Omega)$. Then for all $\phi_h = \{v_0, v_b\} \in V_h$, we have

$$|(\boldsymbol{\Pi}_h \nabla v - \nabla_w R_h v, \nabla_w \phi_h)| \lesssim h^{n+\frac{1}{2}} |\ln h| \|v\|_{W^{n+2,\infty}(\Omega)} \|\phi_h\|_{0,h}. \tag{45}$$

(ii) Let $j \ge 1$, $1 \le n \le j$, and $v \in H_0^1(\Omega) \cap W^{n+2,\infty}(\Omega)$. Then, for all $\phi_h = \{v_0, v_b\} \in V_h$ we have

$$|(\boldsymbol{\Pi}_h \nabla v - \nabla_w R_h v, \nabla_w \phi_h)| \lesssim h^{n+\frac{1}{2}} \|v\|_{W^{n+2,\infty}(\Omega)} \|\phi_h\|_{0,h}. \tag{46}$$

Proof. We first prove part (i). Denote by $\mathcal{E}_{\partial\Omega}$ the set of all edges in $\mathcal{E}_h \cap \partial\Omega$. For any $e \in \mathcal{E}_{\partial\Omega}$, let K_e be the only triangle in $\mathcal{T}_h$ that has e as an edge. Denote by $\mathcal{T}_{\partial\Omega}$ the set of all K_e, for $e \in \mathcal{E}_{\partial\Omega}$. For simplicity of notation, denote $\mathbf{q}_h = \boldsymbol{\Pi}_h \nabla v - \nabla_w R_h v$. Since $(\boldsymbol{\Pi}_h \nabla v - \nabla_w R_h v, \nabla_w \psi_h) = 0$ for all $\psi_h \in V_{0,h}$, without loss of generality, we only need to consider ϕ_h that vanishes on the interior of all triangles and all internal edges. Then by the definition of ϕ_h and ∇_w, the scaling argument, and the Schwarz inequality,

$$\begin{aligned}
|(\boldsymbol{\Pi}_h \nabla v - \nabla_w R_h v, \nabla_w \phi_h)| &= \left| \sum_{K_e \in \mathcal{T}_{\partial\Omega}} (\mathbf{q}_h, \nabla_w (\phi_b|_e))_{K_e} \right| \\
&= \left| \sum_{e \in \mathcal{E}_{\partial\Omega}} (\phi_b, \mathbf{q}_h \cdot \mathbf{n})_e \right| \\
&\lesssim \sum_{e \in \mathcal{E}_{\partial\Omega}} h \|\phi_b\|_{L^\infty(e)} \|\mathbf{q}_h\|_{L^\infty(e)} \\
&\lesssim \|\mathbf{q}_h\|_{L^\infty(\Omega)} \sum_{e \in \mathcal{E}_{\partial\Omega}} h \left(\|\phi_0\|_{L^\infty(K_e)} + \|\phi_0 - \phi_b\|_{L^\infty(e)} \right) \\
&\lesssim \|\mathbf{q}_h\|_{L^\infty(\Omega)} \sum_{K_e \in \mathcal{T}_{\partial\Omega}} \|\phi_h\|_{0,h,K_e} \\
&\lesssim \|\mathbf{q}_h\|_{L^\infty(\Omega)} \left(\sum_{K_e \in \mathcal{T}_{\partial\Omega}} \|\phi_h\|_{0,h,K_e}^2 \right)^{\frac{1}{2}} \left(\sum_{K_e \in \mathcal{T}_{\partial\Omega}} 1 \right)^{\frac{1}{2}} \\
&\lesssim h^{-\frac{1}{2}} \|\mathbf{q}_h\|_{L^\infty(\Omega)} \|\phi_h\|_{0,h}.
\end{aligned}$$

Now, by inequalities (8) and (43), we have

$$\begin{aligned}\|\mathbf{q}_h\|_{L^\infty(\Omega)} &\le \|\nabla v - \mathbf{\Pi}_h \nabla v\|_{L^\infty(\Omega)} + \|\nabla v - \nabla_w R_h v\|_{L^\infty(\Omega)} \\ &\lesssim h^{n+1}\|v\|_{W^{n+2,\infty}(\Omega)} + h^{n+1}|\ln h|\|\Delta v\|_{W^{n,\infty}(\Omega)},\end{aligned}$$

for $0 \le n \le j$. This completes the proof of part (i).

The proof for part (ii) is similar. One simply needs to replace inequality (43) by (44) in the estimation of $\|\mathbf{q}_h\|_{L^\infty(\Omega)}$. □

4 Error Analysis

The main purpose of this section is to analyze the approximation error of the weak Galerkin formulation (6). For simplicity, in this section, we assume that the solution of (6) satisfies $u \in H^{3+\gamma}(\Omega)$ and $w \in H^{1+\gamma}(\Omega)$, where $\gamma > \frac{1}{2}$. This is not an unreasonable assumption, as we know from (4), the solution u can have up to H^4 regularity as long as Ω satisfies certain conditions. However, our assumption does not include all the possible cases for the biharmonic equation.

Testing $w = -\Delta u$ with $\phi_h = \{\phi_0, \phi_b\} \in V_h$, and then by using (13), we have

$$((w, \phi_h)) = (w, \phi_0) = -(\nabla \cdot \nabla u, \phi_0) = (\mathbf{\Pi}_h \nabla u, \nabla_w \phi_h). \tag{47}$$

Similarly, testing $-\Delta w = f$ with $\psi_h = \{\psi_0, \psi_b\} \in V_{0,h}$ gives

$$(\mathbf{\Pi}_h \nabla w, \nabla_w \psi_h) = (f, \psi_0). \tag{48}$$

Comparing (47)–(48) with the weak Galerkin form (6), one immediately sees that there is a consistency error between them. Indeed, since V_h and $V_{0,h}$ are not subspaces of $H^1(\Omega)$ and $H_0^1(\Omega)$, respectively, the weak Galerkin method is nonconforming. Therefore, we would like to first rewrite (47)–(48) into a form that is more compatible with (6). By using (32) and (33), Eqs. (47)–(48) can be rewritten as

$$\begin{cases} ((N_h w, \phi_h)) - (\nabla_w R_h u, \nabla_h \phi_h) = E(w, u, \phi_h), \\ (\nabla_w N_h w, \nabla_w \psi_h) = (f, \psi_0), \end{cases} \tag{49}$$

where

$$E(w, u, \phi_h) = ((N_h w - w, \phi_h)) + (\mathbf{\Pi}_h \nabla u - \nabla_w R_h u, \nabla_w \phi_h).$$

Define $\varepsilon_u = R_h u - u_h \in V_{0,h}$ and $\varepsilon_w = N_h w - w_h \in V_h$. By subtracting (49) from (6), we have

$$\begin{cases} ((\varepsilon_w, \phi_h)) - (\nabla_w \varepsilon_u, \nabla_h \phi_h) = E(w, u, \phi_h) & \text{for all } \phi_h \in V_h, \\ (\nabla_w \varepsilon_w, \nabla_w \psi_h) = 0 & \text{for all } \psi_h \in V_{0,h}. \end{cases} \tag{50}$$

Notice here $(\nabla_w \varepsilon_w, \nabla_w \psi_h) = 0$ does not necessarily imply $\varepsilon_w = 0$, since the equation only holds for all $\psi_h \in V_{0,h}$ while ε_w is in V_h.

Lemma 4.1. *The consistency error $E(w,u,\phi_h)$ is small in the sense that*

$$|E(w,u,\phi_h)| \lesssim h^m \|w\|_{m+1} \|\phi_h\|_{0,h} + h^{n+\frac{1}{2}} |\ln h| \|u\|_{W^{n+2,\infty}(\Omega)} \|\phi_h\|_{0,h},$$

where $\frac{1}{2} < m \le j+1$ and $0 \le n \le j$. Moreover, for $j \ge 1$, we have the improved estimate

$$|E(w,u,\phi_h)| \lesssim h^m \|w\|_{m+1} \|\phi_h\|_{0,h} + h^{n+\frac{1}{2}} \|u\|_{W^{n+2,\infty}(\Omega)} \|\phi_h\|_{0,h},$$

where $\frac{1}{2} < m \le j+1$ and $1 \le n \le j$.

Proof. The proof is straightforward by using the Schwarz inequality, Lemma 3.6, Remark 3.4, and Lemma 3.9. □

To derive an error estimate from (50), let us recall the standard theory for mixed finite element methods. Given two bounded bilinear forms $a(\cdot,\cdot)$ defined on $X \times X$ and $b(\cdot,\cdot)$ defined on $X \times M$, where X and M are finite dimensional spaces. Denote $X_0 \subset X$ by

$$X_0 = \{\phi \in X : b(\phi,\psi) = 0 \text{ for all } \psi \in M\}.$$

Then for all $\chi \in X$ and $\xi \in M$,

$$\sup_{\phi \in X,\, \psi \in M} \frac{a(\chi,\phi) + b(\phi,\xi) + b(\chi,\psi)}{\|\phi\|_X + \|\psi\|_M} \gtrsim \|\chi\|_X + \|\xi\|_M,$$

if and only if

$$\begin{aligned} \sup_{\phi \in X_0} \frac{a(\chi,\phi)}{\|\phi\|_X} &\gtrsim \|\chi\|_X, \qquad \text{for all } \chi \in X_0, \\ \sup_{\phi \in X} \frac{b(\phi,\xi)}{\|\phi\|_X} &\gtrsim \|\xi\|_M, \qquad \text{for all } \xi \in M. \end{aligned} \tag{51}$$

In our formulation, we set $X = V_h$ with norm $\|\cdot\|_{0,h}$ and $M = V_{0,h}$ with norm $|||\cdot|||$. Define

$$a(\chi,\phi) = ((\chi,\phi)), \qquad b(\phi,\xi) = -(\nabla_w \phi, \nabla_w \xi).$$

It is not hard to check that both of these bilinear forms are bounded under the given norms. In particular, the boundedness of $b(\cdot,\cdot)$ has been given in (27). It is also clear that the first inequality in (51) follows from the definition of $a(\cdot,\cdot)$ and $\|\cdot\|_{0,h}$, and the second inequality follows directly from (28). Combine the above, we have for all $\chi \in V_h$ and $\xi \in V_{0,h}$:

$$\sup_{\phi \in V_h,\, \psi \in V_{0,h}} \frac{((\chi,\phi)) - (\nabla_w \phi, \nabla_w \xi) - (\nabla_w \chi, \nabla_w \psi)}{\|\phi\|_{0,h} + |||\psi|||} \gtrsim \|\chi\|_{0,h} + |||\xi|||. \tag{52}$$

Theorem 4.2. *The weak Galerkin formulation (6) for the biharmonic problem (1) has the following error estimate:*

$$\|\varepsilon_w\|_{0,h} + |||\varepsilon_u||| \lesssim h^m \|w\|_{m+1} + h^{n+\frac{1}{2}} |\ln h| \|u\|_{W^{n+2,\infty}(\Omega)},$$

where $\frac{1}{2} < m \leq j+1$ *and* $0 \leq n \leq j$*. Moreover, for* $j \geq 1$*, we have the improved estimate*

$$\|\varepsilon_w\|_{0,h} + |||\varepsilon_u||| \lesssim h^m \|w\|_{m+1} + h^{n+\frac{1}{2}} \|u\|_{W^{n+2,\infty}(\Omega)},$$

where $\frac{1}{2} < m \leq j+1$ *and* $1 \leq n \leq j$*.*

Proof. By (50) and (52),

$$\begin{aligned}\|\varepsilon_w\|_{0,h} + |||\varepsilon_u||| &\lesssim \sup_{\phi_h \in V_h,\, \psi_h \in V_{0,h}} \frac{(\!(\varepsilon_w, \phi_h)\!) - (\nabla_w \phi_h, \nabla_w \varepsilon_u) - (\nabla_w \varepsilon_w, \nabla_w \psi_h)}{\|\phi_h\|_{0,h} + |||\psi_h|||} \\ &= \sup_{\phi_h \in V_h,\, \psi_h \in V_{0,h}} \frac{E(w,u,\phi_h)}{\|\phi_h\|_{0,h} + |||\psi_h|||}.\end{aligned}$$

Combining this with Lemma 4.1, this completes the proof of the theorem. □

Remark 4.1. Assume that the exact solution w and u are sufficiently smooth. It follows from the above theorem that the following convergence holds true:

$$\|\varepsilon_w\|_{0,h} + |||\varepsilon_u||| \lesssim \begin{cases} O(h^{\frac{1}{2}}|\ln h|) & \text{for } j = 0, \\ O(h^{j+\frac{1}{2}}) & \text{for } j \geq 1, \end{cases}$$

where j is the order of the finite element space, i.e., order of polynomials on each element.

At this stage, it is standard to use the duality argument and derive an error estimation for the L^2 norm of ε_u. However, estimating $\|\varepsilon_u\|_{0,h}$ is not an easy task, as is similar to the case of Poisson equations. For simplicity, we only consider $\|\varepsilon_{u,0}\|$, where ε_u is conveniently expressed as $\varepsilon_u = \{\varepsilon_{u,0}, \varepsilon_{u,b}\}$. Define

$$\begin{cases} \xi + \Delta\eta = 0, \\ -\Delta\xi = \varepsilon_{u,0}, \end{cases} \tag{53}$$

where $\eta = 0$ and $\frac{\partial \eta}{\partial \mathbf{n}} = 0$ on $\partial\Omega$. We assume that all internal angles of Ω are less than $126.283696\cdots^\circ$. Then, according to (4), the solution to (53) has H^4 regularity:

$$\|\xi\|_2 + \|\eta\|_4 \lesssim \|\varepsilon_{u,0}\|.$$

Furthermore, since such a domain Ω is convex, the Poisson equation with either the homogeneous Dirichlet boundary condition or the homogeneous Neumann boundary condition has H^2 regularity.

Clearly, Eq. (53) can be written into the following form:

$$\begin{cases} ((N_h\xi,\, \phi_h)) - (\nabla_w R_h\eta,\, \nabla_w\phi_h) = E(\xi,\eta,\phi_h) & \text{for all } \phi_h = \{\phi_0,\, \phi_b\} \in V_h, \\ (\nabla_w N_h\xi,\, \nabla_w\psi_h) = (\varepsilon_{u,0},\, \psi_0) & \text{for all } \psi_h = \{\psi_0,\, \psi_b\} \in V_{0,h}. \end{cases} \tag{54}$$

For simplicity of the notation, denote

$$\Lambda(N_h\xi, R_h\eta_h;\, \phi_h, \psi_h) = ((N_h\xi,\, \phi_h)) - (\nabla_w R_h\eta,\, \nabla_w\phi_h) - (\nabla_w N_h\xi,\, \nabla_w\psi_h).$$

Note that Λ is a symmetric bilinear form. By setting $\phi_h = \varepsilon_w$ and $\psi_h = \varepsilon_u$ in (54) and then subtract these two equations, one get

$$\begin{aligned} \|\varepsilon_{u,0}\|^2 &= E(\xi,\eta,\varepsilon_w) - \Lambda(N_h\xi, R_h\eta;\, \varepsilon_w, \varepsilon_u) \\ &= E(\xi,\eta,\varepsilon_w) - \Lambda(\varepsilon_w, \varepsilon_u;\, N_h\xi, R_h\eta) \\ &= E(\xi,\eta,\varepsilon_w) - E(w,u,N_h\xi). \end{aligned} \tag{55}$$

Here we have used the symmetry of $\Lambda(\cdot,\cdot)$ and Eq. (50).

The two terms, $E(\xi,\eta,\varepsilon_w)$ and $E(w,u,N_h\xi)$, in the right-hand side of Eq. (55) will be estimated one by one. We start from $E(\xi,\eta,\varepsilon_w)$. By using Lemma 4.1, it follows that

(i) When $j = 0$,

$$\begin{aligned} E(\xi,\eta,\varepsilon_w) &\lesssim \left(h\|\xi\|_2 + h^{\frac{1}{2}}|\ln h|\|\eta\|_{W^{2,\infty}(\Omega)} \right) \|\varepsilon_w\|_{0,h} \\ &\lesssim h^{1/2}|\ln h|\,(\|\xi\|_2 + \|\eta\|_4)\,\|\varepsilon_w\|_{0,h}. \end{aligned} \tag{56}$$

(ii) When $j \geq 1$, let $\delta > 0$ be an infinitely small number which ensures the Sobolev embedding from $W^{4,2}(\Omega)$ to $W^{3-\delta,\infty}(\Omega)$. Then

$$\begin{aligned} E(\xi,\eta,\varepsilon_w) &\lesssim \left(h\|\xi\|_2 + h^{\frac{3}{2}-\delta}|\ln h|\|\eta\|_{W^{3-\delta,\infty}(\Omega)} \right) \|\varepsilon_w\|_{0,h} \\ &\lesssim h\,(\|\xi\|_2 + \|\eta\|_4)\,\|\varepsilon_w\|_{0,h}. \end{aligned} \tag{57}$$

Next, we give an estimate for $E(w,u,N_h\xi)$.

Lemma 4.3. *Assume all internal angles of Ω are less than $126.283696\cdots^\circ$, which means the biharmonic problem with clamped boundary condition in Ω has H^4 regularity. Then*

(i) For $j=0$,

$$E(w,u,N_h\xi)\lesssim\left(h^{m+\frac{1}{2}}\|w\|_{m+1}+h^2\|(I-Q_0)f\|+h^{n+1}\|u\|_{n+1}\right)\|\xi\|_2,$$

where $\frac{1}{2}<m\le 1$ *and* $1/2<n\le 1$.

(ii) For $j\ge 1$,

$$E(w,u,N_h\xi)\lesssim\left(h^{m+1}\|w\|_{m+1}+h^2\|(I-Q_0)f\|+h^{n+1}\|u\|_{n+1}\right)\|\xi\|_2,$$

where $\frac{1}{2}<m\le j+1$ *and* $1/2<n\le j+1$.

Proof. By definition,

$$E(w,u,N_h\xi)=(\!(N_hw-w,N_h\xi)\!)+(\mathbf{\Pi}_h\nabla u-\nabla_wR_hu,\nabla_wN_h\xi). \tag{58}$$

First, by the definition of $(\!(\cdot,\cdot)\!)$, the Schwarz inequality, Remarks 3.3 and 3.4, we have

$$\begin{aligned}
&(\!(N_hw-w,N_h\xi)\!)\\
&=(N_0w-Q_0w,N_0\xi)+\sum_{K\in\mathcal{T}_h}h(N_0w-N_bw,N_0\xi-N_b\xi)_{\partial K}\\
&\lesssim\|N_0w-Q_0w\|\|N_0\xi\|+\|N_hw-w\|_{0,h}\|N_h\xi-\xi\|_{0,h}\\
&\lesssim\begin{cases}(h^{m+\frac{1}{2}}\|w\|_{m+1}+h^2\|(I-Q_0)\Delta w\|)\|\xi\|_2 & \text{for } j=0,\ \frac{1}{2}<m\le 1\\ (h^{m+1}\|w\|_{m+1}+h^2\|(I-Q_0)\Delta w\|)\|\xi\|_2 & \text{for } j\ge 1,\ \frac{1}{2}<m\le j+1\end{cases}.
\end{aligned} \tag{59}$$

Next, by using inequalities (11), (33), (13), (10), (37), and (38) one after one, we get

$$\begin{aligned}
&(\mathbf{\Pi}_h\nabla u-\nabla_wR_hu,\nabla_wN_h\xi)\\
&=((\mathbf{\Pi}_h-\mathbf{P}_h)\nabla u,\nabla_wN_h\xi)+(\nabla_w(Q_hu-R_hu),\nabla_wN_h\xi)\\
&=((\mathbf{\Pi}_h-\mathbf{P}_h)\nabla u,\nabla_wN_h\xi)+(\nabla_w(Q_hu-R_hu),\mathbf{\Pi}_h\nabla\xi)\\
&=((\mathbf{\Pi}_h-\mathbf{P}_h)\nabla u,\nabla_w(N_h\xi-Q_h\xi))+((\mathbf{\Pi}_h-\mathbf{P}_h)\nabla u,\mathbf{P}_h\nabla\xi)-(Q_0u-R_0u,\Delta\xi)\\
&\lesssim h^{n+1}\|u\|_{n+1}\|\xi\|_2+((\mathbf{\Pi}_h-I)\nabla u,\mathbf{P}_h\nabla\xi)+h^2\|(I-Q_0)\Delta u\|\|\xi\|_2,
\end{aligned}$$

for $\frac{1}{2}<n\le j+1$. The estimation for $((\mathbf{\Pi}_h-I)\nabla u,\mathbf{P}_h\nabla\xi)$ follows the same technique used in inequality (40). By the definition of $\mathbf{\Pi}_h$ and since $\frac{\partial u}{\partial\mathbf{n}}=0$ on $\partial\Omega$, we know that $(\mathbf{\Pi}_h-I)\nabla u\cdot\mathbf{n}$ also vanishes on $\partial\Omega$. Therefore, using the same argument as in (40), one has

$$((\mathbf{\Pi}_h-I)\nabla u,\mathbf{P}_h\nabla\xi)\lesssim h^{n+1}\|u\|_{n+1}\|\xi\|_2+h^2\|(I-Q_0)\Delta u\|\|\xi\|_2$$

for $\frac{1}{2}<n\le j+1$. Combining the above gives

$$(\mathbf{\Pi}_h \nabla u - \nabla_w R_h u, \nabla_w N_h \xi) \lesssim \left(h^{n+1}\|u\|_{n+1} + h^2\|(I-Q_0)\Delta u\|\right)\|\xi\|_2. \tag{60}$$

for $\frac{1}{2} < n \leq j+1$.

Notice that

$$\begin{aligned} h^2\|(I-Q_0)\Delta u\| &= h^2\|(I-Q_0)w\| \lesssim h^{m+2}\|w\|_m \qquad \text{for } 0 \leq m \leq j+1, \\ h^2\|(I-Q_0)\Delta w\| &= h^2\|(I-Q_0)f\|. \end{aligned} \tag{61}$$

The lemma follows immediately from (58)–(61). □

Finally, combining Theorem 4.2, inequalities (55), (56)–(57), and Lemma 4.3, we get the following L^2 error estimation:

Theorem 4.4. *Assume all internal angles of* Ω *are less than* $126.283696\cdots^\circ$, *which means the biharmonic problem with clamped boundary condition in* Ω *has* H^4 *regularity. Then*

(i) For $j = 0$,

$$\begin{aligned} \|\varepsilon_{u,0}\| \lesssim h^{m+\frac{1}{2}}|\ln h|\|w\|_{m+1} + h|\ln h|^2\|u\|_{W^{2,\infty}(\Omega)} \\ + h^2\|(I-Q_0)f\| + h^{n+1}\|u\|_{n+1}, \end{aligned}$$

where $\frac{1}{2} < m \leq 1$ *and* $\frac{1}{2} < n \leq 1$.

(ii) For $j \geq 1$,

$$\|\varepsilon_{u,0}\| \lesssim h^{m+1}\|w\|_{m+1} + h^{l+\frac{3}{2}}\|u\|_{W^{l+2,\infty}(\Omega)} + h^2\|(I-Q_0)f\| + h^{n+1}\|u\|_{n+1},$$

where $\frac{1}{2} < m \leq j+1$, $\frac{1}{2} < n \leq j+1$ *and* $1 \leq l \leq j$.

Remark 4.2. If u, w, and f are sufficiently smooth, then we get

$$\|\varepsilon_{u,0}\| \lesssim \begin{cases} O(h|\ln h|^2) & \text{for } j = 0, \\ O(h^{j+\frac{3}{2}}) & \text{for } j \geq 1. \end{cases}$$

5 Numerical Results

In this section, we would like to report some numerical results for the weak Galerkin finite element method proposed and analyzed in previous sections. Before doing that, let us briefly review some existing results for H^1-H^1 conforming, equal-order finite element discretization of the Ciarlet–Raviart mixed formulation. As discussed in [5, 32], theoretical error estimates for such schemes are indeed suboptimal due to an effect of $\inf_{\chi_h}\|u-\chi_h\|_2$, where χ_h is taken from the employed H^1 conforming finite element space. For example, when H^1-H^1 conforming quadratic elements are used to approximate both u and w, the error satisfies $\|u-u_h\|_2+\|w-w_h\| \lesssim \inf_{\chi_h}\|u-\chi_h\|_2+\inf_{\chi_h}\|w-\chi_h\| \lesssim O(h)$, while intuitively, one may expect $\|w-w_h\|$ to have an $O(h^2)$ convergence. By using the L^∞ argument, Scholz [32] was able to improve the convergence rate of L^2 norm for w by $h^{\frac{1}{2}}$, and it is known that this theoretical result is indeed sharp. For the weak Galerkin approximation, from the discussing in the previous sections, clearly we are facing the same issue.

However, numerous numerical experiments have illustrated that H^1-H^1 conforming, equal-order Ciarlet–Raviart mixed finite element approximation often demonstrates convergence rates better than the theoretical prediction. Indeed, this has been partly explained theoretically in [33], in which the author proved that optimal order of convergence rates can be recovered in certain fixed subdomains of Ω, when equal-order H^1 conforming elements are used. We point out that similar phenomena have been observed in the numerical experiments using weak Galerkin discretization. This means that numerical results are often better than theoretical predictions.

Another issue in the implementation of the weak Galerkin finite element method is the treatment of nonhomogeneous boundary data:

$$u = g_1 \qquad \text{on } \partial\Omega,$$

$$\frac{\partial u}{\partial \mathbf{n}} = g_2 \qquad \text{on } \partial\Omega.$$

Clearly, both boundary conditions are imposed on u, and $u = g_1$ is the essential boundary condition, while $\frac{\partial u}{\partial \mathbf{n}} = g_2$ is the natural boundary condition. To impose the natural boundary condition, we shall modify the first equation of (6) into

$$(\!(w_h, \phi_h)\!) - (\nabla_w u_h, \nabla_w \phi_h) = -\langle g_2, \phi_b \rangle_{\partial\Omega}.$$

The essential boundary condition should be enforced by taking the L^2 projection of the corresponding boundary data.

Consider three test problems defined on $\Omega = [0,1]\times[0,1]$ with exact solutions

$$u_1 = x^2(1-x)^2y^2(1-y)^2,$$

$$u_2 = \sin(2\pi x)\sin(2\pi y) \qquad \text{and} \qquad u_3 = \sin(2\pi x+\frac{\pi}{2})\sin(2\pi y+\frac{\pi}{2}),$$

respectively. The reason for choosing these three exact solutions is that they have the following type of boundary conditions:

Table 1 Numerical results for the test problem with exact solution u_1 and lowest order of WG elements

h	$\|\|\nabla_w e_u\|\|$	$\|\|e_{u,0}\|\|$	$\|\|e_{u,b}\|\|$	$\|\|\nabla_w e_w\|\|$	$\|\|e_{w,0}\|\|$	$\|\|e_{w,b}\|\|$
0.1	1.33e−03	2.40e−04	4.59e−04	5.66e−02	2.96e−03	6.91e−03
0.05	4.69e−04	6.18e−05	1.17e−04	2.80e−02	9.14e−04	1.99e−03
0.025	2.00e−04	1.55e−05	2.97e−05	1.60e−02	2.64e−04	5.70e−04
0.0125	9.56e−05	3.90e−06	7.44e−06	1.21e−02	8.33e−05	1.89e−04
0.00625	4.72e−05	9.77e−07	1.86e−06	1.13e−02	3.26e−05	7.91e−05
Asym. order $O(h^k)$, $k=$	1.1930	1.9876	1.9877	0.5864	1.6461	1.6298

$$u_1|_{\partial\Omega} = 0 \qquad \left.\frac{\partial u_1}{\partial \mathbf{n}}\right|_{\partial\Omega} = 0,$$

$$u_2|_{\partial\Omega} = 0 \qquad \left.\frac{\partial u_2}{\partial \mathbf{n}}\right|_{\partial\Omega} \neq 0,$$

$$u_3|_{\partial\Omega} \neq 0 \qquad \left.\frac{\partial u_3}{\partial \mathbf{n}}\right|_{\partial\Omega} = 0.$$

This allows us to test the effect of different boundary data on convergence rates. Although the theoretical error estimates are given for $\varepsilon_u = R_h u - u_h$ and $\varepsilon_w = N_h w - w_h$, by using the triangle inequality and the approximation properties of R_h, N_h and Q_h, it is clear that they have at least the same order as $e_u = Q_h u - u_h$ and $e_w = Q_h w - w_h$, provided that the exact solution is smooth enough. Thus for convenience, we only compute different norms for e_u and e_w, instead of for ε_u and ε_w.

The tests are performed using an unstructured triangular initial mesh, with characteristic mesh size 0.1. The initial mesh is then refined by dividing every triangle into four sub-triangles, to generate a sequence of nested meshes with various mesh size h. All discretization schemes are formulated by using the lowest order weak Galerkin element, with $j = 0$. For simplicity of notation, for any $v \in V_h$, denote

$$\|v_b\| = \left(\sum_{K \in \mathcal{T}_h} h \|v_b\|_{\partial K}^2 \right)^{1/2}.$$

The results for test problems with exact solutions u_1, u_2, and u_3 are reported in Tables 1, 2, and 3, respectively. The results indicate that u always achieves an optimal order of convergence, while the convergence for w varies with different boundary conditions. It should be pointed out that both of them have outperformed the convergence as predicted by theory.

Our final example is a case where the exact solution has a low regularity in the domain $\Omega = [0,1] \times [0,1]$. More precisely, the exact solution is given by

Table 2 Numerical results for the test problem with exact solution u_2 and lowest order of WG elements

h	$\|\nabla_w e_u\|$	$\|e_{u,0}\|$	$\|e_{u,b}\|$	$\|\nabla_w e_w\|$	$\|e_{w,0}\|$	$\|e_{w,b}\|$
0.1	9.58e−01	8.66e−02	1.65e−01	4.39e+01	6.09e−01	2.01e+00
0.05	3.34e−01	2.18e−02	4.14e−02	2.32e+01	2.78e−01	7.19e−01
0.025	1.43e−01	5.47e−03	1.03e−02	1.37e+01	1.15e−01	2.81e−01
0.0125	6.81e−02	1.37e−03	2.59e−03	1.02e+01	5.12e−02	1.26e−01
0.00625	3.36e−02	3.42e−04	6.49e−04	9.33e+00	2.45e−02	6.12e−02
Asym. order $O(h^k)$, $k=$	1.1958	1.9958	1.9975	0.5649	1.1709	1.2587

Table 3 Numerical results for the test problem with exact solution u_3 and lowest order of WG elements

h	$\|\nabla_w e_u\|$	$\|e_{u,0}\|$	$\|e_{u,b}\|$	$\|\nabla_w e_w\|$	$\|e_{w,0}\|$	$\|e_{w,b}\|$
0.1	8.23e−01	1.18e−01	2.27e−01	5.61e+01	4.25e+00	9.42e+00
0.05	3.07e−01	3.18e−02	6.09e−02	2.43e+01	1.24e+00	2.58e+00
0.025	1.35e−01	8.13e−03	1.55e−02	1.13e+01	3.28e−01	6.61e−01
0.0125	6.49e−02	2.04e−03	3.90e−03	5.58e+00	8.42e−02	1.67e−01
0.00625	3.21e−02	5.11e−04	9.78e−04	2.77e+00	2.14e−02	4.21e−02
Asym. order $O(h^k)$, $k=$	1.1599	1.9679	1.9682	1.0801	1.9157	1.9558

Table 4 Numerical results for the test problem with exact solution u_4 and lowest order of WG elements

h	$\|\nabla_w e_u\|$	$\|e_{u,0}\|$	$\|e_{u,b}\|$	$\|\nabla_w e_w\|$	$\|e_{w,0}\|$	$\|e_{w,b}\|$
0.1	3.73e−02	9.44e−04	2.15e−03	2.88e+01	4.05e−01	1.78e+00
0.05	1.87e−02	2.55e−04	5.73e−04	4.08e+01	2.86e−01	1.26e+00
0.025	9.37e−03	6.60e−05	1.46e−04	5.77e+01	2.02e−01	8.91e−01
0.0125	4.68e−03	1.67e−05	3.69e−05	8.16e+01	1.42e−01	6.30e−01
0.00625	2.34e−03	4.19e−06	9.24e−06	1.15e+02	1.01e−01	4.45e−01
Asym. order $O(h^k)$, $k=$	0.9984	1.9567	1.9690	−0.4998	0.5008	0.5000

$$u_4 = r^{3/2}\left(\sin\frac{3\theta}{2} - 3\sin\frac{\theta}{2}\right),$$

where (r,θ) are the polar coordinates. It is easy to check that $u \in H^{2.5}$. The errors for weak Galerkin finite element approximations are reported in Table 4. Here, u still achieves an optimal order of convergence, while the convergence rates for w is restricted by the fact that $w \in H^{0.5}$. All the results are in consistency with the theory established in this article.

References

1. Adams, R., Fournier, J.: Sobolev Spaces. Academic press, New York (2003)
2. Adini, A., Glough, R.W.: Analysis of plate bending by the finite element method. NSF report G, 7337 (1961)
3. Arnold, D.N., Scott, L.R., Vogelius, M.: Regular inversion of the divergence operator with Dirichlet boundary conditions on a polygon. Ann. Scuola Norm. Sup. Pisa Cl. Sci. **15**(4), 169–192 (1988)
4. Babuška, I.: The finite element method with Lagrange multipliers. Numer. Math. **20**, 179–192 (1973)
5. Babuška, I., Osborn, J., Pitk aranta, J.: Analysis of mixed methods using mesh dependent norms. Math. Comp. **35**, 1039–1062 (1980)
6. Behrens, E.M., Guzmán, J.: A mixed method for the biharmonic problem based on a system of first-order equations. SIAM J. Numer. Anal. **49**, 789–817 (2011)
7. Blum, H., Rannacher, R., Leis, R.: On the boundary value problem of the biharmonic operator on domains with angular corners. Math. Methods Appl. Sci. **2**, 556–581 (1980)
8. Brenner, S.C., Sung, L.-Y.: C0 interior penalty methods for fourth order elliptic boundary value problems on polygonal domains. J. Sci. Comput. **22/23**, 83–118 (2005)
9. Brezzi, F.: On the existence, uniqueness, and approximation of saddle point problems arising from Lagrange multipliers. RAIRO **8**, 129–151 (1974)
10. Brezzi, F., Fortin, M.: Mixed and Hybrid Finite Elements. Springer, New York (1991)
11. Ciarlet, P., Raviart, P.: A mixed finite element for the biharmonic equation. In: de Boor, C. (ed.) Symposium on Mathematical Aspects of Finite Elements in Partial Differential Equations, pp. 125–143. Academic Press, New York (1974)
12. Ciarlet, P.G.: The Finite Element Method for Elliptic Problems. North-Holland, New York (1978)
13. Cockburn, B., Dong, B., Guzmán, J.: A hybridizable and superconvergent discontinuous Galerkin method for biharmonic problems. J. Sci. Comput. **40**, 141–187 (2009)
14. Engel, G., Garikipati, K., Hughes, T.J.R., Larson, M.G., Mazzei, L., Taylor, R.L.: Continuous/discontinuous finite element approximations of fourth-order elliptic problems in structural and continuum mechanics with applications to thin beams and plates, and strain gradient elasticity. Comput. Methods Appl. Mech. Eng. **191**, 3669–3750 (2002)
15. Falk, R.S., Osborn, J.E.: Error estimates for mixed methods. RAIRO. Numer. Anal. **14**, 249–277 (1980)
16. Fraeijs de Veubeke, B.X.: Displacement and equilibrium models in the finite element method. In: Zienkiewicz, O.C., Holister, G. (eds.) "Stress Analysis". Wiley, New York (1965)
17. Fraeijs de Veubeke, B.X.: Stress function approach. In: International Congress on the Finite Element Methods in Structural Mechanics, Bournemouth, 1975
18. Gastaldi, L., Nochetto, R.H.: Sharp maximum norm error estimates for general mixed finite element approximations to second order elliptic equations. **M2AN 23**, 103–128 (1989)
19. Glowinski, R., Pironneau, O.: Numerical methods for the first biharmonic equation and for the two-dimensional Stokes problem. *SIAM Rev.* **21**, 167–212 (1979)
20. Gudi, T., Nataraj, N., Pani, A.K.: Mixed discontinuous Galerkin finite element method for the biharmonic equation. J. Sci. Comput. **37**, 139–161 (2008)
21. Herrmann, L.: A bending analysis for plates. In: Matrix Methods in Structural Mechanics, On Proceedings of the Conference held at Wright-Patterson Air Force Base, Ohio, AFFDL technical report No. AFFDL-TR-66-88 pp. 577–604 (1965)
22. Herrmann, L.: Finite element bending analysis for plates. J. Eng. Mech. Div. ASCE EM5 **93**, 49–83 (1967)
23. Johnson, C.: On the convergence of a mixed finite element method for plate bending problems. Numer. Math. **21**, 43–62 (1973)
24. Johnson, C., Pitkäranta, J.: Analysis of some mixed finite element methods related to reduced integration. Math. Comp. **38**, 375–400 (1982)

25. Lascaux, P., Lesaint, P.: Some nonconforming finite elements for the plate bending problem. RAIRO Anal. Numer. **R-1**, 9–53 (1985)
26. Malkus, D.S., Hughes, T.J.R.: Mixed finite element methods-reduced and selective integration techniques: A unification of concepts. Comput. Methods Appl. Mech. Eng. **15**, 63–81 (1978)
27. Miyoshi, T.: A finite element method for the solution of fourth order partial differential equations. Kunamoto J. Sci. (Math.) **9**, 87–116 (1973)
28. Morley, L.S.D.: The triangular equilibrium element in the solution of plate bending problems. Aero. Q. **19**, 149–169 (1968)
29. Mu, L., Wang, J., Ye, X.: Weak Galerkin finite element methods on polytopal meshes. arXiv:1204.3655v2
30. Mu, L., Wang, J., Wang, Y., Ye, X.: A computational study of the weak Galerkin method for second order elliptic equations. Numer. Algorithm (2012). arXiv:1111.0618v1, DOI:10.1007/s11075-012-9651-1
31. Raviart, P., Thomas, J.: A mixed finite element method for second order elliptic problems. In: Galligani, I., Magenes, E. (eds.) Mathematical Aspects of the Finite Element Method. Lectures Notes in Math. vol. 606. Springer, New York (1977)
32. Scholz, R.: A mixed method for 4th order problems using linear finite elements. R.A.I.R.O. Numer. Anal. **12**, 85–90 (1978)
33. Scholz, R.: Interior error estimates for a mixed finite element method. Numer. Funct. Anal. Optim. **1**, 415–429 (1979)
34. Wang, J.: Asymptotic expansions and maximum norm error estimates for mixed finite element methods for second order elliptic problems. Numer. Math. **55**, 401–430 (1989)
35. Wang, J., Ye, X.: A weak Galerkin finite element method for second-order elliptic problems, arXiv:1104.2897v1 [math.NA]. J. Comput. Appl. Math. **241**, 103-115 (2013)
36. Wang, J., Ye, X.: A weak Galerkin mixed finite element method for second-order elliptic problems, arXiv:1202.3655v1, submitted to Math Comp.

Domain Decomposition Scheme for First-Order Evolution Equations with Nonselfadjoint Operators

Petr Vabishchevich and Petr Zakharov

Abstract Domain decomposition iterative methods and implicit schemes are usually used for solving evolution equations. An alternative approach is based on constructing non-iterative method based on special schemes of splitting into subdomains. Such regional-additive schemes are based on the general theory of additive operator-difference schemes. Domain decomposition analogues of the classical schemes of alternating direction method, locally one-dimensional schemes, factored schemes, and regularized vector-additive schemes are used here. The main results in the literature are obtained for time-dependent problems with selfadjoint second-order elliptic operators. This paper discusses the Cauchy problem for first-order evolution equations with nonnegative nonselfadjoint operators in a finite-dimensional Hilbert space. Based on the partition of unity, we have constructed nonnegativity preserving decomposition operators for the respective operator term in the equation. We construct unconditionally stable additive domain decomposition schemes based on the principle of regularization of operator-difference schemes and vector-additive schemes.

Keywords First-order evolution equations • Parabolic partial differencial equation • Domain decomposition method • Difference scheme.

Mathematics Subject Classification (2010): 65J08, 65M12

P. Vabishchevich (✉)
Nuclear Safety Institute of RAS, 52, B. Tulskaya, Moscow, 115191, Russia
e-mail: vabishchevich@gmail.com

P. Zakharov
North-Eastern Federal University, 58, Belinskogo, Yakutsk, 677000, Russia
e-mail: zapetch@gmail.com

O.P. Iliev et al. (eds.), *Numerical Solution of Partial Differential Equations: Theory, Algorithms, and Their Applications*, Springer Proceedings in Mathematics & Statistics 45, DOI 10.1007/978-1-4614-7172-1_14,

1 Introduction

Domain decomposition methods are often used for the numerical solution of boundary value problems for partial differential equations on parallel computers. The theory of the domain decomposition (DD) methods is mostly developed for stationary problems [11, 12, 24, 25]. Numerous sequential and parallel algorithms for overlapping and nonoverlapping DD methods are developed and analysed in conjunction with such problems.

Domain decomposition methods for unsteady problems are based on two approaches [14]. In the first approach, standard implicit approximation in time is used. After that, domain decomposition methods developed for steady-state problems can be applied for solving the discrete problem on the new time level. In the case of optimal DD iterative methods, the number of iterations does not depend on space and time discretization steps [3, 4]. In the second approach, non-iterative domain decomposition algorithms are constructed for unsteady problems. In some cases, this can be interpreted as performing at each time step only one iteration of the Schwarz alternating method for the approximate solution of boundary value problems for second-order parabolic equation [6, 7]. We also construct a special scheme of splitting into subdomains (regional-additive schemes [26, 27]).

The construction of regional-additive schemes and the investigation of their convergence are based on the general theory of the splitting schemes [10, 13, 34]. Most interesting for the practice is the situation when the operator is split into a sum of three or more noncommutative nonselfadjoint operators. In the case of such a multicomponent splitting, stable additive splitting schemes are constructed based on the concept of additive approximation. Furthermore, additively averaged summarized approximation schemes are interesting, when we focus on parallel computers. In the class of splitting schemes with full approximation [19], we point to the vector-additive schemes, when the original equation is transformed into a system of similar equations [1, 2, 31]. The most suitable approach for constructing additive regularized operator-difference schemes for multicomponent splitting [18, 23] is the one in which the stability is achieved due to perturbations of the operators of the difference scheme.

A domain decomposition scheme is defined by a decomposition of the computational domain and by defining the splitting of the operator. To construct the decomposition operators when solving BVP for PDEs, it is convenient to use a partition of unity for the computational domain [5, 8, 16, 26, 28, 29, 33]. In the overlapping DD methods, a function is associated with each subdomain, and this function takes value between zero and one. Domain decomposition methods for unsteady convection-diffusion problems are studied in the works [17, 20, 30]. In the extreme case, the width of the overlap of the subdomains is equal to the space discretization step. In this case the regionally additive schemes can be interpreted as nonoverlapping domain decomposition schemes, where the exchange is achieved by setting proper boundary conditions for each of the subdomain. Research results on domain decomposition method for unsteady boundary value problems are summarized in the books [14, 19]. From the more recent studies, we mention [32], where DD schemes which are more suitable for computer implementation are presented.

In this paper, we construct a domain decomposition schemes for first-order evolution equations with general nonnegative operator in a finite-dimensional Hilbert space. Decomposition operators are constructed separately for the selfadjoint and for the skew-symmetric part of the operator. The splitting is based on partition of unity in the appropriate spaces. We propose two classes of unconditionally stable regionally additive regularized schemes, and we consider vector-additive operator-difference domain decomposition scheme.

2 The Cauchy Problem for First-Order Evolution Equations

Let H be finite-dimensional real Hilbert space of grid functions, in which the scalar product and the norm are $(\cdot,\cdot)$ $\|\cdot\|$, respectively. Consider a time independent and nonnegative in H grid operator A:

$$A \geq 0, \quad \frac{d}{dt}A = A\frac{d}{dt}. \tag{1}$$

Let us denote by E the identity operator in H. We seek a solution to the Cauchy problem

$$\frac{du}{dt} + Au = f(t), \quad 0 < t \leq T, \tag{2}$$

$$u(0) = u^0. \tag{3}$$

The problem (1)–(3) is obtained after a finite-difference approximation in space of initial boundary value problems (IBVP) for second-order partial differential equations (PDEs). Similar systems of ordinary differential equations arise when finite element method (FEM) or finite volume method (FVM) are used for space discretization.

Let us give a standard a priori estimate for the problem (1)–(3). We take a scalar product in H of the Eq. (2) and u. In view of (1) we arrive at

$$\frac{1}{2}\frac{d}{dt}\|u\|^2 \leq (f,u). \tag{4}$$

Taking into account

$$(f,u) \leq \|f\|\|u\|,$$

from (4) we obtain

$$\frac{d}{dt}\|u\| \leq \|f\|.$$

Using the Gronwall lemma, we obtain the desired estimate

$$\|u\| \leq \|u^0\| + \int_0^t \|f(\theta)\| d\theta, \tag{5}$$

which expresses the stability of the solution to the initial data and right-hand side.

The scope of this work is to present discretizations in time for the Eq. (2). Our discretizations belong to the class of the two-layer schemes. Let τ be the time step and let $y^n = y(t^n)$, $t^n = n\tau$, $n = 0, 1, \ldots, N$, $N\tau = T$. Equation (2) is approximated by a two-level weighted scheme as follows:

$$\frac{y^{n+1} - y^n}{\tau} + A(\sigma y^{n+1} + (1-\sigma)y^n) = \varphi^n, \quad n = 0, 1, \ldots, N-1, \tag{6}$$

where, for example, $\varphi^n = f(\sigma t^{n+1} + (1-\sigma)t^n)$. It is supplemented by the initial condition

$$y^0 = u^0. \tag{7}$$

Difference scheme (6), (7) has approximation error $\mathcal{O}(\tau^2 + (\sigma - 0.5)\tau)$. An analogy of (5) for the discretized in time function reads as follows:

$$\|y^{n+1}\| \leq \|y^n\| + \tau\|\varphi^n\|, \quad n = 0, 1, \ldots, N-1. \tag{8}$$

We prove the following theorem.

Theorem 1. *The difference scheme (1), (6), (7) is unconditionally stable for $\sigma \geq 0.5$, and the estimate (8) holds for the solution of the above difference equation.*

Proof. Let us rewrite (6) in the form

$$y^{n+1} = Sy^n + \tau(E + \sigma\tau A)^{-1}\varphi^n, \tag{9}$$

where

$$S = (E + \sigma\tau A)^{-1}(E - (1-\sigma)\tau A) \tag{10}$$

is the operator of the transition to a new time level. From (9) we have

$$\|y^{n+1}\| = \|S\| \, \|y^n\| + \tau\|(E + \sigma\tau A)^{-1}\varphi^n\|. \tag{11}$$

For the last term on the right side of (11), in the class of operators (1), under natural conditions $\sigma \geq 0$, we have

$$\|(E + \sigma\tau A)^{-1}\varphi^n\| \leq \|\varphi^n\|.$$

Let us show that if $\sigma \geq 0.5$, for nonnegative operator A, it holds

$$\|S\| \leq 1. \tag{12}$$

In real Hilbert space H, the inequality (12) is equivalent to [9] the fulfilment of the operator inequality

$$SS^* \le E.$$

In view of (10), this inequality takes the form

$$(E+\sigma\tau A)^{-1}(E-(1-\sigma)\tau A)(E-(1-\sigma)\tau A^*)(E+\sigma\tau A^*)^{-1} \le E.$$

Multiplying this inequality on the left by $(E+\sigma\tau A)^{-1}$ and on the right by $(E+\sigma\tau A^*)^{-1}$, we obtain

$$(E-(1-\sigma)\tau A)(E-(1-\sigma)\tau A^*) \le (E+\sigma\tau A)(E+\sigma\tau A^*).$$

It follows from here that

$$\tau(A+A^*)+(\sigma^2-(1-\sigma)^2)\tau^2 AA^* \ge 0.$$

This inequality holds for nonnegative operators A with$\sigma \ge 0.5$. In view of (12), from (11), we have obtained the required estimate (8). □

3 Decomposition Operators

To better understand the formal structure of the operators of the domain decomposition, we give a typical example. We consider a model nonstationary convection-diffusion problem with time-independent (but space-dependent) diffusion coefficient and velocity. The convective term below is written in the so-called (see, e.g., [21]) symmetric form. In a bounded domain Ω, the unknown function $u(\boldsymbol{x},t)$ satisfies the following equation:

$$\begin{aligned} &\frac{\partial u}{\partial t} + \frac{1}{2}\sum_{\alpha=1}^{m}\left(v_\alpha(\boldsymbol{x})\frac{\partial u}{\partial x_\alpha} + \frac{\partial}{\partial x_\alpha}(v_\alpha(\boldsymbol{x})u)\right) \\ &\quad - \sum_{\alpha=1}^{m}\frac{\partial}{\partial x_\alpha}\left(k(\boldsymbol{x})\frac{\partial u}{\partial x_\alpha}\right) = f(\boldsymbol{x},t), \quad \boldsymbol{x}\in\Omega, \quad 0<t<T, \end{aligned} \tag{13}$$

in which $k(\boldsymbol{x}) \ge \kappa > 0$, $\boldsymbol{x}\in\Omega$. Equation (13) is supplemented with homogeneous Dirichlet boundary conditions

$$u(\boldsymbol{x},t) = 0, \quad \boldsymbol{x}\in\partial\Omega, \quad 0<t<T. \tag{14}$$

In addition, we define the initial condition

$$u(\boldsymbol{x},0) = u^0(\boldsymbol{x}), \quad \boldsymbol{x}\in\Omega. \tag{15}$$

We will consider the set of functions $u(\boldsymbol{x},t)$, satisfying the boundary conditions (14). Let us write the above unsteady convection-diffusion problem in the form of differential-operator equation

$$\frac{du}{dt} + \mathcal{A}u = f(t), \quad 0 < t < T. \tag{16}$$

We consider the Cauchy problem for the evolution equation (16):

$$u(0) = u^0. \tag{17}$$

Let us explicitly specify the diffusive and convective operators and rewrite (16) in the following form:

$$\mathcal{A} = \mathcal{C} + \mathcal{D}. \tag{18}$$

The diffusion operator stands for

$$\mathcal{D}u = -\sum_{\alpha=1}^{m} \frac{\partial}{\partial x_\alpha}\left(k(\boldsymbol{x})\frac{\partial u}{\partial x_\alpha}\right).$$

On the set of functions (14) in $\mathcal{H} = \mathcal{L}_2(\Omega)$, the diffusion operator $\mathcal{D}$ is selfadjoint and positive definite:

$$\mathcal{D} = \mathcal{D}^* \geq \kappa\delta\mathcal{E}, \quad \delta = \delta(\Omega) > 0, \tag{19}$$

where $\mathcal{E}$ is the identity operator in $\mathcal{H}$.

The convective transport operator $\mathcal{C}$ is defined by the expression

$$\mathcal{C}u = \frac{1}{2}\sum_{\alpha=1}^{m}\left(v_\alpha(\boldsymbol{x})\frac{\partial u}{\partial x_\alpha} + \frac{\partial}{\partial x_\alpha}(v_\alpha(\boldsymbol{x})u)\right).$$

For any $v_\alpha(\boldsymbol{x})$, the operator $\mathcal{C}$ is skew-symmetric in $\mathcal{H}$:

$$\mathcal{C} = -\mathcal{C}^*. \tag{20}$$

Taking into account the representation (18), from (19), (20), it follows that $\mathcal{A}>0$ $\mathcal{H}$.

A domain decomposition scheme for this problem will be associated with the partition of unity of the computational domain Ω. Let the domain Ω consists of p (possibly overlapping) separate subdomains

$$\Omega = \Omega_1 \cup \Omega_2 \cup \ldots \cup \Omega_p.$$

With each separate subdomain Ω_α, $\alpha = 1,2,\ldots,p$, we associate function $\eta_\alpha(\boldsymbol{x})$, $\alpha = 1,2,\ldots,p$, such that

$$\eta_\alpha(\boldsymbol{x}) = \begin{cases} > 0, \ \boldsymbol{x} \in \Omega_\alpha, \\ 0, \ \ \boldsymbol{x} \notin \Omega_\alpha, \end{cases} \quad \alpha = 1,2,\dots,p, \tag{21}$$

where

$$\sum_{\alpha=1}^{p} \eta_\alpha(\boldsymbol{x}) = 1, \quad \boldsymbol{x} \in \Omega. \tag{22}$$

In view of (21), (22) from (18), we obtain the representation

$$\mathcal{A} = \sum_{\alpha=1}^{p} \mathcal{A}_\alpha, \quad \mathcal{A}_\alpha = \mathcal{C}_\alpha + \mathcal{D}_\alpha, \quad \alpha = 1,2,\dots,p, \tag{23}$$

in which

$$\mathcal{D}_\alpha u = -\sum_{\alpha=1}^{m} \frac{\partial}{\partial x_\alpha}\left(k(\boldsymbol{x})\eta_\alpha(\boldsymbol{x})\frac{\partial u}{\partial x_\alpha}\right),$$

$$\mathcal{C}_\alpha u = \frac{1}{2}\sum_{\alpha=1}^{m}\left(v_\alpha(\boldsymbol{x})\eta_\alpha(\boldsymbol{x})\frac{\partial u}{\partial x_\alpha} + \frac{\partial}{\partial x_\alpha}(v_\alpha(\boldsymbol{x})\eta_\alpha(\boldsymbol{x})u)\right).$$

Similarly to (19), (20), it holds for the subdomain operators:

$$\mathcal{D}_\alpha = \mathcal{D}_\alpha^* \geq 0, \quad \mathcal{C}_\alpha = -\mathcal{C}_\alpha^*, \quad \alpha = 1,2,\dots,p. \tag{24}$$

Due to (24), the operators in the splitting (23) satisfy

$$\mathcal{A}_\alpha \geq 0, \quad \alpha = 1,2,\dots,p, \tag{25}$$

and the selfadjoint part of the operator $\mathcal{A}$ splits into sum of nonnegative selfadjoint operators, and the skew-symmetric operator splits into sum of skew-symmetric operators.

The diffusive transport operator $\mathcal{D}$ is conveniently represented as

$$\mathcal{D} = \mathcal{G}^*\mathcal{G}, \quad \mathcal{G} = k^{1/2}\,\mathrm{grad}, \quad \mathcal{G}^* = -\mathrm{div}\,k^{1/2}, \tag{26}$$

with $\mathcal{G}: \mathcal{H} \to \tilde{\mathcal{H}}$, where $\tilde{\mathcal{H}} = (\mathcal{L}_2(\Omega))^p$ is the corresponding Hilbert space of vector functions. Using these notations, operators $\mathcal{D}_\alpha,\ \alpha = 1,2,\dots,p$ can be written as

$$\mathcal{D}_\alpha = \mathcal{G}^*\eta_\alpha\mathcal{G}, \quad \alpha = 1,2,\dots,p. \tag{27}$$

Similarly, each of $\mathcal{C}_\alpha,\ \alpha = 1,2,\dots,p$ has the representation

$$\mathcal{C}_\alpha = \frac{1}{2}(\eta_\alpha\mathcal{C} + \mathcal{C}\eta_\alpha), \quad \alpha = 1,2,\dots,p. \tag{28}$$

The advantage of the notations (27), (28) is that diffusion and convection operators have clearly visible structure in the subdomains defined by the splitting (21), (22), and it is easy to verify if (24) is satisfied.

A similar consideration can be given for the operator of the general problem defined by (2), (3). Let us discuss it with some details. Let us select the selfadjoint and the skew-symmetric part of the operator A:

$$A = C + D, \quad C = \frac{1}{2}(A - A^*), \quad D = \frac{1}{2}(A + A^*). \tag{29}$$

The nonnegative operator D can be written as

$$D = G^* G, \tag{30}$$

in which $G : H \to \tilde{H}$. Let E and $\tilde{E}$ be identity operators in the spaces H and $\tilde{H}$, respectively, and let the following partitions of unity define the decomposition of the domain

$$\sum_{\alpha=1}^{p} \chi_\alpha = E, \quad \chi_\alpha \geq 0, \quad \alpha = 1, 2, \ldots, p, \tag{31}$$

$$\sum_{\alpha=1}^{p} \tilde{\chi}_\alpha = \tilde{E}, \quad \tilde{\chi}_\alpha \geq 0, \quad \alpha = 1, 2, \ldots, p. \tag{32}$$

In analogy with (23)–(25), we use the splitting

$$A = \sum_{\alpha=1}^{p} A_\alpha, \quad A_\alpha \geq 0, \quad \alpha = 1, 2, \ldots, p, \tag{33}$$

in which

$$A_\alpha = C_\alpha + D_\alpha, \quad D_\alpha = D_\alpha^* \geq 0, \quad C_\alpha = -C_\alpha^*, \quad \alpha = 1, 2, \ldots, p. \tag{34}$$

Based on (32), we set

$$D_\alpha = G^* \tilde{\chi}_\alpha G, \quad \alpha = 1, 2, \ldots, p. \tag{35}$$

The presentation of the terms in the antisymmetric part is based on (31):

$$C_\alpha = \frac{1}{2}(\chi_\alpha C + C \chi_\alpha), \quad \alpha = 1, 2, \ldots, p. \tag{36}$$

Such an additive representation is a discrete analogue of (27), (28), and it is interpreted as respective version of the domain decomposition.

4 Regularized Domain Decomposition Schemes

Various splitting schemes can be used solving the Cauchy problem for Eqs. (2), (3). The transition to a new time level is based on the solution p separate subtasks, each of which is based on solving a problem with individual operators A_α, $\alpha = 1,2,\dots,p$. Taking into account the structure of the operators (see (34)–(36)), the presented splitting schemes belong to the class of regionally additive schemes and are based on consistent application of non-iterative domain decomposition schemes.

Currently, the principle of regularization of difference schemes is being considered as a basic methodological principle for improving the difference schemes [13]. The construction of unconditionally stable additive-difference schemes [19], based on the principle of regularization, will be implemented here in the following ways:

1. A simple difference scheme (called here *generating* difference scheme) is constructed for the original problem. This scheme does usually not possess the desired properties. For example, in the construction of additive schemes, the generating scheme can be only conditionally stable or even can be completely unstable.
2. The difference scheme is rewritten in a form for which the stability conditions are known.
3. Quality of the scheme (e.g., its stability) is improved due to perturbations of the operators of the difference scheme, at the same time preserving the possibility for its computational implementation as an additive scheme.

Let us now illustrate the above methodology by a particular case study. Applied to the problem (2), (3), we choose as a generating scheme the simple explicit scheme

$$\frac{y^{n+1} - y^n}{\tau} + Ay^n = \varphi^n, \quad n = 0,1,\dots,N-1, \tag{37}$$

which is complemented by the initial conditions (7). This scheme stable (see the proof of Theorem 1) if the inequality

$$A + A^* - \tau AA^* \geq 0 \tag{38}$$

is fulfilled. The inequality (38) with $D > 0$ imposes restrictions on the time step, i.e., the scheme (29), (37) is conditionally stable. Note also that if $D = 0$, the scheme (29), (37) is absolutely unstable. Taking into account the splitting (33), we refer to the scheme under consideration as to a scheme from the class of additive schemes.

In the construction of additive schemes, we can consider also an alternative variant, using as generating scheme the more general scheme (6), (7), which is not additive, but which is unconditionally stable for $\sigma \geq 0.5$. In this latter case, the perturbation is applied just in order to obtain an additive scheme while preserving the property of unconditional stability.

Regularization of difference schemes for improving the stability range (in the construction of splitting schemes) can be achieved via perturbation of the operator A. Another way is related to perturbation of the finite-difference approximation of the time derivative term. In the construction of additive schemes, it is convenient to work with the transition operator S, writing down the generating scheme (37) as

$$y^{n+1} = Sy^n + \tau\varphi^n, \quad n = 0,1,\ldots,N-1. \tag{39}$$

In the case of (37), we have

$$S = E - \tau A. \tag{40}$$

A regularized scheme based on the perturbation of the operator S has the form

$$y^{n+1} = \tilde{S}y^n + \tau\varphi^n, \quad n = 0,1,\ldots,N-1. \tag{41}$$

Let us formulate general conditions on $\tilde{S}$.

The generating scheme (39), (40) has first-order approximation in time, and to preserve this order of approximation, we impose on $\tilde{S}$ the following condition:

$$\tilde{S} = E - \tau A + \mathcal{O}(\tau^2). \tag{42}$$

The scheme (41) is stable in the sense of the estimate (8) provided that the following inequality holds:

$$\|\tilde{S}\| \le 1. \tag{43}$$

Additionally, it should be noted that we seek for additive regularization scheme, where the transition to a new time level is achieved via solving individual subproblems for the operators $A_\alpha, \ \alpha = 1,2,\ldots,p$ in the decomposition (33).

The first class of regularized splitting schemes considered here is based on the following additive representation of the transition operator of the generating scheme:

$$S = \frac{1}{p}\sum_{\alpha=1}^{p} S_\alpha, \quad S_\alpha = E - p\tau A_\alpha, \quad \alpha = 1,2,\ldots,p.$$

We use a similar additive representation for the transition operator in the regularized scheme

$$\tilde{S} = \frac{1}{p}\sum_{\alpha=1}^{p} \tilde{S}_\alpha, \quad \alpha = 1,2,\ldots,p. \tag{44}$$

The individual terms $\tilde{S}_\alpha, \ \alpha = 1,2,\ldots,p$ are based on perturbations of the operators $A_\alpha, \ \alpha = 1,2,\ldots,p$. In analogy with (10), we set

$$\tilde{S}_\alpha = (E + \sigma p\tau A_\alpha)^{-1}(E - (1-\sigma)p\tau A_\alpha), \quad \alpha = 1,2,\ldots,p. \tag{45}$$

If $\sigma \geq 0.5$ (see proof of Theorem 1) we have

$$\|\tilde{S}_\alpha\| \leq 1, \quad \alpha = 1,2,\ldots,p.$$

In view of (44), this provides fulfilment of the stability conditions (43).

Accounting for

$$\tilde{S}_\alpha = E - p\tau(E + \sigma p\tau A_\alpha)^{-1} A_\alpha, \quad \alpha = 1,2,\ldots,p$$

the regularized additive scheme (41), (44), (45) can be rewritten in the form

$$\frac{y^{n+1} - y^n}{\tau} + \sum_{\alpha=1}^{p} (E + \sigma p\tau A_\alpha)^{-1} A_\alpha y^n = \varphi^n, \quad n = 0,1,\ldots,N-1. \tag{46}$$

Comparing to the generating scheme (33), (37), we see that the regularization in this case is achieved by perturbation of A. The outcome of our consideration is the following theorem.

Theorem 2. *The additive-difference scheme (7), (41), (44), (45) is unconditionally stable for $\sigma \geq 0.5$, and stability estimate (8) holds for its solution.*

The computational implementation of the scheme (7), (46) can be carried out as follows. We set

$$y^{n+1} = \frac{1}{p} \sum_{\alpha=1}^{p} y_\alpha^{n+1}, \quad \varphi^n = \sum_{\alpha=1}^{p} \varphi_\alpha^n.$$

In this case, we obtain

$$\frac{y_\alpha^{n+1} - y^n}{p\tau} + (E + \sigma p\tau A_\alpha)^{-1} A_\alpha y^n = \varphi_\alpha^n, \quad \alpha = 1,2,\ldots,p \tag{47}$$

for the individual components of the approximate solution at the new time level y_α^{n+1}, $\alpha = 1,2,\ldots,p$. The scheme (47) can be rewritten as

$$\frac{y_\alpha^{n+1} - y^n}{p\tau} + A_\alpha y^n (\sigma y_\alpha^{n+1} + (1-\sigma) y^n) = (E + \sigma p\tau A_\alpha)\varphi_\alpha^n.$$

In this form we can interpret the scheme (47) as a variant of the additive-averaged component splitting scheme [19].

Another class of regularized splitting schemes instead of additive (see (44)), exploits multiplicative representation of the transition operator:

$$\tilde{S} = \prod_{\alpha=1}^{p} \tilde{S}_\alpha, \quad \alpha = 1,2,\ldots,p. \tag{48}$$

Taking into account (42), we have

$$S = \prod_{\alpha=1}^{p} S_\alpha + \mathcal{O}(\tau^2), \quad S_\alpha = E - \tau A_\alpha, \quad \alpha = 1,2,\dots,p.$$

Similarly to (45), we set

$$\tilde{S}_\alpha = (E + \sigma\tau A_\alpha)^{-1}(E - (1-\sigma)\tau A_\alpha), \quad \alpha = 1,2,\dots,p. \tag{49}$$

Under the standard restrictions $\sigma \geq 0.5$, the regularized scheme (41), (48), (49) is stable.

Theorem 3. *The additive-difference scheme (7), (41), (48), (49) is unconditionally stable for $\sigma \geq 0.5$, and the stability estimate (8) holds for its solution.*

Let us discuss a possible computer implementation of the constructed regularized scheme. We introduce auxiliary quantities $y^{n+\alpha/p}$, $\alpha = 1,2,\dots,p$. Taking into account (41), (48), these are defined from the equations

$$y^{n+\alpha/p} = \tilde{S}_\alpha y^{n+(\alpha-1)/p}, \quad \alpha = 1,2,\dots,p-1,$$

$$y^{n+1} = \tilde{S}_p y^{n+(p-1)/p} + \tau\varphi^n. \tag{50}$$

Similar to (47), we obtain from (50)

$$\frac{y^{n+\alpha/p} - y^{n+(\alpha-1)/p}}{\tau} + (E + \sigma\tau A_\alpha)^{-1} A_\alpha y^{n+(\alpha-1)/p} = \varphi_\alpha^n, \tag{51}$$

where

$$\varphi_\alpha^n = \begin{cases} 0, & \alpha = 1,2,\dots,p-1, \\ \varphi^n, & \alpha = p. \end{cases}$$

We write the scheme (51) as

$$\frac{y^{n+\alpha/p} - y^{n+(\alpha-1)/p}}{\tau} + A_\alpha(\sigma y^{n+\alpha/p} + (1-\sigma) y^{n+(\alpha-1)/p}) = \tilde{\varphi}_\alpha^n, \tag{52}$$

in which

$$\tilde{\varphi}_\alpha^n = (E + \sigma\tau A_\alpha)\varphi_\alpha^n, \quad \alpha = 1,2,\dots,p.$$

Scheme (52) can be considered as a special version of the standard component-wise splitting scheme [10, 13, 34]. However, those schemes are additive approximation schemes, while the constructed here scheme is a full approximation one. Regularized scheme (41), (44), (45), built on the additive representation (44) of the transition operator, is more suitable for parallel computations, compared to the regularized schemes (41), (48), (49) which is based on the multiplicative representation (48).

5 Vector Schemes for Domain Decomposition

Difference schemes for nonstationary problems can often be regarded as appropriate iterative methods for approximate solution of stationary problems. The introduced above regularized additive schemes are based on perturbation of the operator A in the producing scheme (37). Such schemes, as well as the standard additive componentwise splitting schemes, are not suitable for constructing iterative methods for solving stationary equations. Better opportunities in this direction are provided by the vector-additive schemes [1, 31].

Instead of a single unknown $u(t)$, we consider p unknowns u_α, $\alpha = 1,2,\dots,p$, which are to be determined from the system

$$\frac{du_\alpha}{dt} + \sum_{\beta=1}^{p} A_\beta u_\beta = f(t), \quad \alpha = 1,2,\dots,p, \quad 0 < t \le T. \tag{53}$$

The following initial conditions are used for the system of equations (53)

$$u_\alpha(0) = u^0, \quad \alpha = 1,2,\dots,p, \tag{54}$$

which follow from (2). Obviously, each function is a solution of (2), (3), (33). Approximate solution of (2), (3), (33) will be constructed on the basis of difference schemes for the vector problem (53), (54).

To solve the problem (53), (54), we use the following two-level scheme:

$$\frac{y_\alpha^{n+1} - y_\alpha^n}{\tau} + \sum_{\beta=1}^{\alpha} A_\beta y_\beta^{n+1} + \sum_{\beta=\alpha+1}^{p} A_\beta y_\beta^n = \varphi^n,$$

$$\alpha = 1,2,\dots,p, \quad n = 0,1,\dots,N-1, \tag{55}$$

complemented with the initial conditions

$$y_\alpha(0) = u^0, \quad \alpha = 1,2,\dots,p. \tag{56}$$

The computational implementation of this scheme is connected with a consecutive inversion of operators $E + \tau A_\alpha$, $\alpha = 1,2,\dots,p$.

Theorem 4. *The vector-additive difference scheme (33), (55), (56) is unconditionally stable, and stability estimate holds for its components*

$$\|y_\alpha^{n+1}\| \le \|y_\alpha^n\| + \tau\|\varphi^0 - Au^0\| + \tau \sum_{k=1}^{n} \tau \left\| \frac{\varphi^k - \varphi^{k-1}}{\tau} \right\|,$$

$$\alpha = 1,2,\dots,p, \quad n = 0,1,\dots,N-1, \tag{57}$$

is valid.

Proof. The analysis of the vector scheme (55), (56) will be carried out following the work [22]. □

We emphasize that the above stability estimates (57) are obtained for each individual component y_α^{n+1}, $\alpha = 1,2,\dots,p$. Each of them or their linear combination

$$y^{n+1} = \sum_{\alpha=1}^{p} c_\alpha y_\alpha^{n+1}, \quad c_\alpha = \text{const} \geq 0, \quad \alpha = 1,2,\dots,p$$

can be regarded as an approximate solution to our problem (2), (3), (33) at time $t = t^{n+1}$.

6 Model Problem

The performance of the considered domain decomposition schemes is illustrated considering a simple example for numerical solution of the boundary value problem for parabolic equation. Consider a rectangular domain

$$\Omega = \{\, \boldsymbol{x} \mid \boldsymbol{x} = (x_1,x_2),\ 0 < x_\alpha < l_\alpha,\ \alpha = 1,2\}.$$

The following boundary value problem

$$\frac{\partial u}{\partial t} = \sum_{\alpha=1}^{2} \frac{\partial^2 u}{\partial x_\alpha^2}, \quad \boldsymbol{x} \in \Omega, \quad 0 < t < T, \tag{58}$$

$$u(\boldsymbol{x},t) = 0, \quad \boldsymbol{x} \in \partial\Omega, \quad 0 < t < T, \tag{59}$$

$$u(\boldsymbol{x},0) = u^0(\boldsymbol{x}), \quad \boldsymbol{x} \in \Omega \tag{60}$$

is to be solved in Ω.

We introduce a uniform rectangular grid in Ω:

$$\bar{\omega} = \{\boldsymbol{x} \mid \boldsymbol{x} = (x_1,x_2), \quad x_\alpha = i_\alpha h_\alpha, \quad i_\alpha = 0,1,\dots,N_\alpha, \quad N_\alpha h_\alpha = l_\alpha\}$$

and let ω be the set of internal nodes ($\bar{\omega} = \omega \cup \partial\omega$). For grid functions $y(\boldsymbol{x}) = 0$, $\boldsymbol{x} \in \partial\omega$, we define Hilbert space $H = L_2(\omega)$ with the scalar product and norm

$$(y,w) \equiv \sum_{\boldsymbol{x} \in \omega} y(\boldsymbol{x}) w(\boldsymbol{x}) h_1 h_2, \quad \|y\| \equiv (y,y)^{1/2}.$$

After spatial approximations of the problem (58), (59), we arrive at the differential-difference equation:

$$\frac{dy}{dt} + Ay = 0, \quad \boldsymbol{x} \in \omega, \quad 0 < t < T, \tag{61}$$

in which

$$\begin{aligned} Ay = & -\frac{1}{h_1^2}(y(x_1+h_1,x_2) - 2y(x_1,x_2) + y(x_1-h_1,x_2)) \\ & -\frac{1}{h_2^2}(y(x_1,x_2+h_2) - 2y(x_1,x_2) + y(x_1,x_2-h_2)), \quad \boldsymbol{x} \in \omega. \end{aligned} \tag{62}$$

In the space H the operator A is selfadjoint and positive definite [13, 15]:

$$A = A^* \geq (\delta_1 + \delta_2)E, \quad \delta_\alpha = \frac{4}{h_\alpha^2}\sin^2\frac{\pi h_\alpha}{2l_\alpha}, \quad \alpha = 1,2. \tag{63}$$

Taking into account (60), Eq. (62) is supplemented with the initial condition

$$y(\boldsymbol{x},0) = u^0(\boldsymbol{x}), \quad \boldsymbol{x} \in \omega. \tag{64}$$

For simplicity, the DD operator in the investigated problem (61)–(64) is constructed without the explicit separation of the operator G and G and the space $\tilde{H}$, focusing on the decomposition (21), (22). We set

$$\begin{aligned} A_\alpha y = & -\frac{1}{h_1^2}\eta_\alpha(x_1+0.5h_1,x_2)(y(x_1+h_1,x_2) - y(x_1,x_2)) \\ & + \frac{1}{h_1^2}\eta_\alpha(x_1-0.5h_1,x_2)(y(x_1,x_2) - y(x_1-h_1,x_2)) \\ & - \frac{1}{h_2^2}\eta_\alpha(x_1,x_2+0.5h_2)(y(x_1,x_2+h_2) - y(x_1,x_2)) \\ & + \frac{1}{h_2^2}\eta_\alpha(x_1,x_2-0.5h_2)(y(x_1,x_2) - y(x_1,x_2-h_2)), \\ & \alpha = 1,2,\ldots,p. \end{aligned} \tag{65}$$

In view of (21), (22) we have

$$A = \sum_{\alpha=1}^{p} A_\alpha, \quad A_\alpha = A_\alpha^*, \quad \alpha = 1,2,\ldots,p. \tag{66}$$

Thus, we are in a class of additive schemes (33), for which we construct different additive schemes.

Numerical calculations are carried out for the problem (58)–(60) in the unit square ($l_1 = l_2 = 1$) when the solution has the form

$$u(\boldsymbol{x},t) = \sin(n_1\pi x_1)\sin(n_2\pi x_2)\exp(-\pi^2(n_1^2+n_2^2)t) \tag{67}$$

Fig. 1 Domain decomposition

for natural n_1 and n_2. We use this solution to set the initial conditions (60). The domain is decomposed into four overlapping subdomains (see Fig. 1). The disconnected subdomains can be considered as one subdomain, and the decomposition in Fig. 1 can be considered as a decomposition into two subdomains and described by two functions: $\eta_\alpha = \eta_\alpha(x_1)$, $\alpha = 1,2$.

Overlapping and nonoverlapping domain decomposition methods can be constructed for problems of type (58)–(60). Methods without overlap require formulation of interface conditions at the common boundaries. Here we consider overlapping DD and therefore do not need to formulate such conditions. However, the proposed here schemes have straightforward extension for the case of nonoverlapping DD.

A fundamental question in DD methods, especially in their parallel implementation, is the exchange of calculated data between different subdomains. The usual explicit schemes can serve as reference in order to explain the exchange challenges. In this case, the domain decomposition can be associated with certain subsets of grid nodes: ω_α, $\alpha = 1,2$, where $\omega = \omega_1 \cup \omega_2$. In the case of (58)–(60) (seven point stencil in space), the transition to a new level in time for the explicit scheme is associated with the use of solution values at the boundary nodes (here we mean the boundary of each subdomain). We need to transfer the calculated data volume $\sim \partial\omega_\alpha$, $\alpha = 1,2$. In solving numerically the problem (61)–(64), we can consider two possibilities for minimal overlap of the subdomains. In our case, the first one corresponds to allocating the inter-subdomain boundary along the grid nodes with integer numbers; the second one is allocating interface lines along nodes with non-integer numbering.

The variant with division along integer-numbered nodes is displayed in Fig. 2. Let the decomposition be carried out in the variable x_1, i.e. $\theta = x_1$. Decomposition of the domain held by the node $\theta = \theta_i$. Given this decomposition, the operator (65)

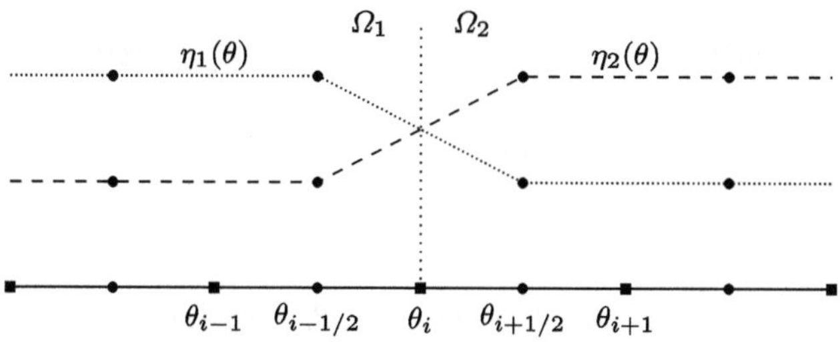

Fig. 2 Decomposition in integer nodes

is written in the form

$$A_1 y = \frac{1}{h_1^2}(y(x_1,x_2) - y(x_1 - h_1,x_2))$$
$$- \frac{1}{2h_2^2}(y(x_1,x_2+h_2) - 2y(x_1,x_2) + y(x_1,x_2-h_2)),$$

$$A_2 y = -\frac{1}{h_1^2}(y(x_1+h_1,x_2) - y(x_1,x_2))$$
$$- \frac{1}{2h_2^2}(y(x_1,x_2+h_2) - 2y(x_1,x_2) + y(x_1,x_2-h_2)), \quad x_1 = \theta_i.$$

This decomposition can be associated with Neumann boundary conditions as exchange boundary conditions. Relationship between the individual subdomains is minimal and they can exchange data with $\theta = \theta_i$. This case can be identified by the decomposition operators (32) as follows:

$$R(\tilde{\chi}_\alpha) = [0,1], \quad \alpha = 1,2,\ldots,p. \tag{68}$$

The values of $\eta_\alpha(x_1 \pm 0.5h_1, x_2)$, $\eta_\alpha(x_1, x_2 \pm 0.5h_1)$, $\alpha = 1,2$ for (65), (67) are equal to 0 or 1.

The second possibility, which is associated with decomposition along the non-integer nodes, is illustrated in Fig. 3. In this case, instead of (68), we have

$$R(\tilde{\chi}_\alpha) = [0,1/2,1], \quad \alpha = 1,2,\ldots,p. \tag{69}$$

In the node $\theta = \theta_i$, difference approximation is used with less twice the flux. With regard to the case in the decomposition of the variable x_1, operators decomposition (65) is

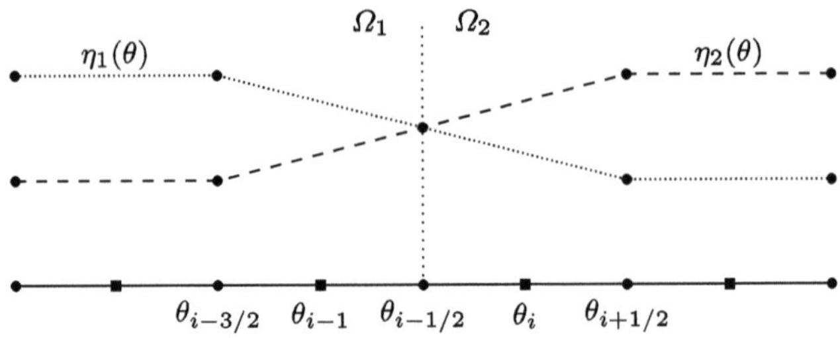

Fig. 3 Decomposition of a half-integer nodes

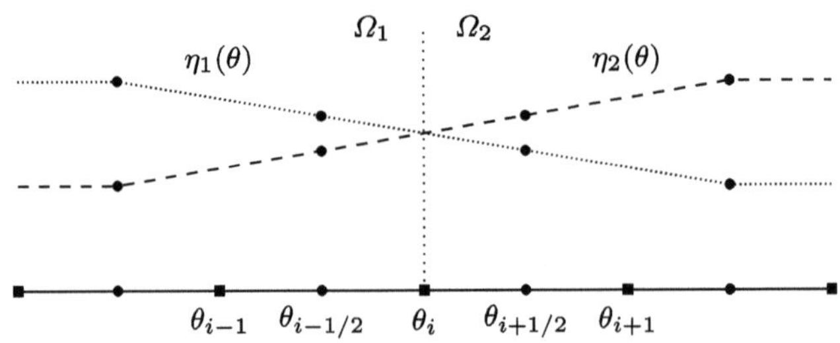

Fig. 4 Decomposition in integer nodes with a width of overlap $3h$

$$
\begin{aligned}
A_1 y = & \frac{1}{2h_1^2}(y(x_1,x_2) - y(x_1 - h_1,x_2)) \\
& - \frac{1}{4h_2^2}(y(x_1,x_2+h_2) - 2y(x_1,x_2) + y(x_1,x_2-h_2)),
\end{aligned}
$$

$$
\begin{aligned}
A_2 y = & -\frac{1}{h_1^2}(y(x_1+h_1,x_2) - y(x_1,x_2)) + \frac{1}{2h_1^2}(y(x_1,x_2) - y(x_1-h_1,x_2)) \\
& - \frac{3}{4h_2^2}(y(x_1,x_2+h_2) - 2y(x_1,x_2) + y(x_1,x_2-h_2)), \quad x_1 = \theta_i.
\end{aligned}
$$

For the calculations in Ω_1 (see Fig. 3), we use half of the flux at the node $\theta = \theta_i$. Thus, when using the domain decomposition method, the exchanges are minimal and coincide with the exchanges in the implementation of the explicit scheme.

The decomposition variants (68), (69) presented above correspond to the case of minimum overlapping of the subdomains. At the discrete level, the width of overlap is determined by the mesh size, h and $2h$, respectively. Similar variants are built for larger overlap of the subdomains. In particular, for the decomposition variant in Fig. 4, we have

$$
R(\tilde{\chi}_\alpha) = [0, 1/3, 2/3, 1], \quad \alpha = 1,2,\dots,p. \tag{70}
$$

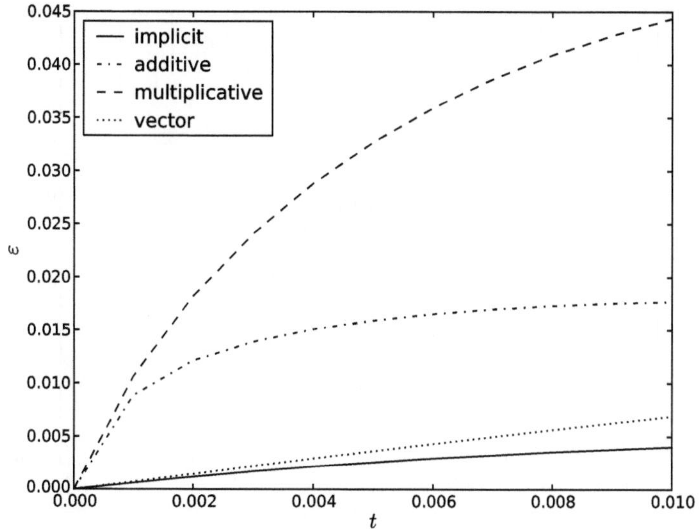

Fig. 5 Accuracy at $N_1 = N_2 = 32$, $N = 10$

In this case the volume of the data exchange is increased, but on the other hand, the transition from one subdomain to another is much smoother. The latter allows us to expect higher accuracy of the approximate solution. Let us present the numerical results obtained in solving (58)–(60). Recall that the exact solution is given by (67) for $n_1 = 2$, $n_2 = 1$ at $T = 0.01$. Square grid $N_1 = N_2$ is used. Regularized fully implicit ($\sigma = 1$) scheme based on additive perturbation (scheme (7), (41), (45), (45)) and based on multiplicative perturbation (scheme (7), (41), (48), (49)) is used, as well as vector-additive scheme (33), (55), (56). The results are compared with the finite-difference solution, which we obtain by using the implicit scheme (1), (6), (7) with $\sigma = 1$ (i.e., scheme without splitting). The errors of the approximate solutions are measured as $\varepsilon(t^n) = \|y^n(\boldsymbol{x}) - u(\boldsymbol{x}, t^n)\|$ on a single time step.

In the case of the decomposition (68) (the width of the overlay is h), the grid space of $N_1 = N_2 = 32$ and grid on time $N = 10$ ($\tau = 0.001$), the error norms of the difference solution using different decomposition schemes are shown in Fig. 5. Figures 6–8 show the local error at the final time. The error is localized in areas of overlap, and for vector decomposition scheme, it is much lower than for the additive and multiplicative versions of regularized additive schemes.

With an increase in the grid space, the error of approximate solution of domain decomposition schemes in comparison with the implicit scheme grows (Fig. 9). In this case, the width of the overlap is reduced by half.

The influence of the width of the overlap is shown in Fig. 10. When using the decomposition (70), there is a substantial increase in the accuracy of the approximate solution compared to the decomposition (68) (compare Figs. 5 and 10).

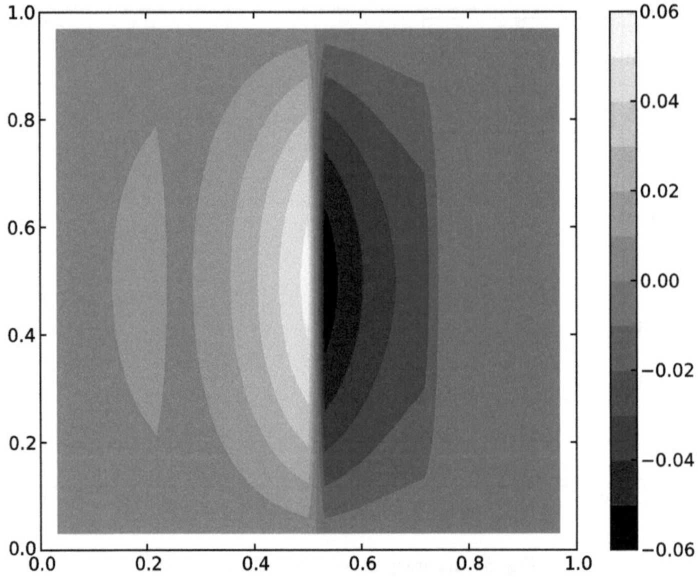

Fig. 6 Error of scheme (7), (41), (48), (49)

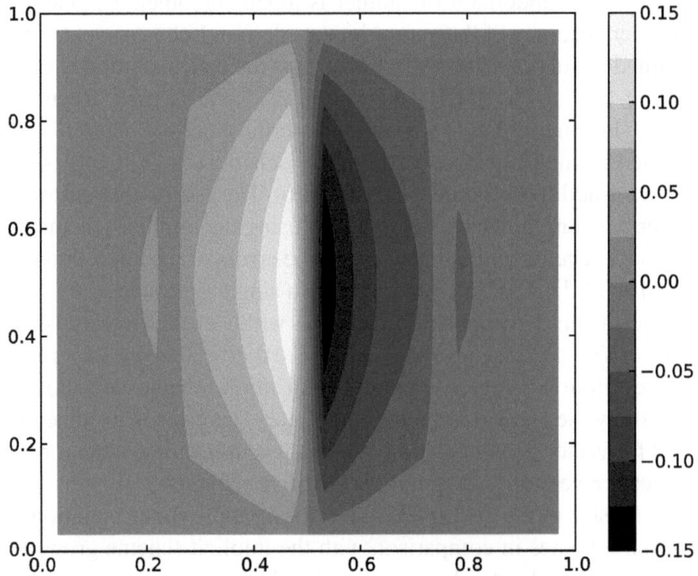

Fig. 7 Error of scheme (7), (41), (45), (45)

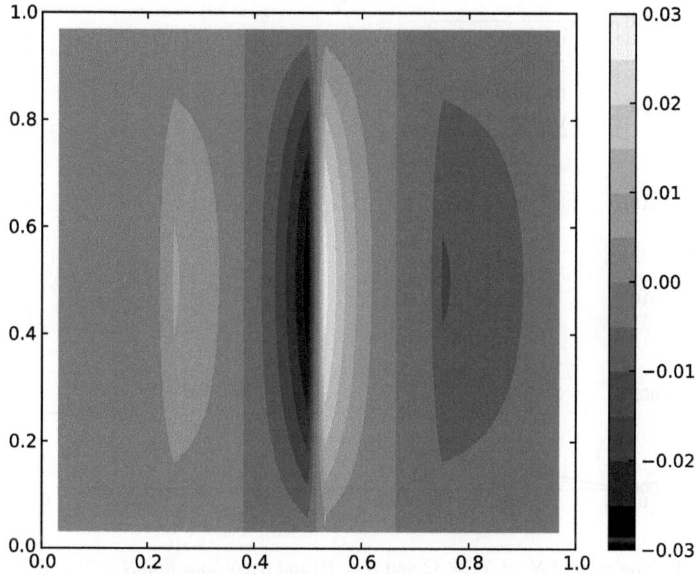

Fig. 8 Error of scheme (33), (55), (56)

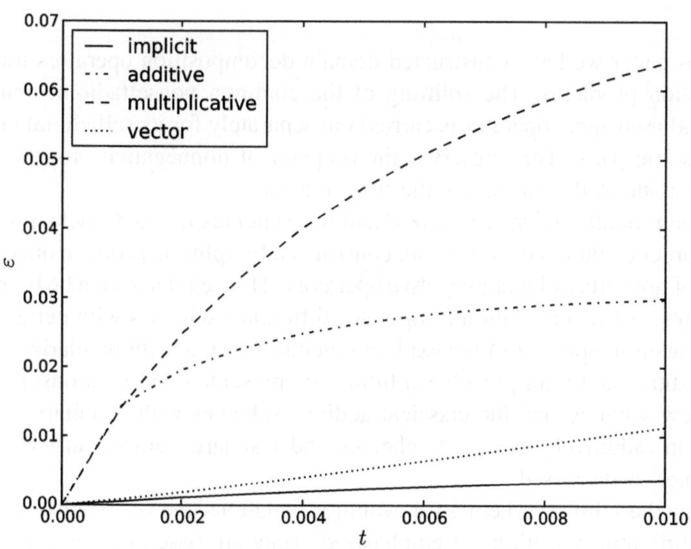

Fig. 9 The error at $N_1 = N_2 = 64$, $N = 10$

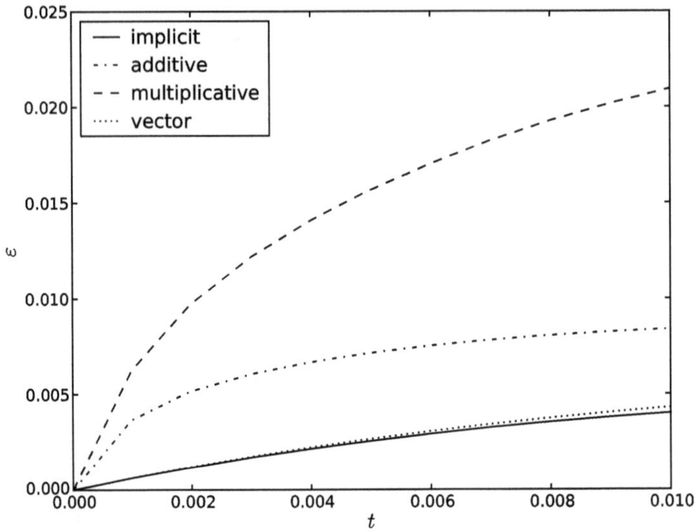

Fig. 10 The error at $N_1 = N_2 = 32$ and $N = 10$ and decomposition $R = [0, 1/3, 2/3, 1]$

7 Conclusions

1. In this paper we have constructed domain decomposition operators for solving evolution problems. The splitting of the common nonselfadjoint nonnegative finite-dimensional operator is carried out separately for its selfadjoint and skew-symmetric parts. This preserves the property of nonnegativity for the operator terms associated with each of the subdomains.
2. Unconditionally stable regularized additive schemes for the Cauchy problem for first-order evolution equations are constructed by splitting problem operators into sum of nonselfadjoint nonnegative operators. This regularization be based on the principles of regularization of operator-difference schemes with perturbation of the transition operator of the explicit scheme. Variants with regularization based on additive and multiplicative splitting are presented, the relationship between the new schemes and the classical additive schemes with summarized approximation (additively averaged schemes and standard component-wise splitting schemes) is discussed.
3. Among the splitting schemes for evolution equations, the vector additive schemes with full approximation are emphasized. They are based on the transition to a system of similar problems in each component with the special organization for computing the approximate solution at the new time level.
4. Numerical simulations for IBVP for a parabolic problem in a rectangular domain are performed. Calculations demonstrate the capabilities of the suggested domain decomposition schemes. The best results in terms of accuracy are demonstrated by the vector-additive scheme of domain decomposition.

Acknowledgements This research was supported by the NEFU Development Program for 2010–2019.

References

1. Abrashin, V.: A variant of the method of variable directions for the solution of multidimensional problems of mathematical-physics. I. Differ. Equat. **26**(2), 243–250 (1990)
2. Abrashin, V., Vabishchevich, P.: Vector additive schemes for second-order evolution equations. Differ. Equat. **34**(12), 1673–1681 (1998)
3. Cai, X.C.: Additive Schwarz algorithms for parabolic convection-diffusion equations. Numer. Math. **60**(1), 41–61 (1991)
4. Cai, X.C.: Multiplicative Schwarz methods for parabolic problems. SIAM J. Sci. Comput. **15**(3), 587–603 (1994)
5. Dryja, M.: Substructuring methods for parabolic problems. In: Glowinski, R., Kuznetsov, Y.A., Meurant, G.A., Périaux, J., Widlund, O. (eds.) Fourth International Symposium on Domain Decomposition Methods for Partial Differential Equations. SIAM, Philadelphia, PA (1991)
6. Kuznetsov, Y.: New algorithms for approximate realization of implicit difference schemes. Sov. J. Numer. Anal. Math. Model. **3**(2), 99–114 (1988)
7. Kuznetsov, Y.: Overlapping domain decomposition methods for FE-problems with elliptic singular perturbed operators. Fourth international symposium on domain decomposition methods for partial differential equations, Proc. Symp., Moscow/Russ. 1990, 223–241 (1991) (1991)
8. Laevsky, Y.: Domain decomposition methods for the solution of two-dimensional parabolic equations. In: Variational-difference methods in problems of numerical analysis, vol. 2, pp. 112–128. Comp. Cent. Sib. Branch, USSR Acad. Sci., Novosibirsk (1987). In Russian
9. Lax, P.D.: Linear algebra and its applications. 2nd edn. Pure and Applied Mathematics. A Wiley-Interscience Series of Texts, Monographs and Tracts, xvi, 376 p. Wiley, New York (2007)
10. Marchuk, G.: Splitting and alternating direction methods. In: Ciarlet, P.G., Lions, J.L. (eds.) Handbook of Numerical Analysis, vol. I, pp. 197–462. North-Holland, Amsterdam (1990)
11. Mathew, T.: Domain decomposition methods for the numerical solution of partial differential equations. Lecture Notes in Computational Science and Engineering, vol. 61, xiii, 764 p. Springer, Berlin (2008)
12. Quarteroni, A., Valli, A.: Domain decomposition methods for partial differential equations. Numerical Mathematics and Scientific Computation, xv, 360 p. Clarendon Press, Oxford (1999)
13. Samarskii, A.: The theory of difference schemes. Pure and Applied Mathematics, Marcel Dekker, vol. 240, 786 p. Marcel Dekker, New York (2001)
14. Samarskii, A., Matus, P., Vabishchevich, P.: Difference schemes with operator factors. Mathematics and Its Applications (Dordrecht), vol. 546, x, 384 p. Kluwer Academic, Dordrecht (2002)
15. Samarskii, A., Nikolaev, E.: Numerical Methods for Grid Equations. Birkhäuser, Basel (1989)
16. Samarskii, A., Vabishchevich, P.: Vector additive schemes of domain decomposition for parabolic problems. Differ. Equat. **31**(9), 1522–1528 (1995)
17. Samarskii, A., Vabishchevich, P.: Factorized finite-difference schemes for the domain decomposition in convection-diffusion problems. Differ. Equat. **33**(7), 972–979 (1997)
18. Samarskii, A., Vabishchevich, P.: Regularized additive full approximation schemes. Doklady. Math. **57**(1), 83–86 (1998)
19. Samarskii, A., Vabishchevich, P.: Additive Schemes for Problems of Mathematical Physics (Additivnye skhemy dlya zadach matematicheskoj fiziki), 320 p. Nauka, Moscow (1999). In Russian

20. Samarskii, A., Vabishchevich, P.: Domain decomposition methods for parabolic problems. In: Lai, C.H., Bjorstad, P., Gross, M., Widlund, O. (eds.) Eleventh International Conference on Domain Decomposition Methods, pp. 341–347. DDM.org (1999)
21. Samarskii, A., Vabishchevich, P.: Numerical Methods for Solution of Convection-Diffusion Problems (Chislennye metody resheniya zadach konvekcii-diffuzii), 247 p. URSS, Moscow (1999). In Russian
22. Samarskii, A., Vabishchevich, P., Matus, P.: Stability of vector additive schemes. Doklady Math. **58**(1), 133–135 (1998)
23. Samarskii, A.A., Vabishchevich, P.N.: Regularized difference schemes for evolutionary second order equations. Math. Model Methods Appl. Sci. **2**(3), 295–315 (1992)
24. Smith, B.: Domain decomposition. Parallel Multilevel Methods for Elliptic Partial Differential Equations, xii, 224 p. Cambridge University Press, Cambridge (1996)
25. Toselli, A., Widlund, O.: Domain decomposition methods – algorithms and theory. Springer Series in Computational Mathematics, vol 34, xv, 450 p. Springer, Berlin (2005)
26. Vabishchevich, P.: Difference schemes with domain decomposition for solving nonstationary problems. U.S.S.R. Comput. Math. Math. Phys. **29**(6), 155–160 (1989)
27. Vabishchevich, P.: Regional-additive difference schemes for nonstationary problems of mathematical physics. Mosc. Univ. Comput. Math. Cybern. (3), 69–72 (1989)
28. Vabishchevich, P.: Parallel domain decomposition algorithms for time-dependent problems of mathematical physics. Advances in Numerical Methods and Applications, pp. 293–299. World Schientific, Singapore (1994)
29. Vabishchevich, P.: Regionally additive difference schemes with a stabilizing correction for parabolic problems. Comput. Math. Math. Phys. **34**(12), 1573–1581 (1994)
30. Vabishchevich, P.: Finite-difference domain decomposition schemes for nonstationary convection-diffusion problems. Differ. Equat. **32**(7), 929–933 (1996)
31. Vabishchevich, P.: Vector additive difference schemes for first-order evolutionary equations. Comput. Math. Math. Phys. **36**(3), 317–322 (1996)
32. Vabishchevich, P.: Domain decomposition methods with overlapping subdomains for the time-dependent problems of mathematical physics. Comput. Methods Appl. Math. **8**(4), 393–405 (2008)
33. Vabishchevich, P., Verakhovskij, V.: Difference schemes for component-wise splitting-decomposition of a domain. Mosc. Univ. Comput. Math. Cybern. **1994**(3), 7–11 (1994)
34. Yanenko, N.: The method of fractional steps. The Solution of Problems of Mathematical Physics in Several Variables, VIII, 160 p. with 15 fig. Springer, Berlin-Heidelberg-New York (1971)

Spectral Coarse Spaces in Robust Two-Level Schwarz Methods

J. Willems

Abstract A survey of recently proposed approaches for the construction of spectral coarse spaces is provided. These coarse spaces are in particular used in two-level preconditioners. At the core of their construction are local generalized eigenvalue problems. It is shown that by means of employing these spectral coarse spaces in two-level additive Schwarz preconditioners one obtains preconditioned systems whose condition numbers are independent of the problem sizes and problem parameters such as (highly) varying coefficients. A unifying analysis of the recently presented approaches is given, pointing out similarities and differences. Some numerical experiments confirm the analytically obtained robustness results.

Keywords Spectral coarse space • Robust preconditioner • Two-level domain decomposition • Additive Schwarz • Multiscale problems

Mathematics Subject Classification (2010): 65-02, 65F10, 65N22, 65N30, 65N35, 65N55

1 Introduction

The robust preconditioning of linear systems of equations resulting from the discretization of partial differential equations is an important objective in the numerical analysis community. The importance arises due to an abundance of applications in the natural and engineering sciences, including, e.g., porous media flows in natural

---

J. Willems (✉)
Radon Institute for Computational and Applied Mathematics (RICAM),
Altenberger Strasse 69, 4040 Linz, Austria
e-mail: joerg.willems@ricam.oeaw.ac.at

O.P. Iliev et al. (eds.), *Numerical Solution of Partial Differential Equations: Theory, Algorithms, and Their Applications*, Springer Proceedings in Mathematics & Statistics 45,
DOI 10.1007/978-1-4614-7172-1_15,

reservoirs or man-made materials and computational solid mechanics. In many practical situations the obtained discrete systems are too large to be solved by direct solvers in acceptable computational time. This leaves the class of iterative solvers as viable alternative. Nevertheless, since the convergence rates of iterative solvers generally depend on the condition numbers of the systems to be solved, suitable preconditioners are necessary to speed up convergence.

More precisely, one is typically faced with a situation where the condition number of the discrete system increases with the size of the problem (or equivalently with decreasing the mesh parameter) and may additionally deteriorate with specific problem parameters. Instances of such problem parameters are, e.g., (highly) varying coefficients or otherwise degenerate parameters. The latter may for instance be observed in linear elasticity in the almost incompressible case, i.e., when the Poisson ratio is close to $1/2$. In view of these two aspects one is therefore interested in the design of preconditioners that yield condition numbers of the preconditioned systems that are independent of mesh and problem parameters. In the following we refer to these preconditioners as *robust* with respect to mesh and problem parameters.

In the absence of degenerate problem parameters obtaining robust preconditioners with respect to the problem size has been successfully addressed for a variety of settings. Here we in particular mention various multilevel and multigrid methods (see e.g. [3, 18, 24, 26] and references therein) and domain decomposition methods (see e.g. [21, 23] and references therein). For problems with varying coefficients these methods remain to work robustly provided the coefficient variations are resolved by the coarsest grid.

However, even for two-level methods the situation is more complicated if the coarse mesh does not resolve the coefficient discontinuities. For certain classes of coefficients robustness of two-level preconditioners could be established by using a coarse space spanned by specially designed multiscale finite element functions (see e.g. [10, 17, 19]) or energy minimizing functions (see e.g. [25, 29]). The dimensions of these "exotic" coarse spaces are essentially given by the dimensions of corresponding standard coarse spaces. While this is desirable from the point of view of computational complexity, it can be shown that for general coefficient configurations the obtained coarse spaces cannot be rich enough to maintain robustness in all situations.

A two-level preconditioner for the scalar elliptic equation with highly varying coefficients that is robust for general coefficient configurations was presented in [15]. Here the authors use local generalized eigenvalue problems in the coarse space construction. More precisely, they consider a family of overlapping subdomains. On each of the subdomains a generalized eigenvalue problem is posed. The eigenfunctions corresponding to eigenvalues below a predefined threshold are then used for constructing the coarse space in the two-level preconditioner. The analysis of this preconditioner then shows that the condition number of the preconditioned system only depends on this predefined threshold, and is thus in particular independent of problem and mesh parameters. The approach of [15] is furthermore refined in [16] where multiscale partition of unity functions are used to reduce the dimension of the coarse space while preserving the robustness of the preconditioner.

Here it should be noted that the idea of using local eigenvalue problems for the coarse space construction has previously been used in [6–8] leading to spectral element-based algebraic multigrid (ρAMGe) methods. More recently, in the framework of ρAMGe and focussing on the robustness with respect to coefficient variations, local generalized eigenvalue problems have been used to construct a tentative coarse space (see [5]). The actual coarse space used in [5] is then obtained from this tentative coarse space after a smoothed aggregation construction (see also [6]). A two-grid method similar to that of [5] is discussed in [20], where additionally advanced polynomial smoothers based on the best uniform polynomial approximation to x^{-1} are considered.

The concept of using local generalized eigenvalue problems in the coarse space construction of robust two-level preconditioners for the scalar elliptic equation with highly varying coefficients is put into a more general framework in [22]. Here the local generalized eigenfunctions corresponding to eigenvalues below a predefined threshold are employed to define functionals. These functionals are in turn used to specify constraints for minimization problems whose solutions are taken as coarse space basis functions. The framework of [22] is a generalization, since it allows for functional constraints not only originating from local generalized eigenproblems. In fact, it is shown that an alternative way for choosing the functional constraints is by specifying averages over suitably chosen, i.e., coefficient dependent, subdomains.

Another generalization of [15, 16] is the use of local generalized eigenvalue problems for the construction of robust preconditioners for abstract symmetric positive definite bilinear forms, which was considered in [12] and later on in [9]. The idea is to formulate the generalized eigenvalue problems only in terms of the abstract bilinear form. This generality makes the theory applicable to a variety of problems such as the scalar elliptic equation with isotropic or anisotropic coefficients, the stream function formulations of Stokes' and Brinkman's problem, the equations of linear elasticity, as well as equations arising in the solution of Maxwell's equations.

The main objective of the chapter at hand is to put the derivations of [12] and [9] in a common perspective, to emphasize their similarities and differences, and to relate them to the original works in [15, 16]. For this we restrict to analyzing the scalar elliptic equation with highly varying isotropic coefficients to keep the argument as simple as possible.

We remark that rather recently the approaches of [15, 16] and [12] have been generalized to multiple levels in [13] and [27], respectively. We note that due to the high computational cost involved in solving generalized eigenvalue problems the generalization to multiple levels provides an important step for keeping the sizes of these eigenvalue problems manageable for overall problem sizes that could hardly be coped with in a two-level framework. Nevertheless, for the sake of simplicity we refrain from including the analysis of these multilevel methods in our present exposition. Finally, for the sake of completeness, we note that these concepts of robust preconditioners have been applied to multiscale anisotropic problems (see [11] for a two-level and [28] for a multilevel method).

The remainder of this chapter is organized as follows. In Sect. 2 we outline the problem setting and formulate the abstract overlapping Schwarz preconditioner. Section 3 is devoted to different related approaches for constructing suitable coarse spaces resulting in robust preconditioners. In Sect. 4 we analyze the coarse space dimension and clarify differences and similarities between the various methods. In Sect. 5 we present some numerical results exemplifying the robustness of the obtained preconditioners before ending with some conclusions.

2 Problem Setting

In order to make our presentation as accessible as possible, we restrict to the following model problem posed in a bounded polyhedral domain $\Omega \subset \mathbb{R}^d$, $d = 2, 3$:

$$-\nabla \cdot (\kappa(\boldsymbol{x})\nabla u) = f \quad \text{in } \Omega, \qquad u = 0 \quad \text{on } \partial\Omega, \tag{1}$$

where $0 < \kappa_{\min} \leq \kappa \leq \kappa_{\max} < \infty$ and $f \in L^2(\Omega)$, with $L^2(\Omega)$ denoting the space of square integrable functions on Ω. It is well-known that the variational formulation of (1) is given by

$$\text{Find } u \in H_0^1(\Omega) \text{ such that } a_\Omega(u, v) = (f, v), \quad \forall v \in H_0^1(\Omega), \tag{2}$$

where $a_\omega(u, v) := \int_\omega \kappa(\boldsymbol{x})\nabla u \cdot \nabla v \, d\boldsymbol{x}$ for any $\omega \subset \Omega$, $(f, v) := \int_\Omega f u \, d\boldsymbol{x}$, and $H_0^1(\Omega)$ denotes the subspace of $L^2(\Omega)$ of functions with square integrable derivatives and zero trace on $\partial\Omega$.

Let $\mathcal{T}_h$ be a quasi-uniform triangulation of Ω with mesh parameter h. Corresponding to $\mathcal{T}_h$ let $\mathscr{V} \subset H_0^1(\Omega)$ be a (possibly higher order) Lagrange finite element space. The finite dimensional problem corresponding to (2) is then given by

$$\text{Find } u \in \mathscr{V} \text{ such that } a_\Omega(u, v) = (f, v), \quad \forall v \in \mathscr{V}. \tag{3a}$$

An equivalent operator notation reads

$$\text{Find } u \in \mathscr{V} \text{ satisfying } Au = F, \tag{3b}$$

where, with $\mathscr{V}'$ denoting the dual space of $\mathscr{V}$, $A : \mathscr{V} \to \mathscr{V}'$ is given by $< Au, v > := a_\Omega(u, v)$, and $F \in \mathscr{V}'$ is defined by $< F, v > := (f, v)$. Here $< \cdot, \cdot >$ denotes the duality pairing of $\mathscr{V}'$ and $\mathscr{V}$.

Our main objective in this chapter is to discuss robust two-level additive Schwarz preconditioners for solving (3). The term "robust" refers to the condition number of the preconditioned system being independent of the mesh parameter h and variations in κ.

Algorithm 1: Additive Schwarz preconditioner $M : \mathscr{V}' \to \mathscr{V}$ corresponding to $\{\mathscr{V}_0(\Omega_j^{(1)})\}_{j=1}^{n_\Omega^{(1)}}$ and $\mathscr{V}_H$.

Let $F \in \mathscr{V}'$.
Set $v \equiv 0 \in \mathscr{V}$.
for $j = 1, \ldots, n_\Omega^{(1)}$ **do**
  Compute $\psi \in \mathscr{V}_0(\Omega_j^{(1)})$ such that

$$a_{\Omega_j}(\psi, w) = F(w), \qquad \forall w \in \mathscr{V}_0(\Omega_j^{(1)}).$$

  $v \leftarrow v + \psi$
end for
Compute $\psi \in \mathscr{V}_H$ such that

$$a(\psi, w) = F(w), \qquad \forall w \in \mathscr{V}_H.$$

$v \leftarrow v + \psi$
return $MF := v$

To make this more precise let $\{\Omega_j^{(1)}\}_{j=1}^{n_\Omega^{(1)}}$ be a family of overlapping subdomains of Ω. For any $\omega \subset \Omega$ we define

$$\mathscr{V}(\omega) := \{v|_\omega \,|\, v \in \mathscr{V}\} \quad \text{and} \quad \mathscr{V}_0(\omega) := \{v \in \mathscr{V} \,|\, \mathrm{supp}(v) \subset \overline{\omega}\}.$$

Also, we identify functions in $\mathscr{V}_0(\omega)$ with their restrictions to ω, and we thus in particular have that $\mathscr{V}_0(\omega) \subset \mathscr{V}(\omega)$. Let $\mathscr{V}_H \subset \mathscr{V}$ be a coarse space whose construction is discussed in Sect. 3. The action of the two-level additive Schwarz preconditioner corresponding to $\mathscr{V}_0(\Omega_j^{(1)})$, $j = 1, \ldots, n_\Omega^{(1)}$ and $\mathscr{V}_H$ is given by Algorithm 1.

Applying M to (3b) yields the following preconditioned system

$$MAu = MF. \tag{4}$$

For $j = 1, \ldots, n_\Omega^{(1)}$ let $\mathscr{I}_j^{(1)} := \{i = 1, \ldots, n_\Omega^{(1)} \,|\, \Omega_i^{(1)} \cap \Omega_j^{(1)} \neq \emptyset\}$. Also, we set $n_{\mathscr{I}}^{(1)} := \max_{j=1,\ldots,n_\Omega^{(1)}} \#\mathscr{I}_j^{(1)}$. Using this notation it follows from [21, Lemma 2.51] that

$$\lambda_{\max}(MA) \leq n_{\mathscr{I}}^{(1)} + 1, \tag{5}$$

where $\lambda_{\max}(\cdot)$ denotes the largest eigenvalue. For establishing the robustness of our preconditioner it, therefore, suffices to derive a lower bound for $\lambda_{\min}(MA)$ independent of h and variations in κ. Here $\lambda_{\min}(\cdot)$ denotes the smallest eigenvalue.

Provided that there exists a constant $K > 0$ such that for any $v \in \mathscr{V}$ there exist $v_H \in \mathscr{V}_H$ and $v_j \in \mathscr{V}_0(\Omega_j^{(1)})$, $j = 1, \ldots, n_\Omega^{(1)}$ satisfying

$$v = v_H + \sum_{j=1}^{n_\Omega^{(1)}} v_j \qquad \text{and} \qquad a_\Omega(v_H, v_H) + \sum_{j=1}^{n_\Omega^{(1)}} a_\Omega(v_j, v_j) \le K a_\Omega(v, v) \tag{6}$$

a standard result for abstract alternating Schwarz methods yields that

$$\lambda_{\min}(MA) \ge K^{-1}, \tag{7}$$

which together with (5) in particular implies the following result (see e.g. [21, Theorem 2.52]).

Theorem 2.1. *The condition number of the additive Schwarz preconditioned system* (4) *is bounded by* $K(n_{\mathscr{G}}^{(1)} + 1)$.

In view of Theorem 2.1 it is, therefore, sufficient to establish a stable decomposition (6) with a constant K independent of h and variations in κ. The crucial ingredient for obtaining such a robust bound is the careful design of the coarse space $\mathscr{V}_H$, which is described in the next section.

3 Spectral Coarse Space Construction

First, we need to introduce some further notation. Let $\{\Omega_j^{(2)}\}_{j=1}^{n_\Omega^{(2)}}$ be another overlapping decomposition of Ω, which may coincide with $\{\Omega_j^{(1)}\}_{j=1}^{n_\Omega^{(1)}}$. Let $\{\xi_j^{(1)}\}_{j=1}^{n_\Omega^{(1)}}$ and $\{\xi_j^{(2)}\}_{j=1}^{n_\Omega^{(2)}}$ be partition of unities subordinate to $\{\Omega_j^{(1)}\}_{j=1}^{n_\Omega^{(1)}}$ and $\{\Omega_j^{(2)}\}_{j=1}^{n_\Omega^{(2)}}$, respectively, such that $\operatorname{supp}(\xi_j^{(i)}) = \overline{\Omega}_j^{(i)}$ for $j = 1, \ldots, n_\Omega^{(i)}$. As a starting point of our derivations, we observe that for any $v_{H,j}^{(i)} \in \mathscr{V}(\Omega_j^{(i)})$, $j = 1, \ldots, n_\Omega^{(i)}$ we have the following two variants of a decomposition of v:

$$v = \underbrace{\sum_{j=1}^{n_\Omega^{(1)}} \xi_j^{(1)} v_{H,j}^{(1)}}_{=:v_H^{(1)}} + \sum_{j=1}^{n_\Omega^{(1)}} \underbrace{\xi_j^{(1)}(v - v_{H,j}^{(1)})}_{=:v_j^{(1)}}, \tag{8a}$$

$$v = \underbrace{\sum_{j=1}^{n_\Omega^{(2)}} \xi_j^{(2)} v_{H,j}^{(2)}}_{=:v_H^{(2)}} + \sum_{j=1}^{n_\Omega^{(1)}} \underbrace{\xi_j^{(1)}(v - v_H^{(2)})}_{=:v_j^{(2)}}. \tag{8b}$$

(8b) is the choice considered in [12], whereas (8a) is essentially the variant considered in [9]. Note that in the first variant there appears only one partition of unity, whereas in the second variant one has the freedom to choose two distinct partition of unities (see Remark 3.4). We now aim at choosing $v_{H,j}^{(i)}$ in such a way that the decompositions (8) are also stable, i.e., have a robust constant K in estimate (6). This approach eventually leads to the definition of a suitable coarse space $\mathscr{V}_H$.

Before proceeding with the actual derivations we note that $v_H^{(i)} = v - \sum_{j=1}^{n_\Omega^{(1)}} v_j^{(i)}$. Thus, by the definition of $n_{\mathscr{G}}^{(1)}$ and a strengthened Cauchy–Schwarz inequality we observe that

$$
\begin{aligned}
a_\Omega\left(v_H^{(i)}, v_H^{(i)}\right) &\leq 2a_\Omega(v,v) + 2a_\Omega\left(\sum_{j=1}^{n_\Omega^{(1)}} v_j^{(i)}, \sum_{j=1}^{n_\Omega^{(1)}} v_j^{(i)}\right) \\
&\leq 2a_\Omega(v,v) + 2n_{\mathscr{G}}^{(1)} \sum_{j=1}^{n_\Omega^{(1)}} a_\Omega\left(v_j^{(i)}, v_j^{(i)}\right).
\end{aligned}
\tag{9}
$$

Thus, for establishing the estimate in (6) with a robust constant K, it suffices to derive the following estimate

$$
\sum_{j=1}^{n_\Omega^{(1)}} a_\Omega\left(v_j^{(i)}, v_j^{(i)}\right) \leq C a_\Omega(v,v), \tag{10}
$$

where C is a generic constant independent of h and variations in κ, i.e., we may disregard the term $a_\Omega(v_H, v_H)$ in the estimate of (6).

Considering the definition of $v_j^{(i)}$ in (8) we aim at choosing $v_{H,j}^{(i)}$ in such a way that (10) holds.

Remark 3.1. We would like to point out here that generally, due to the multiplication by partition of unity functions, we have that $v_H^{(i)}, v_j^{(i)} \notin \mathscr{V}$. This problem can be overcome by considering $I_h v_H^{(i)}$ and $I_h v_j^{(i)}$ instead, where I_h denotes the usual nodal interpolation associated with $\mathscr{V}$.

Another possibility which is proposed in [9] is the use of partition of identity operators in (8) instead of partition of unity functions. At the current place this modification indeed makes the argument more elegant. Nevertheless, this modification shifts the difficulty to the analysis relating the dimension of $\mathscr{V}_H$ to the geometry underlying the variations of κ. This issue will be further addressed in Sect. 4.2.

First we consider the case $i = 1$, i.e., (8a). We observe that

$$
\sum_{j=1}^{n_\Omega^{(1)}} a_\Omega\left(v_j^{(1)}, v_j^{(1)}\right) = \sum_{j=1}^{n_\Omega^{(1)}} \underbrace{a_\Omega\left(\xi_j^{(1)}(v - v_{H,j}^{(1)}), \xi_j^{(1)}(v - v_{H,j}^{(1)})\right)}_{=: m_{\Omega_j^{(1)}}^{(1)}\left(v - v_{H,j}^{(1)}, v - v_{H,j}^{(1)}\right)}. \tag{11}
$$

Similarly, but slightly more complicated, we obtain for the case $i=2$, i.e., (8b),

$$\begin{aligned}
\sum_{k=1}^{n_\Omega^{(1)}} a_\Omega\Big(v_k^{(2)}, v_k^{(2)}\Big) &= \sum_{k=1}^{n_\Omega^{(1)}} a_\Omega\Big(\xi_k^{(1)}(v-v_H^{(2)}), \xi_k^{(1)}(v-v_H^{(2)})\Big)\\
&= \sum_{k=1}^{n_\Omega^{(1)}} a_\Omega\left(\xi_k^{(1)}(v-\sum_{j=1}^{n_\Omega^{(2)}}\xi_j^{(2)}v_{H,j}^{(2)}), \xi_k^{(1)}(v-\sum_{j=1}^{n_\Omega^{(2)}}\xi_j^{(2)}v_{H,j}^{(2)})\right)\\
&\le n_{\mathscr{I}}^{(2)}\sum_{k=1}^{n_\Omega^{(1)}}\sum_{j=1}^{n_\Omega^{(2)}} a_\Omega\Big(\xi_k^{(1)}\xi_j^{(2)}(v-v_{H,j}^{(2)}), \xi_k^{(1)}\xi_j^{(2)}(v-v_{H,j}^{(2)})\Big)\\
&= n_{\mathscr{I}}^{(2)}\sum_{j=1}^{n_\Omega^{(2)}} \underbrace{\sum_{k:\Omega_k^{(1)}\cap\Omega_j^{(2)}\neq\emptyset} a_{\Omega_j^{(2)}}\Big(\xi_j^{(2)}\xi_k^{(1)}(v-v_{H,j}^{(2)}), \xi_j^{(2)}\xi_k^{(1)}(v-v_{H,j}^{(2)})\Big)}_{=:m^{(2)}_{\Omega_j^{(2)}}\left(v-v_{H,j}^{(2)},v-v_{H,j}^{(2)}\right)},
\end{aligned} \tag{12}$$

where $n_{\mathscr{I}}^{(2)}$ is defined analogously to $n_{\mathscr{I}}^{(1)}$ corresponding to the decomposition $\{\Omega_j^{(2)}\}_{j=1}^{n_\Omega^{(2)}}$.

In view of (11) and (12) it is therefore sufficient for satisfying (10) to choose $v_{H,j}^{(i)}$ in such a way that

$$m^{(i)}_{\Omega_j^{(i)}}\Big(v-v_{H,j}^{(i)}, v-v_{H,j}^{(i)}\Big) \le C a_{\Omega_j^{(i)}}(v,v). \tag{13}$$

The following proposition (see e.g. [15, Sect. 3.3.1] or [12, Sect. 2]) is crucial for establishing (13) with a robust constant C.

Proposition 3.2. *Consider the following local generalized eigenvalue problem:*

Find $(\varphi_{j,\lambda}^{(i)}, \lambda) \in \mathscr{V}(\Omega_j^{(i)})\times\mathbb{R}_0^+$ *s.t.* $a_{\Omega_j^{(i)}}\Big(w, \varphi_{j,\lambda}^{(i)}\Big) = \lambda\, m^{(i)}_{\Omega_j^{(i)}}\Big(w, \varphi_{j,\lambda}^{(i)}\Big)\ \forall w\in\mathscr{V}(\Omega_j^{(i)})$. (14)

For $v\in\mathscr{V}$ *let* $v_{H,j}^{(i)} := \Pi_j^{(i)}v \in \mathscr{V}(\Omega_j)$ *be the* $a_{\Omega_j^{(i)}}(\cdot,\cdot)$*-orthogonal projection of* $v|_{\Omega_j^{(i)}}$ *onto those eigenfunctions corresponding to eigenvalues below a predefined "threshold"* $\tau_\lambda^{-1} > 0$*, i.e.,* $\Pi_j^{(i)}v \in span\{\varphi_{j,\lambda}^{(i)}\,|\,\lambda<\tau_\lambda^{-1}\}$ *satisfies*

$$a_{\Omega_j^{(i)}}\Big(v-\Pi_j^{(i)}v, \varphi_{j,\lambda}^{(i)}\Big) = 0 \textit{ for all } \lambda < \tau_\lambda^{-1}.$$

Then we have that

$$m^{(i)}_{\Omega_j^{(i)}}\Big(v-v_{H,j}^{(i)}, v-v_{H,j}^{(i)}\Big) \le \tau_\lambda\, a_{\Omega_j^{(i)}}\Big(v-v_{H,j}^{(i)}, v-v_{H,j}^{(i)}\Big) \le \tau_\lambda\, a_{\Omega_j^{(i)}}(v,v). \tag{15}$$

Proof. The second inequality in (15) is obvious, since $\Pi_j^{(i)} v$ is the $a_{\Omega_j^{(i)}}(\cdot,\cdot)$-orthogonal projection of $v|_{\Omega_j^{(i)}}$.

Next, we note that $v|_{\Omega_j^{(i)}} - \Pi_j^{(i)} v = \sum_{\lambda \geq \tau_\lambda^{-1}} a_{\Omega_j^{(i)}}\left(v, \varphi_{j,\lambda}^{(i)}\right) \varphi_{j,\lambda}^{(i)}$. Thus,

$$
\begin{aligned}
m_{\Omega_j^{(i)}}^{(i)}\left(v - \Pi_j^{(i)} v, v - \Pi_j^{(i)} v\right) &= \sum_{\lambda \geq \tau_\lambda^{-1}} a_{\Omega_j^{(i)}}\left(v, \varphi_{j,\lambda}^{(i)}\right) m_{\Omega_j^{(i)}}^{(i)}\left(v - \Pi_j^{(i)} v, \varphi_{j,\lambda}^{(i)}\right) \\
&= \sum_{\lambda \geq \tau_\lambda^{-1}} \lambda^{-1} a_{\Omega_j^{(i)}}\left(v, \varphi_{j,\lambda}^{(i)}\right) a_{\Omega_j^{(i)}}\left(v - \Pi_j^{(i)} v, \varphi_{j,\lambda}^{(i)}\right) \\
&\leq \tau_\lambda\, a_{\Omega_j^{(i)}}\left(v - \Pi_j^{(i)} v, \sum_{\lambda \geq \tau_\lambda^{-1}} a_{\Omega_j^{(i)}}\left(v, \varphi_{j,\lambda}^{(i)}\right) \varphi_{j,\lambda}^{(i)}\right) \\
&= \tau_\lambda\, a_{\Omega_j^{(i)}}\left(v - \Pi_j^{(i)} v, v - \Pi_j^{(i)} v\right). \qquad \square
\end{aligned}
$$

Note that by choosing the threshold τ_λ we can essentially fix the constant C in estimate (13). Thus, C and therefore also K in (6) only depend on τ_λ and $n_{\mathscr{I}}^{(i)}$, $i = 1, 2$, but are in particular independent of h and variations in κ.

For the solvability of (14) it is also important to note that $m_{\Omega_j^{(i)}}^{(i)}(\cdot,\cdot)$ is positive definite on $\mathscr{V}(\Omega_j^{(i)})$, since $\mathrm{supp}(\xi_j^{(i)}) = \overline{\Omega}_j^{(i)}$ by assumption.

The considerations above suggest choosing the coarse space $\mathscr{V}_H^{(i)}$ as $\mathrm{span}\{\xi_j^{(i)} \varphi_{j,\lambda}^{(i)} \,|\, \lambda < \tau_\lambda^{-1},\ j = 1, \ldots, n_\Omega^{(i)}\}$. However, as indicated in Remark 3.1, this choice in general does not yield a subspace of $\mathscr{V}$. The following proposition resolves this issue by means of applying a nodal interpolation.

Proposition 3.3. *For $i = 1, 2$ let*

$$
\mathscr{V}_H^{(i)} := span\{I_h(\xi_j^{(i)} \varphi_{j,\lambda}^{(i)}) \,|\, \lambda < \tau_\lambda^{-1},\ j = 1, \ldots, n_\Omega^{(i)}\}, \tag{16}
$$

where as above I_h denotes the nodal interpolation corresponding to $\mathscr{V}$. With $v_{H,j}^{(i)}$ as defined in Proposition 3.2 we have that

$$
v = \underbrace{\sum_{j=1}^{n_\Omega^{(1)}} I_h(\xi_j^{(1)} v_{H,j}^{(1)})}_{=: v_{H,I}^{(1)}} + \sum_{j=1}^{n_\Omega^{(1)}} \underbrace{I_h(\xi_j^{(1)}(v - v_{H,j}^{(1)}))}_{=: v_{j,I}^{(1)}} \tag{17a}
$$

and

$$v=\underbrace{\sum_{j=1}^{n_\Omega^{(2)}} I_h(\xi_j^{(2)}v_{H,j}^{(2)})}_{=:v_{H,I}^{(2)}}+\sum_{j=1}^{n_\Omega^{(1)}}\underbrace{I_h(\xi_j^{(1)}(v-v_H^{(2)}))}_{=:v_{j,I}^{(2)}}. \tag{17b}$$

Moreover, $v_{H,I}^{(i)} \in \mathscr{V}_H^{(i)}$ *and the decompositions* (17) *satisfy a stable decomposition property* (6) *with a constant K only depending on* $n_{\mathscr{I}}^{(i)}$, τ_λ, *and the shape regularity of* $\mathscr{T}_h$.

Proof. The identities (17) follow by the linearity of I_h and the fact that $I_h v = v$ for all $v \in \mathscr{V}$.

$v_{H,I}^{(i)} \in \mathscr{V}_H^{(i)}$ follows directly from the definitions of $v_{H,j}^{(i)}$ and $\mathscr{V}_H^{(i)}$.

For showing stability we need to reduce decompositions (17) to the case (8). Thus, it suffices to show that for any $v \in \mathscr{V}$ we have that

$$a_\Omega\Big(I_h(\xi_j^{(i)}v), I_h(\xi_j^{(i)}v)\Big) \le C a_\Omega\Big(\xi_j^{(i)}v, \xi_j^{(i)}v\Big), \tag{18}$$

with a constant C only depending on the mesh regularity of $\mathscr{T}_h$. By [15, Proposition 15] (see also [4, Lemma 4.5.3]) we know that (18) is satisfied. □

Remark 3.4. Note that for the additive Schwarz preconditioner corresponding to the second variant of a stable decomposition, i.e., (17b), the local solves are carried out with respect to $\mathscr{V}_0(\Omega_j^{(1)})$, $j = 1,\dots,n_\Omega^{(1)}$, whereas by the definition of $\mathscr{V}_H^{(2)}$ we see that the supports of the coarse basis functions are given by $\Omega_j^{(2)}$, $j = 1,\dots,n_\Omega^{(2)}$. That is, the subdomains of the local solves do not need to coincide with the supports of the coarse basis functions.

This observation is in contrast to the first variant of a stable decomposition, i.e., (17a), where the support of the coarse basis functions is given by $\Omega_j^{(1)}$, $j = 1,\dots,n_\Omega^{(1)}$, corresponding to the spaces of the local solves, i.e., $\mathscr{V}_0(\Omega_j^{(1)})$, $j,\dots,n_\Omega^{(1)}$.

Remark 3.5. For actual numerical computations it is important to have a basis of $\mathscr{V}_H^{(i)}$ available. Definition (16) obviously provides a generating set of our spectral coarse space. Note, however, that even though the generalized eigenfunctions $\varphi_{j,\lambda}^{(i)}$ are mutually $a_{\Omega_j}(\cdot,\cdot)$ orthogonal, it is not clear that the generating set in (16) also constitutes a basis. In fact, in particular for anisotropic problems (see [28]) it is discussed that this generating set may not be minimal. Nevertheless, for simplicity we assume in the following that the set given in (16) constitutes a basis and refer to [28] for the more general situation.

4 Analysis of Spectral Coarse Space Dimensions

After establishing the robustness of the additive Schwarz preconditioner given by Algorithm 1 and utilizing the coarse spaces $\mathscr{V}_H^{(i)}$ it is important to analyze the dimension of these coarse spaces. This is in particular crucial for the overall computational complexity of the method, and it is generally desirable to keep the dimension of $\mathscr{V}_H^{(i)}$ as small as possible.

By construction the dimension of $\mathscr{V}_H^{(i)}$ is determined by the number of generalized eigenvalues below the threshold τ_λ^{-1} (see Proposition 3.2). We now investigate the number of these "small" eigenvalues for binary geometries and for different choices of subdomains and partition of unities. For this we first recall the well-known min–max/Courant–Fischer principle (see e.g. [14, Theorem 7.36]), which states that

$$\lambda_{j,k}^{(i)} = \min_{\mathscr{V}_k(\Omega_j^{(i)}) \subset \mathscr{V}(\Omega_j^{(i)})} \max_{v \in \mathscr{V}_k(\Omega_j^{(i)})} \frac{a_{\Omega_j^{(i)}}(v,v)}{m_{\Omega_j^{(i)}}^{(i)}(v,v)}, \tag{19}$$

where $\mathscr{V}_k(\Omega_j^{(i)})$ for $k \geq 1$ is a k-dimensional subspace of $\mathscr{V}(\Omega_j^{(i)})$ and $\lambda_{j,k}^{(i)}$ denotes the k-th eigenvalue of (14) sorted in increasing order accounting for multiplicity.

By our assumption of having a binary medium we know that $\overline{\Omega} = \overline{\Omega^p} \cup \overline{\Omega^s}$ such that

$$\kappa(\boldsymbol{x}) = \begin{cases} \kappa_{\max}, & \boldsymbol{x} \in \Omega^s \\ \kappa_{\min}, & \boldsymbol{x} \in \Omega^p, \end{cases}$$

which in particular means that the contrast $\kappa_{\max}/\kappa_{\min}$ is the problem parameter of interest.

For simplicity of the exposition we restrict to the case when $\{\Omega_j^{(2)}\}_{j=1}^{n_\Omega^{(2)}} = \{\Omega_j^{(1)}\}_{j=1}^{n_\Omega^{(1)}}$ and $\{\xi_j^{(2)}\}_{j=1}^{n_\Omega^{(2)}} = \{\xi_j^{(1)}\}_{j=1}^{n_\Omega^{(1)}}$. Thus, without any danger of confusion we may drop the superindices $^{(1)}$ and $^{(2)}$ distinguishing different families of subdomains and partition of unity functions. Note, however, that even with this simplification the bilinear forms $m_{\Omega_j}^{(1)}(\cdot,\cdot)$ and $m_{\Omega_j}^{(2)}(\cdot,\cdot)$ are not identical.

Furthermore, let $\Omega_j^p := \Omega^p \cap \Omega_j$ and similarly $\Omega_j^s := \Omega^s \cap \Omega_j$. Besides, we set

$$\Omega_j^{\text{int}} := \Omega_j \backslash (\bigcup_{k \neq j} \overline{\Omega}_k).$$

Note that we do not exclude the possibility that $\Omega_j^{\text{int}} = \emptyset$. Additionally, let $\Omega_{j,k}^s$, $k = 1, \ldots, L_j$ be the path-connected components of Ω_j^s, where we assume an ordering such that $\Omega_{j,k}^s \backslash \Omega_j^{\text{int}} \neq \emptyset$ for $k = 1, \ldots, \tilde{L}_j$, where $\tilde{L}_j \leq L_j$ is suitably chosen. If $\Omega_{j,k}^s \backslash \Omega_j^{\text{int}} = \emptyset$ for $k = 1, \ldots, L_j$ we set $\tilde{L}_j = 1$ and $\Omega_{j,1}^s = \Omega_j \backslash \Omega_j^{\text{int}}$. The diameter of the subdomains $\{\Omega_j\}_{j=1}^{n_\Omega}$ is assumed to be $\mathcal{O}(H)$, and the width of the overlaps of intersecting subdomains is assumed to be $\mathcal{O}(\delta)$. For a better understanding of these definitions we refer to Fig. 1.

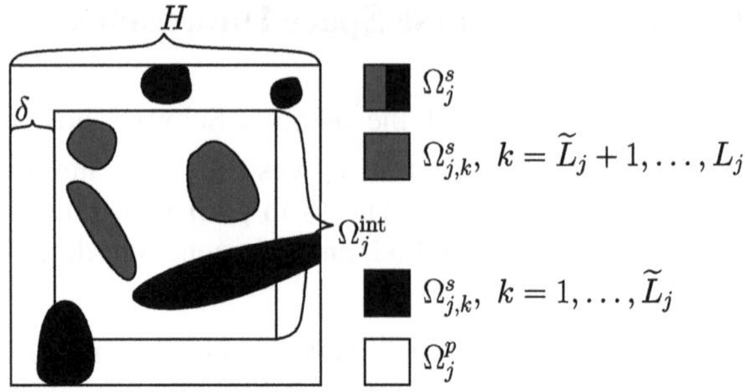

Fig. 1 Subdomain Ω_j with connected components of Ω_j^s and non-overlapping part Ω_j^{int}. In the present configuration $L_j = 7$ and $\tilde{L}_j = 4$

Let

$$\mathscr{V}^c_{\tilde{L}_j}(\Omega_j) := \{v \in \mathscr{V}(\Omega_j) \,|\, \int_{\Omega^s_{j,k}} v\, d\boldsymbol{x} = 0 \text{ for } k = 1,\ldots,\tilde{L}_j\}.$$

Obviously, any $\tilde{L}_j + 1$-dimensional subspace of $\mathscr{V}(\Omega_j)$ has a nontrivial intersection with $\mathscr{V}^c_{\tilde{L}_j}(\Omega_j)$. Thus, by (19) we see that there exists a $w \in \mathscr{V}^c_{\tilde{L}_j}(\Omega_j)$ such that

$$\lambda^{(i)}_{j,\tilde{L}_j+1} \geq \frac{a_{\Omega_j}(w,w)}{m^{(i)}_{\Omega_j}(w,w)}. \tag{20}$$

We first consider the case $(i) = (1)$ and note that by Schwarz' inequality

$$m^{(1)}_{\Omega_j}(w,w) = \int_{\Omega_j} \kappa(\nabla(\xi_j w))^2\, d\boldsymbol{x} \leq 2\int_{\Omega_j} \kappa w^2 (\nabla\xi_j)^2\, d\boldsymbol{x} + 2\underbrace{\int_{\Omega_j} \kappa \xi_j^2 (\nabla w)^2\, d\boldsymbol{x}}_{\leq a_{\Omega_j}(w,w)}. \tag{21}$$

Since $\xi_j \equiv 1$ in Ω_j^{int} we have that

$$\begin{aligned}
\int_{\Omega_j} \kappa w^2 (\nabla\xi_j)^2\, d\boldsymbol{x} &= \int_{\Omega_j\setminus\Omega_j^{\text{int}}} \kappa w^2 (\nabla\xi_j)^2\, d\boldsymbol{x} \\
&\leq C\delta^{-2} \int_{\Omega_j\setminus\Omega_j^{\text{int}}} \kappa w^2\, d\boldsymbol{x} \\
&\leq C\delta^{-2} \left(\sum_{k=1}^{\tilde{L}_j} \int_{\Omega^s_{j,k}} \kappa_{\max} w^2\, d\boldsymbol{x} + \int_{\Omega_j\setminus\Omega_j^{\text{int}}} \kappa_{\min} w^2\, d\boldsymbol{x} \right) \\
&\leq C\left(\frac{H}{\delta}\right)^2 \left(\sum_{k=1}^{\tilde{L}_j} \int_{\Omega^s_{j,k}} \kappa_{\max} (\nabla w)^2\, d\boldsymbol{x} + \int_{\Omega_j\setminus\Omega_j^{\text{int}}} \kappa_{\min} (\nabla w)^2\, d\boldsymbol{x} \right) \\
&\leq C\left(\frac{H}{\delta}\right)^2 a_{\Omega_j}(w,w),
\end{aligned} \tag{22}$$

where we have used Poincaré's inequality, which is possible since $w \in \mathscr{V}^c_{\tilde{L}_j}(\Omega_j)$, and where C is independent of H, h, δ, and $\kappa_{\max}/\kappa_{\min}$.

Similarly, but again slightly more complicated, we obtain for $(i) = (2)$

$$
\begin{aligned}
m^{(2)}_{\Omega_j}(w, w) &= \sum_{k:\,\Omega_k \cap \Omega_j \neq \emptyset} \int_{\Omega_j} \kappa (\nabla(\xi_j \xi_k w))^2 \, d\boldsymbol{x} \\
&\leq 2 \sum_{k:\,\Omega_k \cap \Omega_j \neq \emptyset} \int_{\Omega_j} \kappa w^2 (\nabla(\xi_j \xi_k))^2 + \kappa (\nabla w)^2 (\xi_j \xi_k)^2 \, d\boldsymbol{x} \\
&\leq 4 \int_{\Omega_j} \kappa w^2 (\nabla \xi_j)^2 \, d\boldsymbol{x} + 4 \sum_{k:\,\Omega_k \cap \Omega_j \neq \emptyset} \int_{\Omega_j \cap \Omega_k} \kappa w^2 (\nabla \xi_k)^2 \, d\boldsymbol{x} + 2 a_{\Omega_j}(w, w) \,.
\end{aligned}
$$

Noting that

$$
\begin{aligned}
\int_{\Omega_j \cap \Omega_k} \kappa w^2 (\nabla \xi_k)^2 \, d\boldsymbol{x} &\leq C \delta^{-2} \begin{cases} \displaystyle\int_{\Omega_j \setminus \Omega_j^{\text{int}}} \kappa w^2 \, d\boldsymbol{x}, \text{ if } j = k \\ \displaystyle\int_{\Omega_j \cap \Omega_k} \kappa w^2 \, d\boldsymbol{x}, \text{ if } j \neq k \end{cases} \\
&\leq C \delta^{-2} \int_{\Omega_j \setminus \Omega_j^{\text{int}}} \kappa w^2 \, d\boldsymbol{x}
\end{aligned}
$$

we thus obtain by (22) that

$$
m^{(2)}_{\Omega_j}(w, w) \leq C \left(\frac{H}{\delta} \right)^2 a_{\Omega_j}(w, w) \,. \tag{23}
$$

where C is again independent of H, h, δ, and $\kappa_{\max}/\kappa_{\min}$.

Combining (20), (21), and (22) on the one hand and (20) and (23) on the other hand we thus obtain

$$
\lambda^{(i)}_{j,\tilde{L}_j+1} \geq C \left(\frac{\delta}{H} \right)^2 . \tag{24}
$$

Hence, choosing $\delta = \mathcal{O}(H)$ yields a lower bound of $\lambda^{(i)}_{j,\tilde{L}_j+1}$, which is independent of mesh parameters H and h as well as of the contrast $\kappa_{\max}/\kappa_{\min}$, which is our problem parameter of interest.

4.1 Choice of the Bilinear Form $m_{\Omega_j}(\cdot, \cdot)$

So far, we have carried out our analysis for the bilinear forms $m^{(i)}_{\Omega_j}(\cdot, \cdot)$, $i = 1, 2$, defined by (11) and (12), respectively. Now, we generalize this choice to any bilinear form $\overline{m}_{\Omega_j}(\cdot, \cdot)$ satisfying

$$
m^{(i)}_{\Omega_j}(v, v) \leq C a_{\Omega_j}(v, v) + \overline{m}_{\Omega_j}(v, v) \quad \text{for any } v \in \mathscr{V}(\Omega_j) \tag{25}
$$

for $i=1$ or $i=2$. Now, analogously to (14) consider the corresponding generalized eigenvalue problem

$$\text{Find } (\overline{\varphi}_{j,\lambda}, \lambda) \text{ such that } a_{\Omega_j^{(i)}}\left(w, \overline{\varphi}_{j,\lambda}\right) = \lambda\, \overline{m}_{\Omega_j^{(i)}}\left(w, \overline{\varphi}_{j,\lambda}\right) \text{ for all } w \in \mathscr{V}(\Omega_j^{(i)}).$$

In exactly the same way as (15) in Proposition 3.2 we then obtain

$$\overline{m}_{\Omega_j^{(i)}}(v - \overline{v}_{H,j}, v - \overline{v}_{H,j}) \le \tau_\lambda\, a_{\Omega_j^{(i)}}(v - \overline{v}_{H,j}, v - \overline{v}_{H,j}) \le \tau_\lambda\, a_{\Omega_j^{(i)}}(v, v),$$

where $\overline{v}_{H,j} \in \mathscr{V}(\Omega_j)$ denotes the $a_{\Omega_j}(\cdot,\cdot)$-orthogonal projection of $v|_{\Omega_j}$ onto the span of those eigenfunctions $\overline{\varphi}_{j,\lambda}$ for which $\lambda \le \tau_\lambda^{-1}$. Using (25) together with this estimate we therefore obtain

$$\begin{aligned} m_{\Omega_j}^{(i)}(v - \overline{v}_{H,j}, v - \overline{v}_{H,j}) &\le C a_{\Omega_j}(v - \overline{v}_{H,j}, v - \overline{v}_{H,j}) + \overline{m}_{\Omega_j}(v - \overline{v}_{H,j}, v - \overline{v}_{H,j}) \\ &\le (C + \tau_\lambda)\, a_{\Omega_j}(v - \overline{v}_{H,j}, v - \overline{v}_{H,j}) \\ &\le (C + \tau_\lambda)\, a_{\Omega_j}(v, v). \end{aligned}$$

That is, up to a change in the constant we obtain the same estimate as (15), which implies that in the coarse space construction of our robust preconditioner we may use $\overline{\varphi}_{j,\lambda}$ instead of $\varphi_{j,\lambda}^{(i)}$ in the definition of $\mathscr{V}_H^{(i)}$ (see (16)).

Looking at (21) we see that (25) is satisfied for $(i) = (1)$ and

$$\overline{m}_{\Omega_j}(v, w) := 2 \int_{\Omega_j} \kappa (\nabla \xi_j)^2 v w \, d\boldsymbol{x},$$

which is essentially the choice made in [15, 16].

4.2 *Partition of Unity vs. Partition of Identity*

As indicated in Remark 3.1 the authors of [9] advocate the use of partition of identity operators $\{\Xi_j\}_{j=1}^{n_\Omega}$ instead of partition of unity functions $\{\xi_j\}_{j=1}^{n_\Omega}$. We now elaborate on the changes that this modification necessitates in the analysis of the coarse space dimension.

In the following we consider the case when

$$\Xi_j v := I_h(\xi_j v),$$

where v is either an element of $\mathscr{V}$ or $\mathscr{V}(\Omega_j)$. Instead of $m_{\Omega_j}^{(1)}(\cdot,\cdot)$ given by (12) we then consider $a_{\Omega_j}(\Xi_j w, \Xi_j w)$ for which we obtain

$$a_{\Omega_j}(\Xi_j w, \Xi_j w) = a_{\Omega_j}(I_h(\xi_j w), I_h(\xi_j w)) \le C a_{\Omega_j}(\xi_j w, \xi_j w) = C m_{\Omega_j}^{(1)}(w, w),$$

where we have used estimate (18). The remainder of the analysis proceeds along the lines of (21) and (22).

We note that when using partition of unity functions estimate (18) is needed for establishing the stable decomposition property when employing a coarse space $\mathscr{V}_H^{(i)}$ defined in (16) (see Proposition 3.3). When using partition of identity operators the same estimate is necessary for analyzing the number of asymptotically small (w.r.t. the contrast $\kappa_{\max}/\kappa_{\min}$) generalized eigenvalues and thus the dimension of $\mathscr{V}_H^{(i)}$.

4.3 Choice of the Subdomains

Concerning the choice of the subdomains estimate (24) admits several observations. First of all we see that regardless of the choice of δ the lower bound for $\lambda_{j,\tilde{L}_j+1}^{(i)}$ is independent of the contrast $\kappa_{\max}/\kappa_{\min}$. Our computational experience confirms that this bound on the eigenvalue index is sharp in the sense that $\lambda_{j,k}^{(i)} \to 0$ as $\kappa_{\max}/\kappa_{\min} \to \infty$ for $k = 1, \ldots, \tilde{L}_j$. For (very) high-contrast problems one may therefore expect a "gap" in the spectrum and to recover $\tilde{L}_j$ "small" eigenvalues below τ_λ^{-1} provided this threshold is chosen to lie within this spectral gap.

These considerations imply the following tradeoff regarding the choice of δ. On the one hand one would like to choose δ small, e.g., $\delta = \mathcal{O}(h)$, in order to have Ω_j^{int} as large and thus $\tilde{L}_j$ as small as possible. The latter is desirable, since one is generally interested in a small dimensional coarse space $\mathscr{V}_H$.

On the other hand choosing δ (very) small leads to a (very) small lower bound in (24). In particular choosing a minimal overlap of one layer of fine cells $T \in \mathcal{T}_h$— or more generally $\delta = \mathcal{O}(h)$—results in a lower bound for $\lambda_{j,\tilde{L}_j+1}^{(i)}$ that depends on the mesh parameters and degenerates as $H/h \to \infty$. The occurrence of a spectral gap therefore depends on the relation of H/h and $\kappa_{\max}/\kappa_{\min}$. Hence for a given threshold τ_λ^{-1} and H/h sufficiently large one may in fact recover more "small" eigenvalues than $\tilde{L}_j$, which may ultimately result in a larger dimensional coarse space.

4.4 Eigenvalue Problems in Overlaps of Subdomains

For the case of $\Omega_j^{\text{int}} \neq \emptyset$ a further modification is suggested in [9], which results in a reduction of the number of degrees of freedom involved in the solution of the generalized eigenvalue problem (14). To achieve this one may proceed as follows:

Let

$$\tilde{\mathscr{V}}(\Omega_j) := \{v \in \mathscr{V}(\Omega_j) \,|\, a_{\Omega_j}(v, w) = 0 \; \forall w \in \mathscr{V}_0(\Omega_j^{\text{int}})\}.$$

Note that by construction we have that

$$\mathscr{V}(\Omega_j) = \mathscr{V}_0(\Omega_j^{\text{int}}) \oplus \tilde{\mathscr{V}}(\Omega_j) \quad \text{and} \quad \mathscr{V}_0(\Omega_j^{\text{int}}) \perp_a \tilde{\mathscr{V}}(\Omega_j) \tag{26}$$

and that for small overlaps δ the dimension of $\tilde{\mathscr{V}}(\Omega_j)$ may be much smaller than the dimension of $\mathscr{V}(\Omega_j)$.

Now, consider the following modification of the generalized eigenvalue problem (14) posed with respect to $\tilde{\mathscr{V}}(\Omega_j)$ instead of $\mathscr{V}(\Omega_j)$, i.e.,

$$\text{Find } (\tilde{\varphi}^{(i)}_{j,\lambda}, \lambda) \in \tilde{\mathscr{V}}(\Omega_j) \times \mathbb{R}^+_0 \text{ s.t. } a_{\Omega_j}\left(w, \tilde{\varphi}^{(i)}_{j,\lambda}\right) = \lambda\, m^{(i)}_{\Omega_j \setminus \Omega_j^{\mathrm{int}}}\left(w, \tilde{\varphi}^{(i)}_{j,\lambda}\right) \;\forall w \in \tilde{\mathscr{V}}(\Omega_j), \tag{27}$$

where

$$m^{(1)}_{\Omega_j \setminus \Omega_j^{\mathrm{int}}}(v, w) := a_{\Omega \setminus \Omega_j^{\mathrm{int}}}(\xi_j v, \xi_j w)$$

and

$$m^{(2)}_{\Omega_j \setminus \Omega_j^{\mathrm{int}}}(v, w) := \sum_{k:\Omega_k \cap \Omega_j \neq \emptyset} a_{\Omega_j \setminus \Omega_j^{\mathrm{int}}}(\xi_j \xi_k v, \xi_j \xi_k w)\,,$$

respectively. Note that with these definitions we have

$$m^{(i)}_{\Omega_j}(v, w) = m^{(i)}_{\Omega_j \setminus \Omega_j^{\mathrm{int}}}(v, w) + a_{\Omega_j^{\mathrm{int}}}(v, w)\,. \tag{28}$$

Furthermore, for the solvability of (27) it is again important to note that $m^{(i)}_{\Omega_j \setminus \Omega_j^{\mathrm{int}}}(\cdot, \cdot)$ is positive definite on $\tilde{\mathscr{V}}(\Omega_j)$. This follows from the fact that $\mathrm{supp}(\xi_j) = \overline{\Omega}_j$ by assumption and $v|_{\Omega_j \setminus \Omega_j^{\mathrm{int}}} \not\equiv 0$ for all $v \in \tilde{\mathscr{V}}(\Omega_j) \setminus \{0\}$ by construction.

According to the analysis in Sect. 3 we need to prove a statement analogous to that of Proposition 3.2.

Proposition 4.1. *For $v \in \mathscr{V}$ let $\tilde{v}^{(i)}_{H,j} := \tilde{\Pi}^{(i)}_j v \in \tilde{\mathscr{V}}(\Omega_j)$ be the $a_{\Omega_j}(\cdot, \cdot)$-orthogonal projection of $v|_{\Omega_j}$ onto those eigenfunctions of (27) corresponding to eigenvalues below $\tau_\lambda^{-1} > 0$, i.e., $\tilde{\Pi}^{(i)}_j v \in span\{\tilde{\varphi}^{(i)}_{j,\lambda} \,|\, \lambda < \tau_\lambda^{-1}\}$ satisfies*

$$a_{\Omega_j}\left(v - \tilde{\Pi}^{(i)}_j v, \tilde{\varphi}^{(i)}_{j,\lambda}\right) = 0 \text{ for all } \lambda < \tau_\lambda^{-1}.$$

Then we have that

$$m^{(i)}_{\Omega_j}\left(v - \tilde{v}^{(i)}_{H,j}, v - \tilde{v}_{H,j}\right) \leq (1 + \tau_\lambda)\, a_{\Omega_j}\left(v - \tilde{v}^{(i)}_{H,j}, v - \tilde{v}^{(i)}_{H,j}\right) \leq (1 + \tau_\lambda)\, a_{\Omega_j}(v, v)\,. \tag{29}$$

Proof. The second inequality in (29) is obvious for the same reason as the second inequality in (15).

By (28) we have that

$$\begin{aligned}
& m^{(i)}_{\Omega_j}\left(v - \tilde{\Pi}^{(i)}_j v, v - \tilde{\Pi}^{(i)}_j\right) \\
&= m^{(i)}_{\Omega_j \setminus \Omega_j^{\mathrm{int}}}\left(v - \tilde{\Pi}^{(i)}_j v, v - \tilde{\Pi}^{(i)}_j\right) + \underbrace{a_{\Omega_j^{\mathrm{int}}}\left(v - \tilde{\Pi}^{(i)}_j v, v - \tilde{\Pi}^{(i)}_j v\right)}_{\leq a_{\Omega_j}\left(v - \tilde{\Pi}^{(i)}_j v, v - \tilde{\Pi}^{(i)}_j v\right)}.
\end{aligned}$$

Thus, it remains to show that

$$m^{(i)}_{\Omega_j\setminus\Omega_j^{\text{int}}}\left(v-\tilde{\Pi}_j^{(i)}v,\,v-\tilde{\Pi}_j^{(i)}v\right) \le \tau_\lambda\, a_{\Omega_j}\left(v-\tilde{\Pi}_j^{(i)}v,\,v-\tilde{\Pi}_j^{(i)}v\right). \tag{30}$$

By a reasoning identical to that of Proposition 3.2 it follows that (30) holds for all $v\in\tilde{\mathscr{V}}(\Omega_j)$.

For general $v\in\mathscr{V}(\Omega_j)$ consider the unique $a_{\Omega_j}(\cdot,\cdot)$-orthogonal decomposition $v=v^{\text{int}}+\tilde{v}$ with $v^{\text{int}}\in\mathscr{V}_0(\Omega_j^{\text{int}})$ and $\tilde{v}\in\tilde{\mathscr{V}}(\Omega_j)$. By the $a_{\Omega_j}(\cdot,\cdot)$-orthogonality of $\mathscr{V}_0(\Omega_j^{\text{int}})$ and $\tilde{\mathscr{V}}(\Omega_j)$ and since $\tilde{\Pi}_j^{(i)}$ is an $a_{\Omega_j}(\cdot,\cdot)$-orthogonal projection onto a subspace of $\tilde{\mathscr{V}}(\Omega_j)$ it easily follows that $\tilde{\Pi}_j^{(i)}v=\tilde{\Pi}_j^{(i)}\tilde{v}$. Thus, and since $\operatorname{supp}(v^{\text{int}})\subset\overline{\Omega}_j^{\text{int}}$ we have that

$$\begin{aligned}
m^{(i)}_{\Omega_j\setminus\Omega_j^{\text{int}}}\left(v-\tilde{\Pi}_j^{(i)}v,\,v-\tilde{\Pi}_j^{(i)}v\right) &= m^{(i)}_{\Omega_j\setminus\Omega_j^{\text{int}}}\left(v^{\text{int}}+\tilde{v}-\tilde{\Pi}_j^{(i)}\tilde{v},\,v^{\text{int}}+\tilde{v}-\tilde{\Pi}_j^{(i)}\tilde{v}\right)\\
&= m^{(i)}_{\Omega_j\setminus\Omega_j^{\text{int}}}\left(\tilde{v}-\tilde{\Pi}_j^{(i)}\tilde{v},\,\tilde{v}-\tilde{\Pi}_j^{(i)}\tilde{v}\right)\\
&\le \tau_\lambda\, a_{\Omega_j}\left(\tilde{v}-\tilde{\Pi}_j^{(i)}\tilde{v},\,\tilde{v}-\tilde{\Pi}_j^{(i)}\tilde{v}\right)\\
&\le \tau_\lambda\, a_{\Omega_j}\left(v^{\text{int}}+\tilde{v}-\tilde{\Pi}_j^{(i)}v,\,v^{\text{int}}+\tilde{v}-\tilde{\Pi}_j^{(i)}v\right)\\
&= \tau_\lambda\, a_{\Omega_j}\left(v-\tilde{\Pi}_j^{(i)}v,\,v-\tilde{\Pi}_j^{(i)}v\right),
\end{aligned}$$

where the first inequality holds, since (30) is satisfied for $v\in\tilde{\mathscr{V}}(\Omega_j)$, and the second inequality follows by $a_{\Omega_j}(\cdot,\cdot)$-orthogonality. □

In view of Proposition 4.1 we may perform the same reasoning as in Sect. 3 with $v^{(i)}_{H,j}$ replaced by $\tilde{v}^{(i)}_{H,j}$, and we thus obtain an additive Schwarz preconditioner with a coarse space given by $\tilde{\mathscr{V}}_H^{(i)}:=\operatorname{span}\{I_h(\xi_j^{(i)}\tilde{\varphi}^{(i)}_{j,\lambda})\,|\,\lambda<\tau_\lambda^{-1},\ j=1,\ldots,n_\Omega^{(i)}\}$ yielding a condition number independent of problem and mesh parameters.

4.5 Choice of the Partition of Unity

So far, in the derivations of this section we have tacitly assumed that the choice of our partition of unity functions only depends on the subdomains $\{\Omega_j\}_{j=1}^{n_\Omega}$. According to estimate (24) this choice is certainly viable. As a matter of fact, it is necessary to have $\xi_j\equiv 1$ in Ω_j^{int}.

Nevertheless, in particular for large overlaps δ one may consider to choose $\{\xi_j\}_{j=1}^{n_\Omega}$ in a problem, i.e., κ, dependent way. The objective of such an approach, which was first considered in [16], is to reduce the number of asymptotically small (w.r.t. $\kappa_{\max}/\kappa_{\min}$) eigenvalues without introducing a degeneracy due to an increasingly smaller overlap δ.

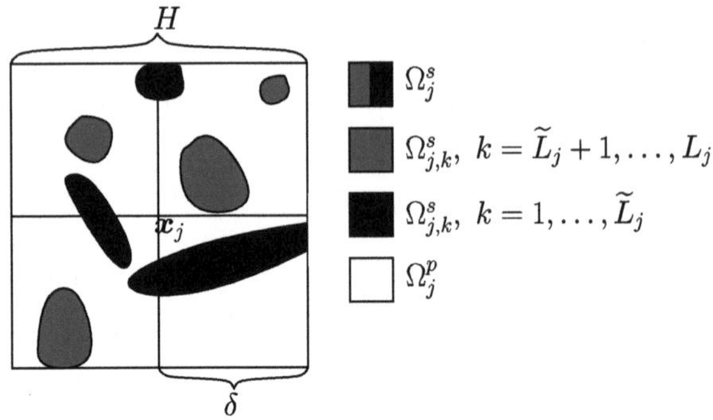

Fig. 2 Subdomain Ω_j with connected components of Ω_j^s. Due to the large overlap $\Omega_j^{\text{int}} = \emptyset$. In the present configuration $L_j = 7$ and $\tilde{L}_j = 3$

Let us consider a coarse grid $\mathcal{T}_H$ of cells obtained by agglomerating fine cells in $\mathcal{T}_h$ (cf. [26, Sect. 1.9] for a description of an agglomeration procedure). The agglomerate coarse cells are assumed to have diameters $\mathcal{O}(H)$. We consider an overlapping decomposition $\{\Omega_j\}_{j=1}^{n_\Omega}$ of Ω, where each subdomain Ω_j is associated with a coarse node $\boldsymbol{x}_j$ and is given by $\Omega_j := interior(\cup\{T \in \mathcal{T}_H \,|\, \boldsymbol{x}_j \in T\})$, i.e., the union of all cells $T \in \mathcal{T}_H$ containing this coarse node. Thus, we obviously have that $\delta = \mathcal{O}(H)$ and $\Omega_j^{\text{int}} = \emptyset$.

In the following we outline the construction of a multiscale partition of unity—henceforth denoted by $\{\xi_j^{\text{ms}}\}_{j=1}^{n_\Omega}$. Let ξ_j^{ms} satisfy $\nabla \cdot (\kappa \nabla \xi_j^{\text{ms}}) = 0$ in those $T{\in}\mathcal{T}_H$ for which $T \subset \overline{\Omega}_j$. Here we assume that $\xi_j^{\text{ms}}|_T$ satisfies suitable boundary conditions on ∂T, which are chosen in such a way that $\sum_{j=1}^{n_\Omega} \xi_j^{\text{ms}} \equiv 1$. One may for instance think of the boundary conditions as being given by the solutions of lower dimensional problems along the agglomerate edges constituting the boundary of T. Here we suppose that ξ_j^{ms} constructed in this way satisfies $0 \le \xi_j^{\text{ms}} \le 1$, which is guaranteed if the validity of a discrete maximum principle is assumed. For a more general situation we refer to [27, Sect. 5].

As above we denote by $\Omega_{j,k}^s$, $k = 1,\dots,L_j$, the path-connected components of Ω_j^s. This time we assume an ordering such that those $\Omega_{j,k}^s$ are ordered first for which it holds that $\Omega_{j,k}^s \cap (\partial T \backslash \partial \Omega_j) \neq \emptyset$ for some $T \in \mathcal{T}_H$ with $T \subset \overline{\Omega}_j$. The number of these path-connected components of Ω_j^s is denoted by $\tilde{L}_j \le L_j$. We refer to Fig. 2 for a better understanding of the current setting. The idea of this construction is that $(\kappa \nabla \xi_j^{\text{ms}})|_{\Omega_{j,k}^s}$, $k = \tilde{L}_j + 1,\dots,L_j$ is small, which seems desirable when looking at the definition of $\overline{m}_{\Omega_j}(\cdot,\cdot)$. More precisely, it is shown in [12, Sect. 5] that provided

$$\|\nabla \xi_j^{\text{ms}}\|_{L^\infty(\Omega_j)} \le CH^{-1} \quad \text{and} \quad \|\kappa_{\max} \nabla \xi_j\|_{L^\infty(\Omega_{j,k}^s)} \le CH^{-1},\ \forall k = \tilde{L}_j + 1,\dots,L_j \tag{31}$$

we have that

$$\lambda^{(i)}_{j,\tilde{L}_j+1} \geq C > 0,$$

where C is independent of $\kappa_{\max}/\kappa_{\min}$, δ, H, and h. Although a rigorous analysis clarifying the question when (31) can be expected to hold is still a largely unsolved problem for general coefficients κ, the computational practice shows that using multiscale partition of unity functions as opposed to standard ones may significantly reduce the coarse space dimension, while maintaining the robustness of the overall preconditioner.

Remark 4.2. It should be noted here that the analysis above generalizes to different symmetric positive definite bilinear forms corresponding, e.g., to the equations of linear elasticity or the *curl–curl* equation with a positive L^2-term arising in the solution of Maxwell's equations (see [27]). The major difficulty in a rigorous, fully discrete analysis is the establishment of an estimate analogous to (18). Also, the construction of a suitable (multiscale) partition of unity/identity resulting in small dimensional coarse spaces has not been addressed in the literature, so far.

5 Numerical Experiments

We now turn to some numerical experiments to exemplify the robustness of two-level additive Schwarz preconditioners using spectral coarse spaces. To demonstrate the necessity of employing this spectral coarse space we also report numerical results for two-level additive Schwarz preconditioners using standard coarse spaces and coarse spaces spanned by multiscale finite element functions. More precisely, we consider the following four different cases

- $\mathscr{V}_H^{\mathrm{st}} := \mathrm{span}\{\xi_j \,|\, j = 1,\ldots,n_\Omega\}$ (cf. [21, Sect. 2.5.3]).
- $\mathscr{V}_H^{\mathrm{ms,st}} := \mathrm{span}\{\xi_j^{\mathrm{ms}} \,|\, j = 1,\ldots,n_\Omega\}$ (cf. [17]).
- $\mathscr{V}_H := \mathrm{span}\{I_h(\xi_j \varphi_{j,\lambda}) \,|\, \lambda < \tau_\lambda^{-1},\ j = 1,\ldots,n_\Omega\}$ (see (16)).
- $\mathscr{V}_H^{\mathrm{ms}} := \mathrm{span}\{I_h(\xi_j^{\mathrm{ms}} \varphi_{j,\lambda}) \,|\, \lambda < \tau_\lambda^{-1},\ j = 1,\ldots,n_\Omega\}$ (see Sect. 4.5),

with $\{\xi_j\}_{j=1}^{n_\Omega}$ a standard partition of unity and $\{\xi_j^{\mathrm{ms}}\}_{j=1}^{n_\Omega}$ as in Sect. 4.5. The subdomains are chosen as described in Sect. 4.5, and the bilinear form $m_{\Omega_j}(\cdot,\cdot)$ is chosen as $1/2\overline{m}_{\Omega_j}(\cdot,\cdot)$ in Sect. 4.1 with the standard partition of unity ξ_j and the multiscale partition of unity ξ_j^{ms}, respectively. The eigenvalue threshold is fixed by setting $\tau_\lambda = 2$.

On our computational domain $\Omega := (0,1)^2$ we use a 256×256 fine and a 16×16 coarse tensor grid. The problems under consideration are discretized using bilinear Lagrange finite elements. We emphasize that there is no essential difficulty in treating more realistic settings. In particular one can consider the case of an unstructured two- or three-dimensional fine grid and a corresponding coarse grid resulting from an agglomeration procedure as, e.g., outlined in [26, Sect. 1.9].

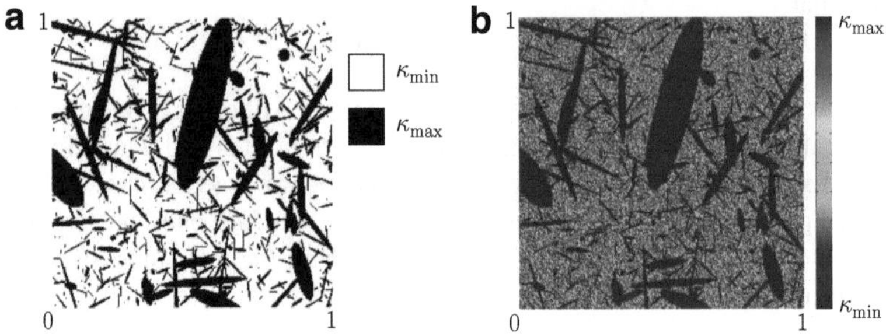

Fig. 3 κ for two random geometries. (**a**) κ for a binary random multiscale geometry. (**b**) Logarithmic plot of κ for a non-binary random multiscale geometry

We choose two different configurations for κ. The first geometry depicted in Fig. 3a is a binary one, i.e., κ only takes two values. As opposed to this, Fig. 3b shows a coefficient κ which assumes a multitude of values between $\kappa_{\min}$ and $\kappa_{\max}$. Although both geometries are artificial in the sense that they do not represent any concrete real life application, we consider them to be "hard" test problems. In particular the (highly) varying coefficients represent multiscale features, which is common in, e.g., reservoir simulations.

In our numerical experiments below we consider the cases $\kappa_{\min} = 1$ and $\kappa_{\max} = 1e1, \ldots, 1e6$ to test our preconditioner for robustness. We would also like to point out that the coefficient variations are not aligned with the coarse 16×16 grid.

For completeness we remark that our implementations are carried out in C++ using the deal.II finite element library (cf. [2]), which in turn uses the LAPACK software package (cf. [1]) for solving all appearing direct and eigenvalue problems.

In Table 1(1) we report the results obtained for the binary geometry shown in Fig. 3a. The table shows the condition numbers of the additive Schwarz preconditioned systems, where we employ the different choices of coarse spaces listed at the beginning of this section. The numbers reported in parentheses are the respective coarse space dimensions. As we can see, the condition numbers corresponding to the spaces $\mathscr{V}_H^{\mathrm{st}}$ and $\mathscr{V}_H^{\mathrm{ms,st}}$ increase quite substantially with increasing the contrast $\kappa_{\max}/\kappa_{\min}$. As opposed to this the preconditioners with the spectral coarse spaces $\mathscr{V}_H$ and $\mathscr{V}_H^{\mathrm{ms}}$ yield condition numbers which are robust with respect to the contrast. This robustness comes at the expense of having to solve local generalized eigenvalue problems, which of course can be done completely in parallel, and of having a larger dimensional coarse space, which is in particular pronounced for higher contrasts. We emphasize, however, that this increase in complexity can be significantly reduced by multiscale partition of unity functions, i.e., by using $\mathscr{V}_H^{\mathrm{ms}}$ instead of $\mathscr{V}_H$. For the highest considered contrast the dimension of the former is less than 3 times as large as the dimension of $\mathscr{V}_H^{\mathrm{st}}$ and $\mathscr{V}_H^{\mathrm{ms,st}}$, whereas for the latter the factor is close to 10. As indicated above the dimension of the spectral coarse spaces changes with increasing the contrast. Nevertheless, in coherence with our theory in Sect. 4 this increase appears to reach some saturation for very high contrasts.

Table 1 Condition numbers of the additive Schwarz preconditioned systems for the geometries shown in Fig. 3 with different contrasts $\kappa_{\max}/\kappa_{\min}$. In parentheses we report the coarse space dimension.

(1) Results for Fig. 3a

$\frac{\kappa_{\max}}{\kappa_{\min}}$	$\mathscr{V}_H^{st}$	$\mathscr{V}_H^{ms,st}$	$\mathscr{V}_H$	$\mathscr{V}_H^{ms}$
1e1	4.7e0(225)	4.7e0(225)	4.7e0(279)	4.7e0(276)
1e2	1.2e1(225)	8.2e0(225)	4.9e0(570)	5.3e0(340)
1e3	7.6e1(225)	3.6e1(225)	4.6e0(1477)	5.2e0(547)
1e4	7.2e2(225)	3.4e2(225)	4.7e0(1995)	5.2e0(669)
1e5	6.1e3(225)	3.1e3(225)	4.8e0(2081)	5.2e0(668)
1e6	4.4e4(225)	2.8e4(225)	4.8e0(2093)	5.2e0(665)

(2) Results for Fig. 3b

$\frac{\kappa_{\max}}{\kappa_{\min}}$	$\mathscr{V}_H^{ms}$
1e1	4.6e0(276)
1e2	4.7e0(273)
1e3	4.9e0(275)
1e4	4.9e0(306)
1e5	5.3e0(380)
1e6	5.4e0(461)

In order to not only test our theory for binary geometries we also consider the coefficient depicted in Fig. 3b. For this geometry we only report the results corresponding to the coarse space $\mathscr{V}_H^{ms}$ in Table 1(2). As we can see, the condition numbers also behave robustly in this situation. Also, similarly to the binary geometry, we can observe the trend that increasing the contrast tends to increase the dimension of the coarse space. Nevertheless, even for the highest contrast $1e6$ the size of the coarse space is still rather manageable and in particular smaller than the corresponding one for the binary geometry.

We close this section by some comments regarding the computational complexity of the discussed domain decomposition methods using spectral coarse spaces. These remarks apply not only to the considered two-dimensional examples but also to the three-dimensional case.

The bottleneck of the discussed methods is the coarse space construction and in particular the solution of the local generalized eigenvalue problems. As indicated above these eigenvalue problems are solved using LAPACK. The algorithm implemented in the subroutine DSYGVX first reduces the generalized eigenvalue problems to standard ones by performing Cholesky decompositions. The resulting matrices are then reduced to Hessenberg tridiagonal form, which can be done by Householder transformations. A QR-algorithm employing Givens rotations can then be used to compute the actual eigenpairs. The overall complexity of this algorithm is cubic in the number of unknowns.

Even though the generalized eigenvalue problems can be solved in parallel, it may be unreasonably costly to construct a spectral coarse space, if one is only interested in solving a single problem on a given geometry. However, if one needs to solve many problems on a single geometry, which, e.g., is the case when computing an approximate solution of a time-dependent problem by an implicit time-stepping scheme, constructing a spectral coarse space may be rather reasonable.

If one wants to solve many problems on a single geometry, it makes sense to distinguish between an offline phase, which in particular includes the construction of the spectral coarse space, and an online phase, which is the actual application of the preconditioner. As the computations in the offline phase are only carried out once, the computational cost of the online phase becomes the major concern. Considering the discussed methods we see that one iteration of a two-level algorithm with a coarse space given by $\mathscr{V}_H$ or $\mathscr{V}_H^{\mathrm{ms}}$ is about as expensive as one iteration of a two-level algorithm with a coarse space given by $\mathscr{V}_H^{\mathrm{st}}$ or $\mathscr{V}_H^{\mathrm{ms,st}}$. The only difference making the former somewhat more expensive than the latter is due to the increased coarse space dimension. Although this space dimension is inherently problem dependent, we note that $\dim(\mathscr{V}_H^{\mathrm{ms}})$ remains rather manageable for the considered examples. In view of drastically reduced condition numbers of the preconditioned systems, this slight increase in computational complexity for one iteration in the online phase seems justified. After all, the number of preconditioned conjugate gradient iterations needed to achieve a prescribed accuracy depend on the condition number of the preconditioned system, and one may therefore expect significant overall computational savings by employing two-level preconditioners using spectral coarse spaces.

6 Conclusions

We have given an overview of several recently proposed approaches for constructing spectral coarse spaces for robust preconditioners. For this we have developed a monolithic framework enabling us to detail the similarities and distinctions of the different methods and to discuss their advantages and shortcomings. In this context we have in particular related the more recent abstract works for general symmetric positive definite bilinear forms to the originally introduced concepts and ideas for the scalar elliptic equation. To show the applicability of the discussed analysis, we have presented some numerical examples to validate the theoretical results.

Acknowledgements The research of J. Willems was supported in parts by NSF Grant DMS-1016525.

References

1. Anderson, E., Bai, Z., Bischof, C., Blackford, S., Demmel, J., Dongarra, J., Du Croz, J., Greenbaum, A., Hammarling, S., McKenney, A., Sorensen, D.: LAPACK Users' Guide, 3rd edn. Society for Industrial and Applied Mathematics, Philadelphia, PA (1999)
2. Bangerth, W., Hartmann, R., Kanschat, G.: deal.II—a general purpose object oriented finite element library. ACM Trans. Math. Software **33**(4), 24/1–24/27 (2007)

3. Bramble, J.H.: Multigrid Methods, 1st edn. Longman Scientific&Technical, Essex (1993)
4. Brenner, S.C., Scott, L.R.: The Mathematical Theory of Finite Element Methods, 2nd edn. Springer, New York (2002)
5. Brezina, M., Vassilevski, P.: Smoothed aggregation spectral element agglomeration AMG: SA-ρAMGe. In: Lirkov, I., Margenov, S., Wasniewski, J. (eds.) Large-scale Scientific Computing. Volume 7116 of Lecture Notes in Computer Science, pp. 3–15. Springer, Heidelberg (2012)
6. Brezina, M., Heberton, C., Mandel, J., Vanek, P.: An iterative method with convergence rate chosen a priori. Technical Report, Denver, CO (1999)
7. Chartier, T., Falgout, R.D., Henson, V.E., Jones, J., Manteuffel, T., McCormick, S., Ruge, J., Vassilevski, V: Spectral AMGe (ρAMGe). SIAM J. Sci. Comput. **25**(1), 1–26 (2003)
8. Chartier, T., Falgout, R.D., Henson, V.E., Jones, J., Manteuffel, T., McCormick, S., Ruge, J., Vassilevski, P.S.: Spectral element agglomerate AMGe. In: Widlund, O.B., Keyes, D.E. (eds.) Domain Decomposition Methods in Science and Engineering XVI. Volume 55 of Lecture Notes in Computational Science and Engineering, pp. 513–521. Springer, Berlin (2007)
9. Dolean, V., Hauret, P., Nataf, F., Pechstein, C., Scheichl, R., Spillane, N.: Abstract robust coarse spaces for systems of PDEs via generalized eigenproblems in the overlaps. Technical Report 2011-07, University of Linz, Institute of Computational Mathematics (2011) (submitted)
10. Efendiev, Y., Hou, T.Y.: Multiscale Finite Element Methods: Theory and Applications. Volume 4 of Surveys and Tutorials in the Applied Mathematical Sciences. Springer, New York (2009)
11. Efendiev, Y., Galvis, J., Lazarov, R.D., Margenov, S., Ren, J.: Multiscale domain decomposition preconditioners for anisotropic high-contrast problems. Technical Report ISC-Preprint-2011-05, Institute for Scientific Computation (2011)
12. Efendiev, Y., Galvis, J., Lazarov, R., Willems, J.: Robust domain decomposition preconditioners for abstract symmetric positive definite bilinear forms. Math. Model. Numer. Anal. **46**, 1175–1199 (electronic) (2012)
13. Efendiev, Y., Galvis, J., Vassilevski, P.S.: Multiscale spectral AMGe solvers for high-contrast flow problems. Technical Report 2012-02, Institute for Scientific Computation (2012)
14. Fuhrmann, P.: A Polynomial Approach to Linear Algebra (Universitext), 2nd edn. Springer, New York (2012)
15. Galvis, J., Efendiev, Y.: Domain decomposition preconditioners for multiscale flows in high-contrast media. Multiscale Model. Simulat. **8**(4), 1461–1483 (2010)
16. Galvis, J., Efendiev, Y.: Domain decomposition preconditioners for multiscale flows in high contrast media: reduced dimension coarse spaces. Multiscale Model. Simulat. **8**(5), 1621–1644 (2010)
17. Graham, I.G., Lechner, P.O., Scheichl, R.: Domain decomposition for multiscale PDEs. Numer. Math. **106**(4), 589–626 (2007)
18. Hackbusch, W.: Multi-Grid Methods and Applications, 2nd edn. Springer Series in Computational Mathematics. Springer, Berlin (2003)
19. Hou, T.Y., Wu, X.-H., Cai, Z.: Convergence of a multiscale finite element method for elliptic problems with rapidly oscillating coefficients. Math. Comput. **68**(227), 913–943 (1999)
20. Kraus, J., Vassilevski, P., Zikatanov, L.: Polynomial of best uniform approximation to 1/x and smoothing in two-level methods. Comput. Meth. Appl. Math. **12**(4), 448–468 (2012)
21. Mathew, T.P.A.: Domain Decomposition Methods for the Numerical Solution of Partial Differential Equations. Lecture Notes in Computational Science and Engineering. Springer, Berlin (2008)
22. Scheichl, R., Vassilevski, P.S., Zikatanov, L.T.: Weak approximation properties of elliptic projections with functional constraints. Multiscale Model. Simulat. **9**(4), 1677–1699 (2011)
23. Toselli, A., Widlund, O.: Domain Decomposition Methods—Algorithms and Theory. Springer Series in Computational Mathematics. Springer, New York (2005)
24. Trottenberg, U., Oosterlee, C. W., Schüller, A.: Multigrid. Academic, San Diego, CA (2001) (With contributions by A. Brandt, P. Oswald and K. Stüben)
25. Van lent, J., Scheichl, R., Graham, I.G.: Energy-minimizing coarse spaces for two-level Schwarz methods for multiscale PDEs. Numer. Linear Algebra Appl. **16**(10), 775–799 (2009)

26. Vassilevski, P.S.: Multilevel Block Factorization Preconditioners: Matrix-based Analysis and Algorithms for Solving Finite Element Equations. Springer, New York (2008)
27. Willems, J.: Robust multilevel methods for general symmetric positive definite operators. Technical Report RICAM-Report 2012-06, Radon Institute for Computational and Applied Mathematics (2012)
28. Willems, J.: Robust multilevel solvers for high-contrast anisotropic multiscale problems. Technical Report RICAM-Report 2012-17, Radon Institute for Computational and Applied Mathematics (2012)
29. Xu, J., Zikatanov, L.T.: On an energy minimizing basis for algebraic multigrid methods. Comput. Vis. Sci. **7**(3–4), 121–127 (2004)

About the Editors

Oleg P. Iliev is a senior scientist in Fraunhofer Institute for Industrial Mathematics in Kaiserslautern, Germany, APL professor in the Faculty of Mathematics in the Technical University of Kaiserslautern, and visiting professor in KAUST. He is working in the area of mathematical modeling and computer simulation of industrial and environmental processes.

Svetozar D. Margenov is a professor of mathematics at the Institute of Information and Communication Technologies of the Bulgarian Academy of Sciences in Sofia, Bulgaria. His research is in the area of numerical methods for partial differential equations and large-scale scientific computing.

Peter D. Minev is a professor of applied mathematics at the University of Alberta in Edmonton, AB, Canada, working in the area of numerical analysis and computational mechanics.

Panayot S. Vassilevski is a computational mathematician at the Center for Applied Scientific Computing of Lawrence Livermore National Laboratory in Livermore, CA, USA, working in the area of multilevel methods for solving large-scale problems typically arising from finite element discretization of partial differential equations.

Ludmil T. Zikatanov is a professor of mathematics at The Pennsylvania State University in University Park, PA, USA. His research is in numerical analysis and numerical solution of partial differential equations.

O.P. Iliev et al. (eds.), *Numerical Solution of Partial Differential Equations: Theory, Algorithms, and Their Applications*, Springer Proceedings in Mathematics & Statistics 45, DOI 10.1007/978-1-4614-7172-1, © Springer Science+Business Media New York 2013

FSC
www.fsc.org
MIX
Papier aus verantwortungsvollen Quellen
Paper from responsible sources
FSC® C105338

If you have any concerns about our products,
you can contact us on
ProductSafety@springernature.com

In case Publisher is established outside the EU,
the EU authorized representative is:
Springer Nature Customer Service Center GmbH
Europaplatz 3, 69115 Heidelberg, Germany

Printed by Libri Plureos GmbH
in Hamburg, Germany